How this book will help you...

It doesn't matter whether you're heading for mocks in Year 11, or in the final run-up to your GCSE exam – **this book will help you to produce your very best.**

Whichever approach you decide to take to revision, this book will provide everything you need:

1 Total revision support
2 Quick revision check-ups
3 Exam practice

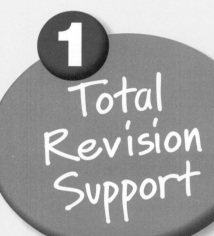

1

Total Revision Support

Everything you need to know

This book contains all the topics you'll be studying at school. **It covers the content of all the double and single-award Science syllabuses produced by the Exam Boards.**

Short, easy-to-use sections

If, when you're revising you need to go over something you haven't understood, you'll have no trouble finding it in this book. We've divided each chapter into a number of **short sections with clear headings**. Just look up the topic you want in the contents list or index. **We've put all the important scientific terms into bold black type** (e.g. **membrane**) so that you can't miss them.

Clear diagrams and charts

There are lots of diagrams in the book. It's often easier to remember facts if they're presented visually. **All the diagrams are in full colour and clearly labelled**. You'll find them really easy to understand.

2 Quick Revision Check-ups

Check yourself questions

It can be really hard knowing where to start when you're revising. Sitting down and wading through pages of facts isn't easy. You're probably asleep before the third page! This book makes it easy to stay awake – **because it makes revising ACTIVE**.

We came up with the idea of putting **'Check yourself'** questions into each chapter. **The questions test your understanding of all the important scientific concepts and ideas** in each section of the chapter. In this way, you can find out quickly and easily just how much you know. You don't need to read through all the text first – just try the questions. If you get all the questions right, you can move straight on to the next section. If you get several of the questions wrong, you know you need to read through the whole section carefully. **This really cuts down on revision time – and helps you focus on where you need to put most effort.**

Answers and Tutorials

If you want the 'Check yourself' questions to be a genuine test of how much you know, then you need to cover up the answers. But, if you'd rather, you can read through a question, then the answer and then the **'tutorial'**. This will still do you a lot of good – and doesn't require quite as much effort!

We've included 'tutorials', as well as answers, to give you even more help with your revision. The tutorials contain extra information, they point out common mistakes that we know candidates make, and give you hints on answering similar questions in exams.

Exam checklists

These are another important revision aid. **The checklists summarise all of the key ideas in each topic.** Use them to tick off the areas that you know about – and to spot the ones where you need to do more work.

3

Exam Practice

Exam technique

Knowing the facts is important. But **it's even more important to know how to use them to answer exam questions properly**. The authors see hundreds of exam scripts a year and students very often lose marks not because they don't know their facts, but because **they haven't understood how to tackle exam questions**.

Questions and sample students' answers

It's often easiest to explain what to do and what not to do by looking at **actual examples of students' answers to exam questions**. This is why we've included sample answers in this book.

These are typical answers, not perfect ones. They highlight the kind of mistakes students often make. **In the examiner's comments, the authors run through these mistakes and show you clearly what you need to do to score full marks on the questions**.

Questions to Answer

We've also included **lots of past exam questions from different Exam Boards for you to have a go at**. The answers are at the back of the book so it's easy not to cheat. Have a go at the questions yourself and then compare them with the answers. **We've provided comments on most of the answers to give you extra help** – and if you're still unsure you can go back to the relevant section in the book.

Three final tips:

1 Work as consistently as you can during your whole GCSE Science course. If you don't understand something, ask your teacher straight away, or look it up in this book. You'll then find revision much easier.

2 Plan your revision carefully and focus on the areas you know you find hard. The 'Check yourself' questions in this book will help you do this.

3 Try to do some exam questions as though you were in the actual exam. Time yourself and don't cheat by looking at the answers until you've really had a good go at working out the answers.

About your GCSE Science course

Does this book match my course?

This book has been written to support all the double and single-award Science GCSE syllabuses produced by the five Exam Boards in England and Wales. These syllabuses include content from the three main areas of Biology, Chemistry and Physics and are based on the National Curriculum Key Stage 4 Programmes of Study. This means that although different syllabuses may arrange the content in different ways, they must all cover the same work.

The double-Science courses lead to the award of two GCSE Science grades. The single-Science courses result in a single GCSE grade. The content of the single award syllabuses is taken from the double award. Ask your teacher whether you are following a single or double-award course.

Foundation and Higher tier papers

In your GCSE Science exam you will be entered for either the Foundation tier exam papers or the Higher tier exam papers. The Foundation exams allow you to obtain grades from G to C. The Higher exams allow you to obtain grades from D to A*.

Higher tier							
A*	A	B	C	D	E	F	G
			Foundation tier				

What will my exam questions be like?

The exam questions will be of a type known as structured questions. Usually these are based on a particular topic and will include related questions. Some of these questions will require short answers involving a single word, phrase or sentence. Other questions will require a longer answer involving extended prose. You will have plenty of practice at both types of questions as you work through the chapters in this book.

Short answer questions

These are used to test a wide range of knowledge and understanding quite quickly.

They are often worth one mark each.

Extended prose questions

They are used to test how well you can link different ideas together. Usually they ask you to explain ideas in some detail. It is important to use the correct scientific terms and to write clearly.

They may be worth four or five marks and sometimes more.

How should I answer exam questions?

- Look at the number of marks. The marks should tell you how long to spend on a question. A rough guide is a minute for every mark. The number of marks will indicate how many different points are required in the answer.

- Look at the space allocated for the answer. If only one line is given, then only a short answer is required, e.g. a single word, a short phrase or a short sentence. Some questions will require more extended writing and so four or more lines will be allocated.

- Read the question carefully. Students frequently answer the question they would like to answer rather than the one that has actually been set! Circle or underline the key words. Make sure you know what you are being asked to do. Are you choosing from a list? Are you using the Periodic Table? Are you completing a table? Have you been given the formula you will need to use in a calculation? Are you describing or explaining?

Biology is the study of living things. An individual living thing is called an **organism**. The two main types of organism are **animals** (which includes us) and **plants**. Although there is a huge variety of organisms, there are many things that they all have in common. This includes the way they are built as well as how they work.

CHARACTERISTICS OF LIVING THINGS

There are seven characteristics, or features, that all living things have.

- **Movement.** This can be reasonably fast like a dog running or much slower like a plant opening and closing its flowers.
- **Respiration.** This means releasing energy from food – don't confuse it with breathing. Plants and animals do this.
- **Sensitivity.** This means organisms can sense and respond to things around them. For example a plant shoot will grow towards the light or a rabbit will run from the sight of a fox.
- **Growth.** Humans, for example, grow from babies into adults, and when adult continue to grow by replacing damaged or worn out parts such as blood cells or growing new skin after a cut.
- **Reproduction.** This can be sexual (involving sex cells) or asexual (involving growth but no sex cells).
- **Excretion.** This means getting rid of substances that the organism has made but does not want. Humans breathing out carbon dioxide is a good example.
- **Nutrition.** This means using food whether or not you have got it by eating something else, like animals, or by making it yourself, like plants.

You can remember this list by remembering that the first letters make up a name:

MRS GREN.

Living things show all these characteristics. There is some argument about viruses. They do not show most of these features therefore most scientists would not describe them as living.

CELLS

Most living things are made up of building blocks called **cells**. Most cells are so small that they can only be seen with a microscope. There are many different types of cells which have different jobs, such as nerve cells, muscle cells and blood cells. However there are certain features that most cells have in common. There are also some features that most plant cells have that are not found in animal cells. The diagrams on the next page show a typical animal cell and a typical plant cell.

The different parts of cells have different jobs.

- **Cell membrane**. This holds the cell together and controls substances entering or leaving the cell.
- **Cytoplasm**. This is more complicated than it looks. This is where many different chemical processes happen.

Animal cell

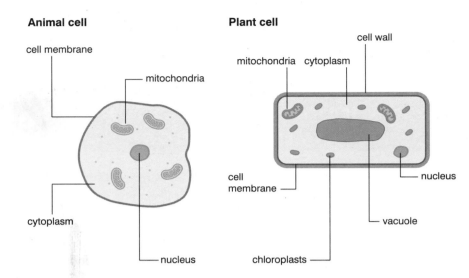

Plant cell

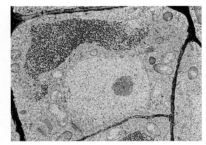

A plant cell. You can see the cell wall (the surrounding dark line), the nucleus (the pale patch containing a red dot), and some mitochondria (red circles and ellipses).

Human cheek cells, a good example of animal cells. You can see the various 'organelles' (the organs of a cell), including the nucleus, which are stained orange.

- **Nucleus**. This contains **chromosomes** and **genes**. They control how a cell grows and works. Genes also control the features that can be passed on to offspring.
- **Mitochondria** (singular: mitochondrion). This is where energy is released from food in the process known as respiration.
- **Cell wall**. Plants have this as well as a cell membrane. It is much more rigid than the cell membrane and gives the cell shape and support.
- **Vacuole**. This contains a liquid called cell sap which is water with various things dissolved in it. In a healthy plant the vacuole is large and helps support the cell.
- **Chloroplasts**. These contain the green pigment **chlorophyll** which absorbs the light energy that plants need to make food in the process known as **photosynthesis**.

A typical animal cell is different from a typical plant cell although they do have things in common.

Similarities	Differences	
Most cells contain	**Animal cells**	**Plant cells**
cell membrane	No cell wall	Have a cell wall
cytoplasm	No large vacuole (although there may be small ones)	Usually have a large vacuole
nucleus	No chloroplasts	Green parts of a plant contain chloroplasts
mitochondria	Many irregular shapes	Usually have a regular shape

So far we have looked at 'typical' cells but many cells are different to this. This is because they are **specialised** which means they have special features that allow them to carry out different jobs. In the same way a child could draw a 'typical' car although there are many different kinds, such as racing cars or people carriers, that are built differently to do different jobs.

One example of a specialised cell is a sperm cell. The sperm cell's job is to swim to an egg cell and fertilise it by joining its nucleus with the egg cell's nucleus. It is specialised for this in several ways:

- It has a tail which moves so it can swim
- It has many mitochondria to provide the energy for swimming
- The acrosome contains digestive **enzymes** that make a hole in the egg cell's membrane so the sperm cell can enter
- The nucleus contains chromosomes which carry the genetic information that will be passed on to the child

There are many other examples of specialised cells. You will find out about some of them in later chapters.

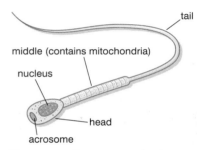

The sperm cell is an example of a specialised cell.

ORGANISATION

Different cells are specialised to carry out different jobs. Each type of cell is organised in a way that helps the body work effectively. Cells of the same type form **tissue**. Different tissues may make up an **organ**. Organs that work together make up an organ **system**. For example:

The circulatory system includes the heart organ which contains muscle tissue and nervous tissue which are made of muscle cells and nerve cells.

You will find out more about different systems in later chapters.

The main systems of organs in your body.

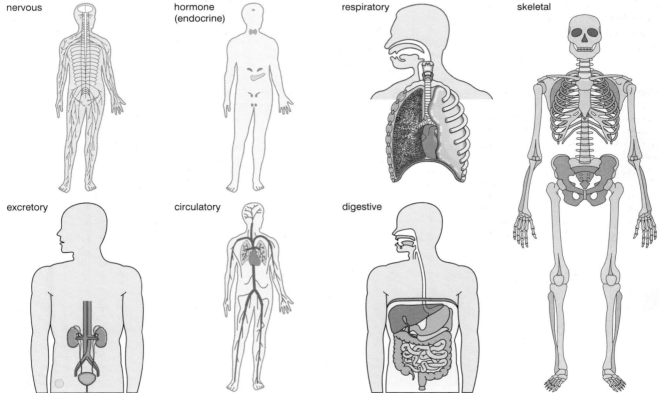

Check yourself

QUESTIONS

Q1 A tree is a living thing but a motor car is not. Explain why.

Q2 Look at the diagram of an onion cell.

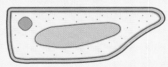

a) How is the onion cell different to a typical plant cell ?
b) Explain the reason for this difference.
c) How is the onion cell different to an animal cell ?

Q3 Name which part of a cell does the following:
a) releases energy from food,
b) allows oxygen to enter,
c) contains the genes,
d) contains cell sap,
e) stops plant cells swelling if they take in a lot of water.

Q4 For each of the following say whether it is a cell, a tissue, an organ, a system or an organism:
(i) oak tree, (ii) human egg,
(iii) brain, spinal cord and nerves (iv) leaf,
(v) stomach lining, (vi) kidney.

REMEMBER! Cover the answers if you want to.

ANSWERS

A1 A tree can carry out all seven of the characteristics of life. A car can not.

A2
a) It does not contain any chloroplasts (and therefore no chlorophyll).
b) Onions grow underground so they will not receive any light. There is no point therefore having chloroplasts as they will not be able to photosynthesise.
c) It has a cell wall, a large vacuole and a regular shape.

A3
a) Mitochondria.
b) Cell membrane.
c) Nucleus or chromosomes. Both are correct.
d) Vacuole.
e) Cell wall.

A4 (i) Organism, (ii) cell, (iii) system,
(iv) organ, (v) tissue, (vi) organ.

TUTORIAL

T1 *You could argue that a car does show some of the characteristics of life: movement, respiration (it releases energy from petrol), sensitivity (if it has an alarm for example), excretion (fumes are released) and nutrition (it takes in petrol). It does not however grow or reproduce so it does not show all seven characteristics.*

T2
a) *Any green parts of a plant will contain chloroplasts.*
b) *By going into detail and mentioning photosynthesis you show clearly that you understand the reason.*
c) *There are some exceptions in that some plant cells may not have large vacuoles or a regular shape. They all have a cell wall though.*

T3
a) *This is a better answer than simply saying cytoplasm.*
b) *Don't forget that everything that goes in or out of a cell goes through the membrane.*
c) *Chromosomes is a more specific answer and shows more understanding.*
d) *Don't forget only plants have large vacuoles.*
e) *The membrane would simply burst if the cell became too big. The cell wall is more rigid and resists this. You will find out more about this in chapter 4.*

T4 *Organisms are whole animals or plants. (Don't forget that there are some organisms, like bacteria, that are only one cell big.) Organs are the separate parts of a body each of which has its own job(s). Organs are usually made of different groups of cells (tissues).*

TRANSPORT INTO AND OUT OF CELLS

Not only do whole organisms take in substances, for example when animals eat and breathe, or give them out, for example when they excrete, but each cell in a body has to take in and give out different things. All of these things will have to pass through the **cell membrane**.

There are three main ways substances enter and leave cells:

- diffusion
- osmosis
- active transport.

DIFFUSION

Substances like water, oxygen, carbon dioxide and food, are made of particles. You will find out more about the different kinds of particles (molecules, atoms and ions) in chapter 8.

In liquids and gases the particles are constantly moving around. This means that they will tend to spread themselves out evenly. For example if you dissolve sugar in a cup of tea, even if you do not stir it, the sugar will eventually spread throughout the tea because the sugar molecules are constantly moving around, colliding with and bouncing off other particles. This is an example of **diffusion**. The sugar molecules have moved from being in a big lump to being spread throughout where there was originally no sugar. In other words:

diffusion is movement from an area of high concentration to an area of low concentration.

This is shown in the diagrams below.

 water molecules

 sugar molecules

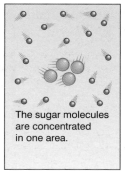
The sugar molecules are concentrated in one area.

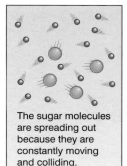
The sugar molecules are spreading out because they are constantly moving and colliding.

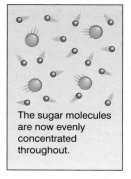
The sugar molecules are now evenly concentrated throughout.

So to summarise:

- Diffusion occurs when there is a difference in concentration.
- Particles diffuse from areas of high concentration to areas of low concentration. This is described as moving along the **concentration gradient**.
- Diffusion stops when the particles are evenly concentrated. But this does not mean that the particles themselves stop moving.
- Diffusion happens because particles are constantly and randomly moving. It does not need an input of energy from a plant or animal.

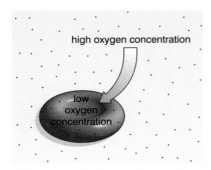

A high concentration gradient leads to diffusion. Red blood cells low in oxygen will soak it up from more oxygen rich surroundings.

Substances can enter and leave cells by diffusion. If there is a higher concentration on one side of the membrane than the other and the substance can move through the membrane, then it will. For example, red blood cells travel to the lungs to collect oxygen. There is a *low* oxygen concentration in the cells (because they have given up their oxygen to other parts of the body) but a *high* oxygen concentration in the air spaces in the lungs. Therefore oxygen diffuses into the red blood cells.

Other examples of diffusion are oxygen and carbon dioxide leaving and entering plant leaves, and parts of digested food in the digestive system entering the blood. You can read more about diffusion in chapter 8.

OSMOSIS

Osmosis is a special example of diffusion where *only water* moves into or out of cells. It occurs because cell membranes are **semi-permeable** or partially permeable – they allow some substances to move through them but not others. Water will diffuse from a place where there is a high concentration of water molecules (such as a dilute sugar solution) to where there is a low concentration of water molecules (such as a concentrated sugar solution).

Stop and think about this a second time. Many people confuse the concentration of the solution with the concentration of the water. Remember it is the water that is moving, so we must think of the amount of *water* in the solution instead of the amount of substance dissolved in it. A low concentration of dissolved substances means a high concentration of water. A high concentration of dissolved substances means a low concentration of water. So the water is still moving from a high concentration (of water) to a low concentration (of water), even though this is often described as water moving from a low concentration *solution* to a high concentration *solution*.

The water is less concentrated on the right and more concentrated on the left. The water diffuses to the right. This sort of diffusion is known as osmosis.

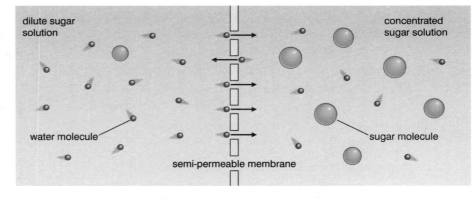

One important example of osmosis is water entering the roots of plants. If animal cells such as blood cells are placed in different strength solutions they will swell up or shrink depending on whether they gain or lose water via osmosis.

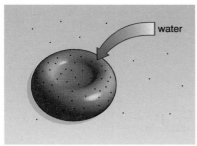

A blood cell in pure water.

ACTIVE TRANSPORT

There are occasions when cells need to absorb substances *against* a concentration gradient. In other words from an area of low concentration to an area of high concentration. This is the *opposite* direction to the way the substances would diffuse. For example, root hair cells on a plant's roots may take in nitrate ions from the soil even though there is a higher concentration of them in the plant than in the soil. The way that the nitrate ions are absorbed is called **active transport**. Another example of active transport is sugar being quickly absorbed from the small intestine into the blood.

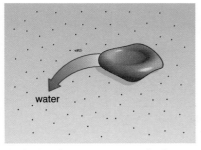

A blood cell in a concentrated solution.

Active transport occurs when special 'carrier proteins' on the surface of a cell pick up particles from one side of the membrane and transport them to the other side. You can see this happening in the diagram below.

Active transport uses energy that the cells release during respiration.

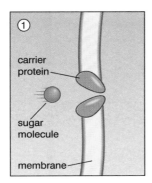
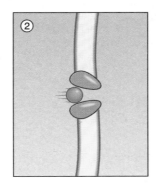
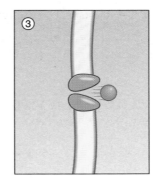

Check yourself

QUESTIONS

Q1 During the day, leaf cells photosynthesise and produce oxygen. The oxygen passes out of the leaf in to the air. Explain why the oxygen passes out of the leaf.

Q2 A plant cell wall is permeable but the plant cell membrane is semi-permeable. What is the difference?

Q3 An animal cell placed in pure water may swell up and burst but a plant cell will not. Explain why.

Q4 An old fashioned way of killing slugs in the garden is to sprinkle salt on them. This kills them by drying them out. Explain why this would dry them out.

Q5 Which of the following are examples of: diffusion, osmosis, or neither?
a) Carbon dioxide entering a leaf when it is photosynthesising.
b) Food entering your stomach when you swallow.
c) Tears leaving your tear ducts when you cry.
d) A dried out piece of celery swelling up when placed in a bowl of water.

Q6 Why would you expect plant root-hair cells to contain more mitochondria than other plant cells ?

REMEMBER! Cover the answers if you want to.

ANSWERS

A1 As oxygen is being made in photosynthesis there will be a higher concentration inside the leaf compared with the outside of the leaf. Therefore oxygen will diffuse out of the leaf.

A2 The cell wall will let any substance through but the cell membrane will only let some substances through.

TUTORIAL

T1 *In diffusion, substances move from areas of high concentration to areas of lower concentration. As oxygen is continually being made inside the leaf there will always be a higher concentration there than outside.*

T2 *Don't get mixed up between cell walls and membranes. Only plant cells have cell walls and they are there to provide support not to act as a barrier.*

ANSWERS

A3 Both cells take in water by osmosis as they contain more concentrated solutions than the water. The animal cell will continue to take in water until it bursts but in the plant cell the rigid cell wall holds the cell together so it does not burst.

A4 The salt forms a very concentrated solution on the slug's surface so water leaves its body by osmosis.

A5 a) Diffusion.
b) Neither.
c) Neither.
d) Osmosis.

A6 To provide energy for active transport of minerals like nitrates into the roots.

TUTORIAL

T3 *Eventually the plant cell will stop taking in water even though it still contains a more concentrated solution because the cell wall will not expand very much. This gives the plant cells the 'firmness' they need to support themselves even though they do not contain a skeleton. You can read more about this in chapter 4.*

T4 *This happens with slugs because their skin is semi-permeable. It would not happen with us because our skin is not.*

T5 *a) There is a lower concentration of carbon dioxide inside the leaf as it is continually being used up.*
b) The food is squeezed along by muscles in the gullet.
c) Again water is being forced out.
d) As the celery is dried up the cells contain very concentrated solutions.

T6 *The concentration of minerals in soils is always lower than the concentration of minerals in plant cells. Plants therefore can't rely on diffusion to get more of these minerals and must expend energy on actively 'pulling' them in.*

EXAMINATION CHECKLIST

The facts and ideas that you should understand by studying this topic.

To be successful at Foundation GCSE Tier you should be able to:

- list the seven life processes common to all living things
- describe the differences and similarities between plant and animal cells
- recall the functions of the parts of cells: nucleus, membrane, cytoplasm, mitochondria, cell wall, vacuole and chloroplasts
- relate the structure of a cell to its functions
- recall that cells work together in tissues which work together in organs which work together in organ systems
- explain how molecules enter and leave cells by diffusion through the cell membrane from high to low concentration.

In addition, to be successful at Higher GCSE Tier you should be able to:

- explain how osmosis occurs due to differences in concentration across a semi-permeable membrane
- predict the direction of water movement in osmosis
- recall that active transport uses energy to move substances from low concentrations to high concentrations.

Key Words

Tick each word when you are sure of its meaning

active transport	excretion	respiration
cell membrane	gene	semi-permeable
cell wall	mitochondria	specialised
chlorophyll	nucleus	system
chloroplast	nutrition	tissue
chromosome	organ	vacuole
concentration gradient	organism	
cytoplasm	osmosis	
diffusion	reproduction	

EXAM PRACTICE

Sample Student's Answers & Examiner's Comments

a) *This is the correct answer. The question says 'name the process' so only a short answer is required.*

b) *No marks are gained for explaining how water gets in to the soil as this was not asked for. Neither are marks gained for repeating information already given in the question about water entering through the root hair cells. The process involved is osmosis and one mark has been gained for identifying this. However the candidate has not **described** the process as the question asked. Marks would have been gained for describing the movement of the water from the high concentration of water to the low concentration of water (one mark) which happens through a semi-permeable membrane (one mark).*

c) *Plants wilt because they have lost too much water so one mark was gained for explaining that they were 'drying out'. However this mark has only just been achieved. An answer like 'the roots lost water into the soil' would have been better. The question asked the candidate to **explain** so the second mark was for explaining that water left the plants because the sea water was a more concentrated solution (with the dissolved salt) than the plants, or alternatively that the sea water had a lower concentration of water than the plants.*

● *With only 3 points, this is a grade C/D answer. Grade B candidates would expect to get 5 points.*

---

1 The drawing shows a root-hair cell from near the tip of a young root.

soil   root

a) This cell needs oxygen. Name the process by which oxygen enters the cell from the air in the soil.

diffusion ✓

(1)

b) Describe the process by which water enters the root hair cells.

when it rains water soaks into the soil. Then it goes through the cell wall into the root hair cell. This is called osmosis ✓

(3)

c) A seaside garden is flooded by the sea in an exceptionally severe storm. Several of the plants wilt and die.

Explain why flooding with sea water caused the plants to wilt.

The salt kills the plants by drying them out ✓

3/6

(2)

Northern Examination and Assessment Board

Questions to Answer

Answers to questions 2 – 5 can be found on page 357.

2 a) Inherited characteristics are carried in sperm and egg cells.

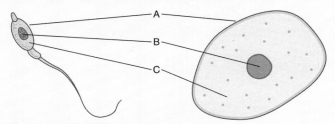

Name the parts labelled A, B and C. (3)

b) Which part of these cells carries information on inherited characteristics? (1)

Midland Examining Group

3 The following parts may be found in a cell:

A cell wall

B cell membrane

C cytoplasm

D nucleus

a) Which of the parts controls the activities of the cell? (1)

b) In which of the parts do most of the chemical reactions take place? (1)

c) Which of the parts controls the passage of substances into the cell? (1)

Northern Examination and Assessment Board

4 Use words from the list to label the plant cell.

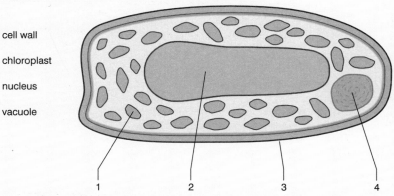

cell wall

chloroplast

nucleus

vacuole

(4)

Northern Examination and Assessment Board

11

5 In an osmosis experiment, 10 cylinders of potato each EXACTLY 50 mm long were used.

Five of the cylinders were placed in water
Five of the cylinders were placed in 25% sugar solution.

After one hour the cylinders were removed from the solution and their surfaces were dried. They were placed on the graph paper as shown in the diagram below.

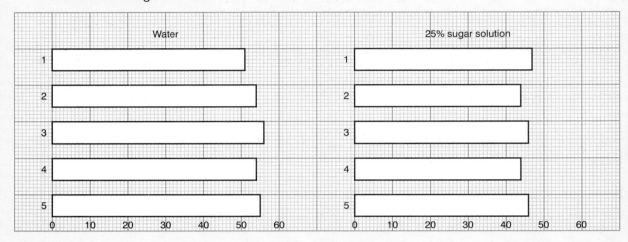

a) Calculate the average length of the cylinders that were in the water. (1)

b) The cylinders that were in the water and those which were in the sugar solution had different average lengths. Explain how this happened. (3)

c) Which other measurements could have been taken to show that the cylinders had changed during the experiment? (2)

London Examinations

Our bodies need energy to keep us alive. We release energy from our food by a process called **respiration**. Oxygen is often needed to help release the energy. In this chapter you will find out more about respiration as well as how **oxygen** and **food** are taken in and transported around our bodies.

Our bodies need energy to keep us alive. We release energy from our food by a process called **respiration**. Oxygen is often needed to help release the energy. In this chapter you will find out more about respiration as well as how **oxygen** and **food** are taken in and transported around our bodies.

RESPIRATION

The energy that our bodies need is released from our food. Releasing the energy is called **respiration** and happens in *every cell* of our body. The food is usually **glucose** (sugar) but other kinds of food can be used if enough glucose is not available. Respiration usually involves oxygen. This kind of respiration is called **aerobic respiration**. Water and carbon dioxide are produced as waste products. Aerobic respiration can be summarised by a word equation:

glucose + oxygen → water + carbon dioxide + energy

It can also be written as a chemical equation:

$$C_6H_{12}O_6 + 6O_2 \rightarrow 6H_2O + 6CO_2 + energy$$

The energy released is used to keep warm and for all the processes characteristic of life: movement, sensitivity, growth, reproduction, excretion and nutrition (feeding). Eventually most of the energy is lost as heat.

KEY POINTS

Digestion gets food into the body.

Breathing gets oxygen into the body.

Blood gets food and oxygen to every cell.

Respiration releases energy from food.

ANAEROBIC RESPIRATION

Aerobic respiration provides most of the energy we need. If we need to release more, for example if we are exercising, then aerobic respiration increases. For this to happen the cells involved, in this case the muscle cells, need an increased supply of glucose and oxygen. For this reason breathing gets faster and deeper to take in more oxygen and the heart beats faster to get the oxygen and glucose to the muscles more quickly.

However there is a limit to how fast we can breathe and how fast the heart can beat. This means the muscles may need to release more energy but can't get enough oxygen. In this case another kind of respiration is used that does not need oxygen. This is called **anaerobic respiration** and is shown by the word equation:

glucose → **lactic acid** + energy

Although anaerobic respiration may be happening to a small extent a lot of the time, it really only happens to a great extent when aerobic respiration can not provide all the energy needed.

Differences between aerobic and anaerobic respiration.

Aerobic respiration	Anaerobic respiration
Uses oxygen	Does not use oxygen
Does not make lactic acid	Makes lactic acid
Makes carbon dioxide	Does not make carbon dioxide
Makes water	Does not make water
Releases a large amount of energy	Releases a small amount of energy

Athletes are all too familiar with the effects of lactic acid.

There are other kinds of anaerobic respiration in other organisms that produce different substances. For example, yeast respires anaerobically converting sugar directly into alcohol, carbon dioxide and energy.

Oxygen debt

The lactic acid that builds up during anaerobic respiration in humans is poisonous. It is lactic acid that causes muscle fatigue (tiredness) and makes muscles ache. It has to be broken down and oxygen is needed to do this. This is why you continue to breathe quickly even after you have finished exercising. You are taking in the extra oxygen you need to remove the lactic acid. This is sometimes called 'repaying the oxygen debt'. Only when all the lactic acid has been broken down does your heart rate and breathing return to normal.

Check yourself

QUESTIONS

Q1 Why do living things respire?

Q2 Where does respiration happen?

Q3 Complete the word equation for aerobic respiration:

.... + oxygen → carbon dioxide + ... + energy

Q4
a) Why is aerobic respiration better than anaerobic respiration?
b) If aerobic respiration is better why does anaerobic respiration sometimes happen?

REMEMBER! Cover the answers if you want to.

ANSWERS

A1 To provide energy for life processes like movement and to generate warmth.

A2 In every cell of the body.

A3
a) Glucose.
b) Water.

A4
a) Aerobic respiration provides more energy (for the same amount of glucose) and does not produce lactic acid (which will have to be broken down).
b) If more energy is needed and the oxygen necessary cannot be provided.

TUTORIAL

T1 *Don't give vague answers like 'to stay alive'. If a question in an examination has several marks then give the correct number of points in your answer.*

T2 *You could be even more specific and point out that respiration occurs in the cytoplasm or in the mitochondria. Many examination candidates confuse respiration and breathing and would have said 'lungs'.*

T3
a) *Sugar or even food might be acceptable answers but it depends on the level of the question. Other foods such as fat or protein can be used in respiration if necessary but are not used in preference to glucose.*
b) *A substantial amount of water can be produced in respiration. (For some desert animals this may be their major source of water.)*

T4
a) *Given sufficient oxygen, respiration is normally aerobic.*
b) *Anaerobic respiration is not as efficient as aerobic respiration. It is only used as a 'top up' to aerobic respiration.*

BLOOD AND THE CIRCULATORY SYSTEM

Blood is the body's transport system, carrying materials from one part of the body to another. Some of the substances transported are shown in the table below:

Substance	Carried from	Carried to
food (glucose, amino acids, fat)	small intestine	all parts of the body
water	intestines	all parts of the body
oxygen	lungs	all parts of the body
carbon dioxide	all parts of the body	lungs
urea (waste)	liver	kidneys
hormones	glands	all parts of the body (different hormones affect different parts)

The blood also plays a part in fighting disease and in controlling body temperature. You can find out more about these in chapter 3.

Parts of the blood	Job
Plasma (pale yellow liquid making up most of the blood)	Transports food, carbon dioxide, urea, hormones and other substances all dissolved in water Heat is also redistributed around the body
Red blood cells	Carry oxygen (and some carbon dioxide)
White blood cells	Defend body against disease. (See chapter 3)
Platelets	Involved in blood clotting

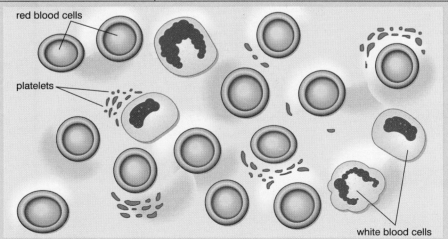

red blood cells

platelets

white blood cells

Blood is mostly water, containing cells and many dissolved substances.

RED BLOOD CELLS

Feature of red blood cells	How it helps
'Biconcave' disc shape (flattened with a dimple in each side)	Large surface area for oxygen to enter and leave
No nucleus	More room to carry oxygen
Contains **haemoglobin** (red pigment)	Haemoglobin combines with oxygen to form **oxyhaemoglobin**. The oxygen is released when the cells reach tissues that need it.
Small	Can fit inside the smallest blood capillaries. Small cells can quickly 'fill up' with oxygen as it is not far for the oxygen to travel right to the centre.
Flexible	Can squeeze into the smallest capillary
Large number	Can carry a lot of oxygen

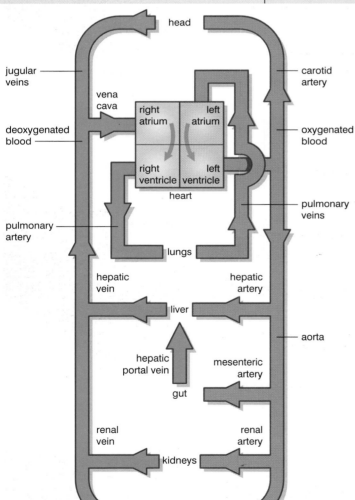

Red blood cells are specialised to carry oxygen.

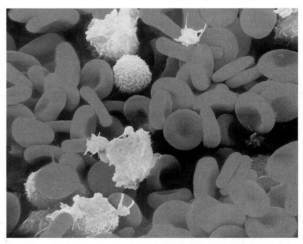

In this micrograph you can see the distinctive 'biconcave' shape of red blood cells.

CIRCULATORY SYSTEM

Blood flows around the body through **arteries**, **veins** and **capillaries**. The **heart** pumps to keep the blood flowing.

THE HEART

The heart is a muscular bag that pumps blood by expanding in size, filling with blood, and then contracting, forcing the blood on its way. The heart is two pumps in one. The right side pumps blood to the lungs to collect oxygen. The left side then pumps the **oxygenated** blood around the rest of the body. The **deoxygenated** (without oxygen) blood then returns to the right side to be sent to the lungs again.

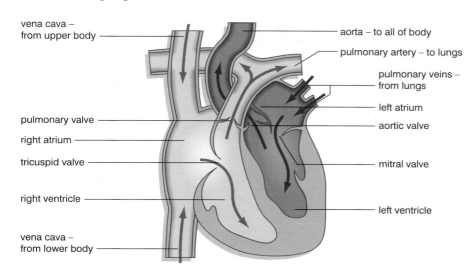

vena cava – from upper body

aorta – to all of body

pulmonary artery – to lungs

pulmonary veins – from lungs

left atrium

aortic valve

pulmonary valve

right atrium

tricuspid valve

mitral valve

right ventricle

left ventricle

vena cava – from lower body

artery:
thick-walled carrying blood at high pressure

The heart contains several valves and four chambers, two called atria (singular: atrium) and two called ventricles.

- The **atria** have thin walls. They collect blood before it enters the ventricles.
- The **ventricles** have thick muscular walls that contract, forcing the blood out.
- The **valves** allow the blood to only flow one way preventing it flowing back the way it has come from.

vein:
thin-walled carrying blood at low pressure

BLOOD VESSELS

Blood leaves the heart through arteries and returns through veins. Capillaries connect the two. (Remember **A** for **A**rteries that travel **A**way from the heart. Ve**IN**s carry blood **IN**to the heart.)

Arteries, veins and capillaries are adapted to carry out their different jobs.

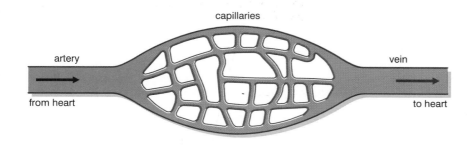

capillaries

artery

vein

from heart

to heart

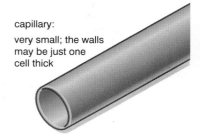

capillary:
very small; the walls may be just one cell thick

Blood Vessel	Job	Adaptations	Explanation
Arteries	Carry blood away from heart.	Thick muscular and elastic wall.	Blood leaves the heart under high pressure. The thick wall is needed to withstand and maintain the pressure. The elastic wall gradually reduces the harsh surge of pumped blood to a steadier flow.
Veins	Carry blood back to the heart.	Thinner walls than arteries.	Blood is now at a lower pressure so there is no need to withstand it.
		Large lumen (space in middle).	Provides less resistance to blood flow.
		Valves.	Prevent back flow which could happen because of the reduced pressure.
Capillaries	Exchange substances with body tissues.	Thin, permeable wall (may only be one cell thick).	Substances, such as oxygen and food, can enter and leave the blood through the capillary walls.
		Small size.	Can reach inside body tissues and between cells.

The valves in the veins prevent back flow.

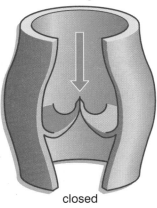

normal blood flow

veins have valves to stop the blood flowing backwards

open

closed

The heart pumps blood to the lungs and back, then to the body and back. This is a double circulation.

DOUBLE CIRCULATION

In humans (but not all animals) the blood travels through the heart twice on each complete journey around the body. This is a **double circulation**.

Once blood has been pushed through a system of capillaries (in either the lungs or the rest of the body) the pressure drops dramatically. The pressure required to push the blood through the lungs and then the rest of the body in one go would be enormous and would destroy the blood vessels. A double circulation system maintains the high blood pressure needed for efficient transport of materials around the body.

The double circulation also allows for the fact that the pressure needed to push blood through the lungs (a relatively short round trip) is significantlly smaller than the pressure needed to push blood around the rest of the body. This is why the left half of the heart is much more muscular than the right half.

Check yourself

QUESTIONS

Q1 Small organisms like A*moeba*, a single-celled animal, do not need a transport system. So why do bigger organisms need them?

Q2 This equation shows the reversible reaction between oxygen and haemoglobin.

oxygen + haemoglobin $\rightleftharpoons$ oxyhaemoglobin

a) Where in the body would oxyhaemoglobin form?
b) Where in the body would oxyhaemoglobin break down?
c) People who live at high altitudes, where there is less oxygen, have more red blood cells per litre of blood than people who live at lower altitudes. Suggest why.

Q3 Most arteries carry oxygenated blood except one. Which one?

Q4
a) In the heart, the ventricles have thicker walls than the atria. Why is this?
b) Why does the left ventricle have a thicker wall than the right ventricle?

Q5
a) List three ways that veins differ from arteries?
b) Substances, such as oxygen and food, enter and leave the blood through the capillary walls. Why not through the walls of arteries and veins?

REMEMBER! Cover the answers if you want to.

ANSWERS

A1 Substances, for example oxygen, will quickly diffuse into the centre of a small organism. For larger organisms, diffusion would take too long so a transport system is needed to ensure quick movement of substances from one part of the organism to another.

A2
a) In the lungs.
b) In respiring tissues around the body.
c) To ensure that their bodies collect enough oxygen.

A3 The pulmonary artery. (Leading to the lungs.)

A4
a) The ventricles pump the blood but the atria simply receive the blood before it enters the ventricles.
b) The left ventricle has to pump blood around the whole body (apart from the lungs). The right ventricle 'only' sends it to the lungs.

A5
a) Veins have valves, thinner walls and a larger lumen.
b) Capillary walls are permeable allowing diffusion. Arteries and veins do not have permeable walls.

TUTORIAL

T1 *Another way of explaining this is to say that smaller organisms have a larger surface area : volume ratio (or a larger surface relative to their size).*

T2
a) *The blood arriving at the lungs contains little oxygen and is there to collect more.*
b) *Every cell will need oxygen. Active cells like those in muscles will need most.*
c) *If the air contains less oxygen then less oxygen will enter the blood unless the body has some way of compensating.*

T3 *Look at the diagram on p.16. All arteries take blood away from the heart.*

T4
a) *The ventricle walls contract and relax to squeeze out blood and take more in. This takes a lot of muscle. The atria, by comparison, do not need to squeeze as hard and must be easily inflated by the incoming low-pressure blood.*
b) *This is why your heart sounds louder on your left side.*

T5
a) *The veins contain blood at a reduced pressure and are built so that they don't resist the flow of blood but rather help it on its way to the lungs.*
b) *The arteries carry blood quickly to each organ or part of the body and the veins bring it back. The capillaries form a branching network inside organs.*

KEY POINTS

Digestion gets food into the body.

Breathing gets oxygen into the body.

Blood gets food and oxygen to every cell.

Respiration releases energy from food.

BREATHING

Breathing is the way that oxygen is taken into our bodies and carbon dioxide removed. Sometimes it is called **ventilation**. Do not confuse this with respiration. Respiration is a chemical process that happens in every cell in the body. Unfortunately, the confusion is not helped when you realise that the parts of the body responsible for breathing are known as the **respiratory system**!

How we breathe

When we breathe, air is moved into and out of the lungs. This involves different parts of the respiratory system inside the **thorax** (chest).

- When we breathe in, air enters though the nose and mouth. In the nose the air is moistened and warmed.
- The air travels down the **trachea** (windpipe) to the lungs. Tiny hairs called **cilia** help to remove dirt and microbes. (You can find out more about the cilia in chapter 3.)
- The air enters the lungs through the **bronchi** (singular: bronchus) which branch and divide to form a network of **bronchioles**.
- At the end of the bronchioles are air sacs called **alveoli** (singular: alveolus) which are covered in tiny blood capillaries. This is where oxygen enters the blood and carbon dioxide leaves the blood.

INHALATION AND EXHALATION

Breathing in and out is known as **inhalation** and **exhalation** (or sometimes inspiration and expiration). Both happen because of changes in the volume of

The respiratory system.

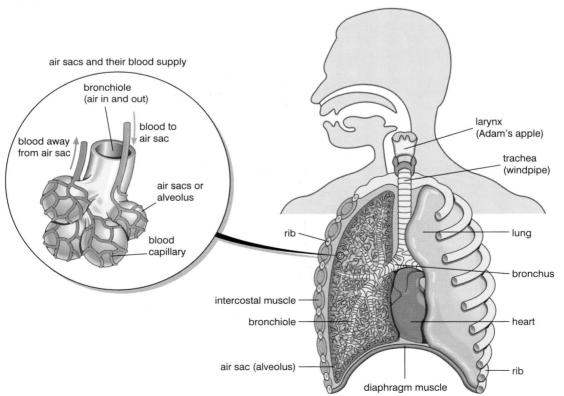

the thorax (chest cavity). This causes pressure changes which in turn cause air to enter or leave the lungs. The changes in thorax volume are caused by:

- the **diaphragm** which is a domed sheet of muscle under the lungs
- the **intercostal muscles** which connect the ribs. There are two sets: the internal intercostal muscles and the external intercostal muscles.

Inhalation

1. The diaphragm contracts and flattens in shape.
2. The external intercostal muscles contract making the ribs move upwards and outwards.
3. These changes cause the volume of the thorax to increase.
4. This causes the air pressure inside the thorax to decrease.
5. This causes air to enter the lungs.

Rings of **cartilage** in the trachea, bronchi and bronchioles keep the air passages open and prevent them from collapsing when the air pressure decreases.

Exhalation

1. The diaphragm relaxes and returns to its domed shape, pushed up by the liver and stomach. This means it pushes up on the lungs.
2. The external intercostal muscles relax allowing the ribs to drop back down. This also presses on the lungs. If you are breathing hard the internal intercostal muscles also contract helping the ribs to move down.
3. These changes cause the volume of the thorax to decrease.
4. This causes the air pressure inside the thorax to increase.
5. This causes air to leave the lungs.

Asthma sufferers use inhalers to calm their symptoms. The inhaler sprays a fine mist into the lungs. The mist contains a drug that opens the bronchioles and makes it easier to breathe.

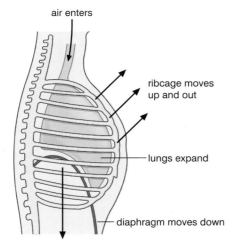

air enters

ribcage moves up and out

lungs expand

diaphragm moves down

Inhalation

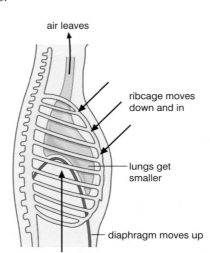

air leaves

ribcage moves down and in

lungs get smaller

diaphragm moves up

Exhalation

Composition of inhaled and exhaled air

The air we breathe contains many gases. Oxygen is taken into the blood. Carbon dioxide and water vapour are added to the air we breathe out. The other gases in the air are breathed out almost unchanged except for being a little warmer.

	In inhaled air	In exhaled air
Oxygen	21%	16%
Carbon dioxide	0.03%	4%
Nitrogen and other gases	79%	79%
Water	variable	high
Temperature	variable	high

ALVEOLI

The alveoli are where oxygen and carbon dioxide diffuse into and out of the blood. For this reason the alveoli are described as the site of **gaseous exchange** or the **respiratory surface**. They are adapted (have special features) to make them efficient at gaseous exchange:

- Thin, permeable walls to allow a short pathway for diffusion.
- Moist lining in which oxygen dissolves first before it diffuses through.
- Large surface area. Lots of alveoli means a very large surface area.
- Good supply of oxygen and good blood supply. This means that a **concentration gradient** is maintained ensuring that oxygen and carbon dioxide rapidly diffuse across.

A concentration gradient is a difference in concentration of substance from one place to another. A sugar lump sitting in a cup of water is an example of a high concentration gradient – sugar is highly concentrated in the lump and lowly concentrated in the water. Such a high concentration gradient encourages diffusion of the particles, the sugar starts to dissolve which lowers the concentration gradient. (See chapter 1 for more on diffusion and concentration gradients.)

Gaseous exchange in an air filled alveolus.

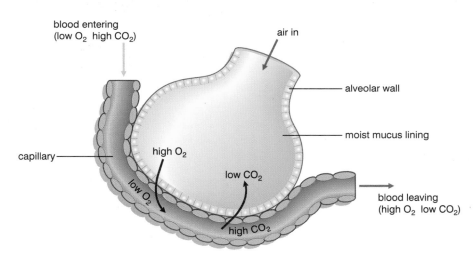

blood entering (low O_2 high CO_2)

air in

alveolar wall

capillary

high O_2

moist mucus lining

low O_2

low CO_2

high CO_2

blood leaving (high O_2 low CO_2)

Check yourself

QUESTIONS

Q1 What is the difference between respiration and breathing?

Q2 Put these in the correct order to show the route that air takes when it enters the body: alveoli, trachea, bronchioles, blood, bronchi.

Q3
a) How many cells does oxygen pass through on its way from the alveoli to the red blood cells?
b) Why is it important for oxygen to have a large concentration gradient between the inside of the alveoli and the blood?

Q4 Inhaled air contains the same amount of nitrogen as exhaled air. Why doesn't it change?

Q5 When you breathe in how do the positions of the diaphragm and ribcage change?

Q6 Why do we need rings of cartilage in the walls of the air passages?

REMEMBER! Cover the answers if you want to.

ANSWERS

A1 Respiration is the release of energy from food which happens in every cell of the body. Breathing is the movement of air into and out of the lungs.

A2 Trachea, bronchi, bronchioles, alveoli, blood

A3
a) Just two, the wall of the alveolus and the wall of the blood capillary.
b) To ensure that oxygen rapidly diffuses from the area of high concentration, in the alveoli, to the area of low concentration, in the blood.

A4 Nitrogen is not used or needed by the body so it is simply breathed out unchanged.

A5 The diaphragm lowers and the ribcage moves upwards and outwards.

A6 To keep them open to ensure that air can move freely.

TUTORIAL

T1 *Aerobic respiration uses up the oxygen taken in by breathing and produces the carbon dioxide which is removed by breathing. But they are not the same thing.*

T2 *Don't forget the whole point of breathing is to get oxygen into the blood so it can then move around the body.*

T3
a) *This short distance is important so gases can easily move between the alveoli and the blood. Another way of looking at it is that the short distance increases the concentration gradient.*
b) *Don't forget that there is also a concentration gradient for carbon dioxide.*

T4 *Don't forget that some oxygen is also breathed out unused because there has not been time for more of it to be absorbed into the blood.*

T5 *Both of these changes increase the volume of the thorax.*

T6 *This is particularly important because when the air pressure drops to take in more air from outside this could otherwise make the air passages close in on themselves.*

FOOD AND DIGESTION

KEY POINTS

Digestion gets food into the body.

Breathing gets oxygen into the body.

Blood gets food and oxygen to every cell.

Respiration releases energy from food.

The food we eat provides us with energy and the raw materials needed for our bodies to grow. The table shows the main types of food in our diet.

Food type	Why we need it	Examples of sources
Carbohydrate. For example sugar and starch.	For energy (released in respiration).	Bread, pasta, biscuits, rice, potatoes, sugar.
Protein	For growth (making new cells) and repairing damaged tissues. (Cell membranes and cytoplasm contain a lot of protein.) Making enzymes and antibodies. Can be used for energy in cases of starvation.	Meat, fish, eggs, milk, beans, peas.
Fat	For energy (although carbohydrates are used in preference). To be stored as a reserve of energy, which is why we put on weight if we eat fatty foods. For insulation.	Red meat, full cream milk, margarine and butter.
Minerals	Iron is needed to make new red blood cells. Calcium is needed for healthy bones and teeth. Minerals are only needed in small amounts. There are many minerals and each has a particular job.	Iron is in red meat, green vegetables. Calcium is in milk and dairy products like cheese.
Vitamins	Lack of vitamins can cause deficiency diseases. Lack of vitamin C can cause scurvy which affects the joints and gums. Lack of vitamin D in children can cause rickets where the bones don't grow properly. Vitamins are only needed in small amounts.	Vitamin C is in fruit and vegetables. Vitamin D is in eggs and dairy products. It can also be made by the skin when exposed to sunlight.
Fibre (roughage)	Helps food move along the digestive system. It is not absorbed by the body and passes out as waste.	Bran, vegetables.
Water	Our bodies are mainly water. Most chemical changes and processes take place in solution.	Almost everything we eat and drink contains water.

It is important that the food we eat contains the right amount of each type of food, in other words a **balanced diet**. You can test foods to see which food types they contain.

Food type	Test	Result
Simple (small molecule) sugars like glucose.	Add Benedict's solution (blue) and heat.	An orange-red colour shows sugar is present.
Starch	Add iodine (yellow-brown).	A blue-black colour shows starch is present.
Protein	Add potassium hydroxide solution followed by a few drops of copper sulphate solution (blue). This is called the Biuret test.	A purple colour shows protein is present.
Fat	Shake the food in some ethanol. Add the ethanol to some water.	A milky colour shows fat is present.
	Rub the food on some paper.	A translucent mark (even when dry) shows the food contains fat.

DIGESTION

The food we eat passes through the digestive system but if it is to be of any use it must enter the blood so that it can travel to every part of the body. Many of the foods we eat are made up of *large* molecules which would not easily enter the blood. This means they have to be broken down into *small* molecules which can easily enter and be carried dissolved in the blood. Breaking down the molecules is called **digestion**. There are two stages:

- **Mechanical digestion** occurs mainly in the mouth where food is broken down into smaller pieces by the teeth and tongue.
- **Chemical digestion** is the breakdown of large food molecules into smaller ones.

Some molecules, like glucose, vitamins, minerals and water are already small enough to pass through the gut wall and do not need to be digested.

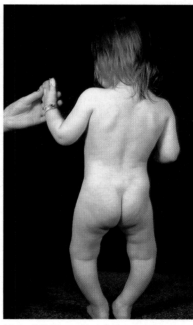

This child is suffering from rickets. The bones have not hardened and are malformed because of a deficiency of vitamin D. Vitamin D enables us to extract calcium from the food we eat.

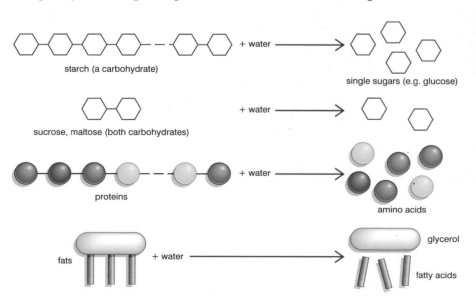

Breaking down food for absorption.

25

Chemical digestion happens because of chemicals called **enzymes**. Enzymes are a type of **catalyst** found in living things. (You can find out more about catalysts in chapter 11.)

Enzymes:

- are proteins
- are produced by cells
- change chemical substances into new products
- are 'specific' which means that each enzyme only works on one substance
- work best at a particular temperature (around 30–40 °C for digestive enzymes) called their 'optimum temperature'
- work best at a particular pH called their 'optimum pH'.

Enzymes work best at an optimum temperature and pH.

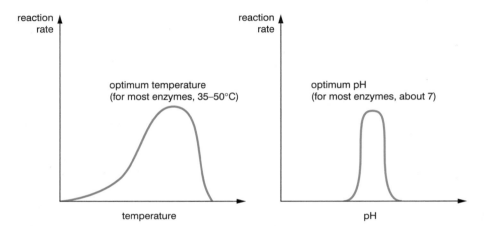

At temperatures that are too high the structure of an enzyme will be changed so that it will not work. This is irreversible and the enzyme is said to be **denatured**. Extremes of pH can also have the same effect.

Every cell contains many enzymes which control the many chemical reactions that happen inside it. Digestive enzymes are only one kind. They are produced in the cells lining parts of the digestive system and are secreted to mix with the food. There are different groups of digestive enzymes such as:

- proteases which break down proteins
- lipases which break down fats
- amylase which breaks down starch
- maltase, sucrase and lactase which break down the different sugars, maltose, sucrose and lactose.

There are some other substances produced in the digestive system that also help digestion:

- Hydrochloric acid is secreted in the stomach. This is important to kill bacteria in food but the enzymes in the stomach also work best at a low pH.
- Sodium hydrogencarbonate is secreted from the pancreas to neutralise the acid to enable the enzymes in the small intestine to work.
- **Bile** is produced in the liver, stored in the gall bladder, and passes along the bile duct into the duodenum. Bile **emulsifies** fats. This means it breaks down large droplets into smaller ones which means that a larger surface area is exposed for the enzymes to work on.

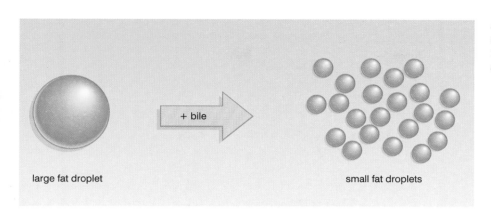

large fat droplet   small fat droplets

Bile lowers the surface tension of large droplets of fat so that they break up. This part of the digestive process is called emulsification.

THE DIGESTIVE SYSTEM

Eating food involves several different processes:

- **ingestion** or taking food into the body
- **digestion** of food into small molecules
- **absorption** of digested food into the blood
- **egestion** or removal of indigestible material (**faeces**) from the body.

Note that egestion is *not* the same as excretion which is the removal of waste substances that have been made in the body. All these different processes take place in different parts of the digestive system or **alimentary canal**.

Part of digestive system	What happens there
Mouth	Teeth and tongue break down food into smaller pieces. Saliva from salivary glands moistens food so it is easily swallowed and contains amylase to begin break down of starch.
Oesophagus or gullet	Each lump of swallowed food, called a **bolus**, is moved along by waves of muscle contraction called **peristalsis**.
Stomach	Food enters through a ring of muscle known as a **sphincter**. Acid and protease are secreted to start protein digestion. Movements of the muscular wall churn up food into a liquid known as **chyme** (pronounced 'kime'). The bulk of the food is stored while a little of the partly digested food at a time passes through another sphincter into the duodenum.
Gall bladder	Stores bile. The bile is passed along the bile duct into the duodenum.
Pancreas	Secretes amylase, lipase and protease as well as sodium hydrogencarbonate into the duodenum.
Small intestine (made up of duodenum and ileum)	Secretions from the gall bladder and pancreas as well as sucrase, maltase, lactase, protease and lipase from the wall of the duodenum complete digestion. Digested food is absorbed into the blood through the **villi**.
Large intestine or colon	Water is absorbed from the remaining material.
Rectum	The remaining material (**faeces**), made up of indigestible food, dead cells from the lining of the alimentary canal and bacteria, is compacted and stored.
Anus	Faeces is egested through a sphincter.

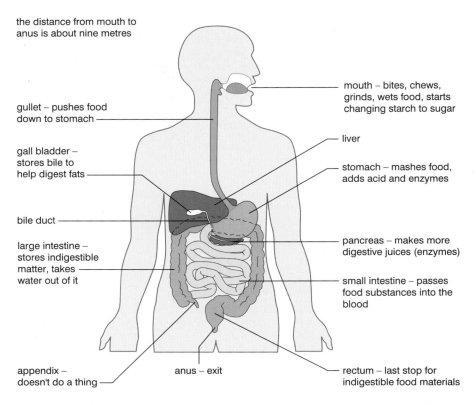

the distance from mouth to anus is about nine metres

mouth – bites, chews, grinds, wets food, starts changing starch to sugar

gullet – pushes food down to stomach

liver

gall bladder – stores bile to help digest fats

stomach – mashes food, adds acid and enzymes

bile duct

pancreas – makes more digestive juices (enzymes)

large intestine – stores indigestible matter, takes water out of it

small intestine – passes food substances into the blood

appendix – doesn't do a thing

anus – exit

rectum – last stop for indigestible food materials

Food moves along the digestive system because of the contractions of the muscles in the walls of the alimentary canal. This is called **peristalsis**.

Peristalsis moves food along the digestive system.

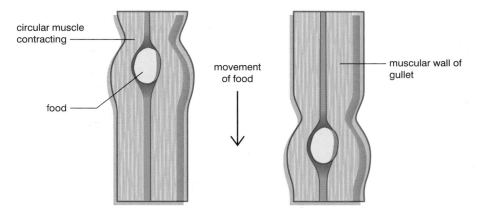

circular muscle contracting

movement of food

muscular wall of gullet

food

ABSORPTION OF FOOD

After food has been digested it can enter the blood. This happens in the main part of the small intestine known as the **ileum**. To help this process, the lining of the ileum is covered in millions of small finger-like projections called villi (singular: villus).

The ileum is adapted for efficient absorption of food by having a large surface area. This is because:

- the ileum is long (6–7 metres in an adult)
- the inside is covered with villi
- the villi are covered in their own **microvilli**.

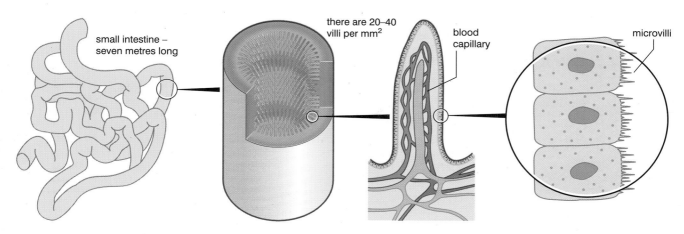

small intestine –
seven metres long

there are 20–40
villi per mm^2

blood
capillary

microvilli

The structure of the ileum.

The villi themselves have other features that also help absorption:

● thin, permeable walls

● good blood supply which maintains a concentration gradient that aids diffusion

● they contain **lymph vessels** (or lacteals) which absorb some of the fat. The lymph vessels eventually drain into the blood system.

Check yourself

QUESTIONS

Q1
a) Which foods give us energy?
b) Which type of food would a body builder need to grow large muscles?
c) Why do you lose weight if you go on a 'calorie controlled' diet?

Q2
a) Protein molecules are long chains of amino acid molecules joined together. Why do proteins need to be digested?
b) What is the difference between mechanical and chemical digestion?
c) What happens to enzymes at high temperatures?
d) Why is the alkaline sodium hydrogencarbonate secreted into the duodenum?

Q3
a) What is the difference between ingestion, egestion and excretion?
b) Where do ingestion and egestion happen?

Q4 Why is it important that the stomach is large enough to store food?

Q5 Where are sphincter muscles found in the digestive system?

Q6 The villi are adapted to absorb food and the alveoli in the lungs are adapted to absorb oxygen. In what ways are they similar?

ANSWERS

A1
a) Carbohydrates, fats and proteins.
b) Protein.
c) The body has to use up reserves of fat in the absence of enough carbohydrates.

A2
a) Proteins are too big to pass through the wall of the ileum. So they have to be broken down into the smaller amino acids.
b) Mechanical digestion is simply breaking food into smaller pieces. This is a *physical* change. Chemical digestion breaks down large molecules into smaller ones. This is a *chemical* change because new substances are formed.
c) They are denatured.
d) To neutralise the acid from the stomach because the enzymes in the small intestine can not work at an acidic pH.

A3
a) Ingestion is taking in food. Egestion is passing out the undigested remains. Excretion is getting rid of things we have made.
b) Ingestion: mouth. Egestion: anus

A4
The intestines are fairly narrow so it is difficult for food to pass along quickly. The stomach allows you to eat a reasonably large meal and store it until there is room for the food to carry on.

A5
At the entrance and exit of the stomach and at the anus. There is also one where the ileum joins the colon.

A6
They provide a large surface area. They have thin permeable walls. They have a good blood supply causing a concentration gradient. They are both moist.

TUTORIAL

T1
a) *Fats actually provide more energy than carbohydrates but fat is more difficult to digest. Most of our energy should come from carbohydrates. If there are not enough carbohydrates available, the body has to use fat reserves. Protein is only used as a last resort.*
b) *New cells need protein: membranes and cytoplasm contain a lot of protein.*
c) *On the other hand if we eat more carbohydrates than we need they are stored as fat.*

T2
a) *Only molecules that are already small do not need to be digested.*
b) *Many candidates if asked to describe digestion lose marks because they do not describe the break down of **molecules**.*
c) *This means they change their shape irreversibly and can not work. In an examination do not say that they are 'killed' because they were never alive.*
d) *Many candidates think that stomach acid itself breaks down food. It is the fact that it allows enzymes in the stomach to work that is important.*

T3
a) *Many candidates lose marks because they confuse egestion and excretion. Passing out faeces is egestion. Getting rid of urea in urine or breathing out carbon dioxide are examples of excretion because these substances are made in the body.*
b) *Excretion occurs in various places. You will find out more in chapter 3.*

T4
This would be particularly important in animals like lions that eat a large meal every few days.

T5
Their job is to control movement of material through the digestive system.

T6
The similarities are not a coincidence. These factors are vital to ensure the efficient absorption of the different substances.

EXAMINATION CHECKLIST

The facts that you should understand by studying this topic.

To be successful at Foundation GCSE Tier you should be able to:

- recall the approximate percentages of oxygen and carbon dioxide in inhaled and exhaled air
- write down the word equation for aerobic respiration:

  glucose + oxygen → carbon dioxide + water + energy
- explain that when we exercise we need more energy so breathing and pulse rate increase to supply the oxygen that is needed
- recall that during hard exercise the oxygen supply is insufficient to meet energy demands, and anaerobic respiration then takes place
- write down the word equation for anaerobic respiration:

  glucose → lactic acid + energy
- recall that anaerobic respiration releases much less energy than aerobic
- explain how the parts of the circulatory system work together to bring about the transport of substances around the body
- recall that the heart is a muscular pump of 4 chambers and 4 valves and be able to name and locate these parts
- recall the function of the constituents of blood
- recall where O_2, CO_2 and food, enter and leave the bloodstreams
- name and locate the main parts of the breathing system
- explain how the parts of the respiratory system work together to bring about gaseous exchange
- describe the changes in position of the ribs and diaphragm that cause inhalation and exhalation
- recall the functions of the main parts of the digestive system and explain how they work together to bring about digestion
- recall that small molecules are absorbed into the blood in the small intestine by diffusion
- recall that the digestive enzymes break down large food molecules into smaller ones, and that stomach acid aids enzyme function.

In addition, to be successful at Higher GCSE Tier you should be able to:

- write down the symbol equation for aerobic respiration:

  $C_6H_{12}O_6 + 6O_2 → 6CO_2 + 6H_2O$ (+ energy)
- explain fatigue in terms of lactic acid build up and how this is removed during recovery (oxygen debt)
- recall that haemoglobin in red blood cells reacts with oxygen in the lungs forming oxyhaemoglobin and the reverse of this reaction in the tissues
- explain how blood vessels are adapted to their functions
- explain the advantage of the double circulatory system
- explain how inhalation and exhalation are brought about
- explain the function of cartilage in bronchioles and bronchi
- explain how the alveoli are adapted for efficient gaseous exchange
- recall how enzyme activity is affected by pH and temperature
- recall that bile improves fat digestion by emulsification
- explain how the small intestine is adapted for absorption.

Key Words

Tick each word when you are sure of its meaning

anaerobic

absorption

bronchus

denature

diaphragm

digestion

double circulation

egestion

emulsify

enzyme

gaseous exchange

haemoglobin

intercostal muscle

lactic acid

oxygen debt

peristalsis

protein

respiration

respiratory surface

ventilation

ventricles

EXAM PRACTICE

Sample Student's Answers & Examiner's Comments

a) Explaining that 'some of the oxygen was used up' goes some way towards gaining one of the marks but not far enough. The **reason** why oxygen is used up is needed so some reference to **respiration** would be needed. No mark has been gained for simply stating that carbon dioxide was breathed out. This is only repeating information given in the question and therefore would not gain credit. To gain the other mark you would need to explain that carbon dioxide is being produced as a waste product of respiration.
The fact that the candidate has been asked to 'explain', that there is more than one mark and that four lines space has been given are all signs that a fairly detailed answer is needed.

b) This is a more straightforward question and the answer is acceptable and gains the mark.

c) This is the correct answer. You may know that yeast can respire anaerobically producing carbon dioxide and alcohol however the question refers to a 'muscle cell'.

d) What the candidate has written may be true (for sprinters) but is not enough to gain any marks as it does not really answer the question. This is a very common mistake that candidates make. Again, the number of marks available (2) and the space provided all point to a more detailed answer. The question is clearly linked to the one before because they both refer to an athlete so look to link the answer to anaerobic respiration. The explanation that would gain full marks is that an athlete wants to release as much energy as possible and there is a limit to how fast aerobic respiration can occur (governed by breathing and heart rates which determine how fast oxygen can get to the muscles). Anaerobic respiration must be used which requires that an oxygen debt is built up.

● This candidate only gained 2 marks. A grade C candidate should have gained at least 3 marks. A grade A candidate should have gained full marks.

---

1 A scientist took some measurements of air breathed in and out by an athlete. The table shows the percentage of gases in the air samples.

gas	air breathed in / %	air breathed out / %
oxygen	20	16
carbon dioxide	0.04	4
nitrogen	79	79

a) Explain the reason for the difference in the amounts of oxygen and carbon dioxide in the two samples.

some of the oxygen was used up and carbon dioxide was

breathed out.

(2)

b) Why does the percentrage of nitrogen remain constant in the two samples?

It wasn't used up. ✓

(1)

When the athlete exercises vigorously her muscles build up an 'oxygen debt'.

c) Complete the word equation for anaerobic respiration in a muscle cell.

glucose → lactic acid + energy ✓

(1)

d) Explain why the ability to build up an oxygen debt is an advantage to an athlete.

so they don't have to breathe during a race, they can hold

their breath until they finish.

2/6

(2)

Midland Examining Group

Questions to Answer

Answers to questions 2 – 6 can be found on pages 357–358.

2 The diagrams below show a section through the left side of a human heart during two stages of a heart beat.

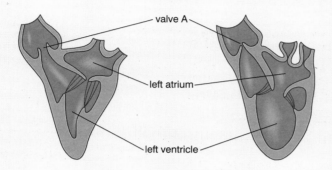

a) Place an arrow on EACH diagram to show the direction of blood flow in the heart during the stages shown. (2)

b) What is the function of valve **A**, labelled on the diagram? (2)

c) Name the vessel bringing blood to the left atrium. (1)

d) Explain why the wall of the left ventricle is more muscular than the wall of the left atrium. (1)

London Examinations

3 This question is about the blood. The blood carries substances to and from the cells of the body.

a) Write down **two** substances carried by the **plasma**. (2)

b) Oxygen is carried in the blood by red blood cells.
Describe how red blood cells carry oxygen. (2)

Midland Examining Group

4 Look at the diagram of the lungs.

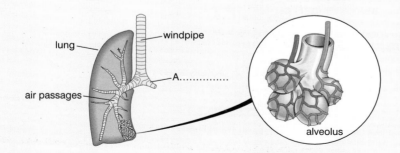

In some illnesses the alveoli become partly filled with fluid.

a) How would this affect someone's breathing rate?

Choose from: **make it slower, make it faster, not affect it at all.** (1)

b) Explain your answer to **a**. (1)

c) Label part A on the diagram. (1)

Midland Examining Group

5 This question is about the digestive system.

Look at this diagram of three types of food molecules found in the small intestine.

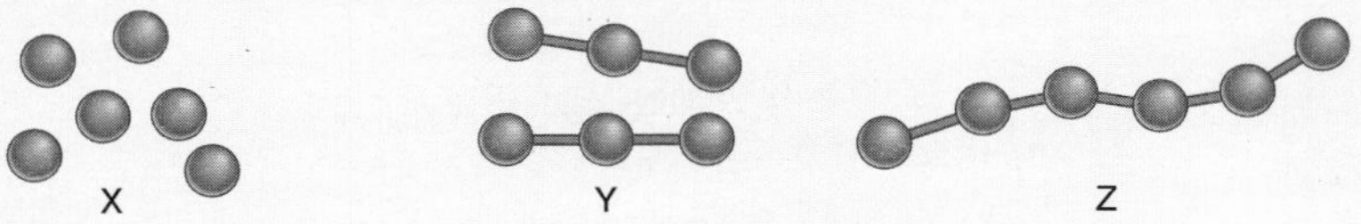

a) Which type of food molecule will get into the blood most easily? (1)

b) How do food molecules get into the blood? Write down the name of the process. (1)

c) Why do food molecules need to get into the blood? (1)

d) Chemical substances produced in the intestine digest food.

(i) Write down the name of these chemical substances. (1)

(ii) Explain how they digest food. (1)

Midland Examining Group

6 The diagram shows a piece of food moving through the intestine.

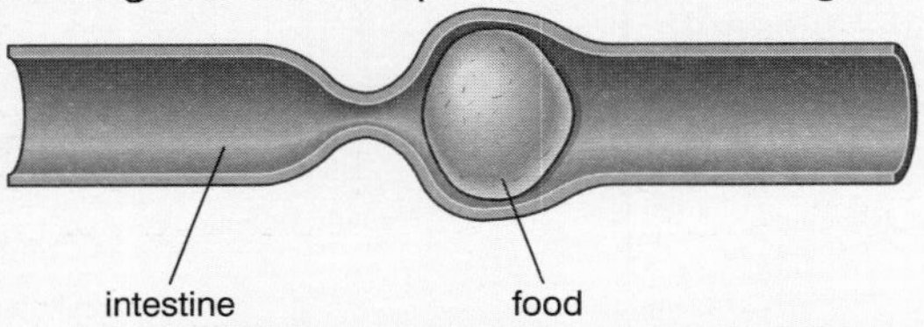

intestine   food

a) Describe how the food is moved and how fibre in the diet may make this process more effective. (3)

b) (i) Bread contains: **starch sugar protein fat calcium**

Which of these food types stains blue-black when iodine solution is added to it? (1)

(ii) Describe how you would test the bread for protein and state the result you would expect if protein was present. (2)

c) Explain why a sample of human saliva could change a 1% starch solution into maltose, but a sample of digestive juice from the stomach would not have this effect. (2)

d) 'Undigested food is egested and not excreted.'

Explain what is meant by this statement. (2)

Midland Examining Group

Under ordinary circumstances, the human body is always trying to work as well as possible. Conditions inside the body, for example body **temperature** and **water balance**, have to be monitored and controlled. **Disease** and **drugs** can interfere with the body's monitoring and control mechanisms as well as having other effects.

The two body systems that regulate how the body works are the **nervous system** and the **hormonal** (or **endocrine**) **system**. These also monitor the surroundings for anything else that could affect the body.

THE NERVOUS SYSTEM

The nervous system collects information about changes inside or outside the body, decides how the body should respond and controls that response.

RECEPTORS

Information is collected by **receptor cells** that are usually grouped together in **sense organs**, also known as **receptors**. Each type of receptor is sensitive to a different kind of change or **stimulus**.

Sense organ	Sense	Stimulus
Skin	Touch	Pressure, pain, hot/cold temperatures
Tongue	Taste	Chemicals in food and drink
Nose	Smell	Chemicals in the air
Eyes	Sight	Light
Ears	Hearing Balance	Sound Movement/position of head

THE EYE

The eye is the receptor that detects light.

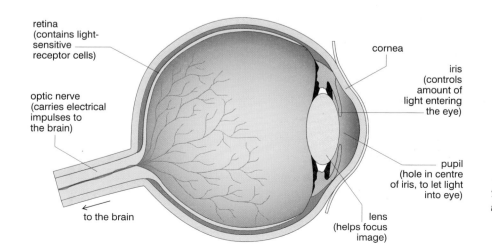

retina (contains light-sensitive receptor cells)

optic nerve (carries electrical impulses to the brain)

to the brain

cornea

iris (controls amount of light entering the eye)

pupil (hole in centre of iris, to let light into eye)

lens (helps focus image)

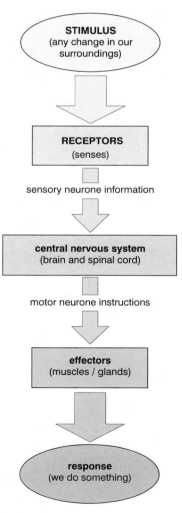

STIMULUS (any change in our surroundings)

RECEPTORS (senses)

sensory neurone information

central nervous system (brain and spinal cord)

motor neurone instructions

effectors (muscles / glands)

response (we do something)

The flow of information from stimulus to response.

Part of eye	Job
Cornea	Transparent cover that does most of the bending of light
Iris	Alters the size of the pupil to control the amount of light entering the eye
Lens	Changes shape to focus light on to the retina
Humour	Clear jelly that fills the inside of the eye
Retina	Contains light-sensitive receptor cells which change the light image into electrical impulses. There are two types of receptor cells: **rods** which are sensitive in dim light but can only sense 'black and white', **cones** which are sensitive in bright light and can detect colour
Optic nerve	Carries electrical impulses to the brain

The iris (the ring shaped, coloured part of the eye) controls the amount of light entering the eye by controlling the size of the hole in the centre, the pupil. The iris contains **circular** and **radial** muscles. In bright light the circular muscles contract (and the radial muscles relax) making the pupil smaller. This reduces how much light enters the eye as too much could do damage. The reverse happens in dim light when the eye has to collect as much light as possible to see clearly.

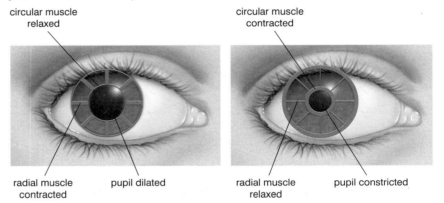

circular muscle relaxed

circular muscle contracted

radial muscle contracted

pupil dilated

radial muscle relaxed

pupil constricted

The lens changes shape in order to focus light from objects which are distant or near. You can find out how glass lenses form images in chapter 17.

NERVES

The sense organs are connected to the rest of the nervous system which is made up of the **brain**, **spinal cord** and **peripheral nerves**. It is in the brain and spinal cord that information is processed and decisions made. The brain and spinal cord together are called the **central nervous system** (CNS). Signals are sent through the nervous system in the form of electro-chemical impulses.

There are three types of nerve cells or **neurones**:

- **sensory neurones** which carry signals to the CNS
- **motor neurones** which carry signals from the CNS controlling how we respond
- **relay** (intermediate or connecting) **neurones** which connect other neurones together.

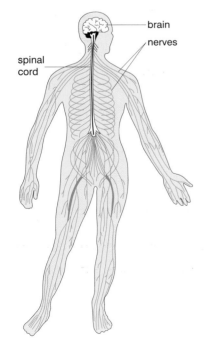

brain

nerves

spinal cord

The nervous system.

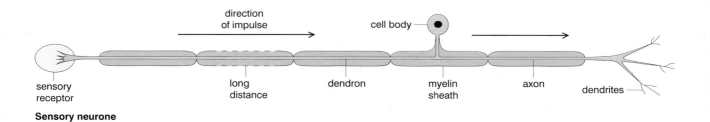

sensory receptor

long distance

dendron

cell body

myelin sheath

axon

dendrites

direction of impulse

Sensory neurone

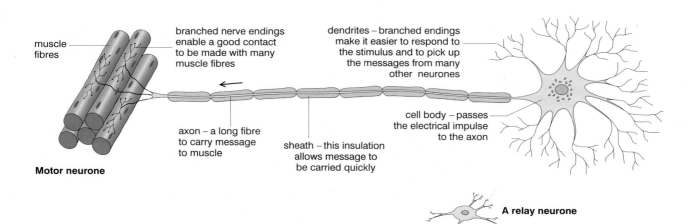

muscle fibres

branched nerve endings enable a good contact to be made with many muscle fibres

dendrites – branched endings make it easier to respond to the stimulus and to pick up the messages from many other neurones

cell body – passes the electrical impulse to the axon

axon – a long fibre to carry message to muscle

sheath – this insulation allows message to be carried quickly

Motor neurone

A relay neurone

Neurones are specialised in several ways:

- They can be very long to carry signals from one part of the body to another.
- They have many branched nerve endings or **dendrites** to collect and pass on signals.
- Many neurones are wrapped in a layer of fat and protein, the **myelin sheath**, which insulates cells from each other and allows the impulses to travel faster.

Neurones are usually grouped together in bundles called **nerves**.

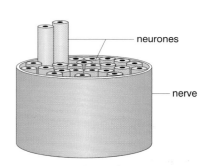

neurones

nerve

Cross section of a nerve.

REFLEXES

The different parts of the nervous system may all be involved when we respond to a stimulus. The simplest type of response is a **reflex**. Reflexes are rapid, automatic responses which often act to protect us in some way. For example, blinking if something gets in your eye or sneezing if you breathe in dust. The pathway that signals travel along during a reflex is called a **reflex arc**:

stimulus → receptor → sensory neurone → CNS → motor neurone → effector → response

For example:

stand on pin → nerve endings → sensory neurone → spinal cord → motor neurone → leg muscles → leg moves

smell of food → nose → sensory neurone → brain → motor neurone → salivary glands → mouth waters

Simple reflexes are usually **spinal reflexes** which means that the signals are processed by the spinal cord, not the brain. The spine sends a signal back to the **effector**. Effectors are the parts of the body that respond – either muscles or glands. Examples of spinal reflexes include standing on a pin or touching a hot object. When the spine sends a signal to the effector, other signals are sent on to the brain so that it is aware of what is happening. There are also reflexes in which the signals are sent straight to the brain, these are called **cranial reflexes**. Examples include blinking when dirt lands in your eye, or salivating at the smell of food.

An example of a spinal reflex.

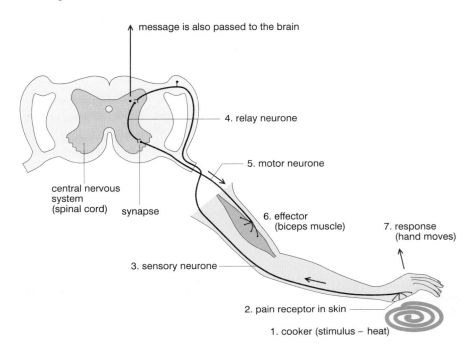

message is also passed to the brain

4. relay neurone

5. motor neurone

central nervous system (spinal cord)

synapse

6. effector (biceps muscle)

7. response (hand moves)

3. sensory neurone

2. pain receptor in skin

1. cooker (stimulus – heat)

SYNAPSES

Signals are passed from one neurone to another even though they do not touch. Between nerve endings there are very small gaps called **synapses**. Signals can be passed across synapses because when an electrical impulse reaches a synapse, chemicals called **transmitter substances** are released from the membrane of one nerve ending and travel across to special **receptor sites** on the membrane of the next nerve ending, triggering off another nerve impulse. Signals are not always passed across synapses, which stops us from responding to every single stimulus. You can find out more about synapses in the section about drugs on pages 51–52.

A synapse.

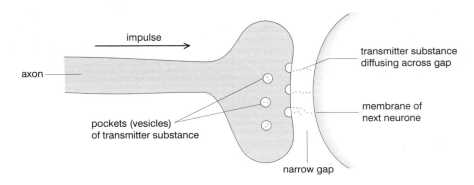

impulse

axon

transmitter substance diffusing across gap

membrane of next neurone

pockets (vesicles) of transmitter substance

narrow gap

THE BRAIN

Human behaviour is more complicated than a simple series of reflexes because we are able to think. The brain is where conscious behaviour as well as much automatic behaviour is controlled.

Part of brain	Job
Cerebrum or cerebral hemispheres	Conscious thought occurs here. Memory, speech, sensations are processed here
Cerebellum	Controls and co-ordinates movement and balance
Medulla	Controls important but automatic actions like breathing and heart rate
Hypothalamus	Controls some basic body functions like water balance and temperature control.

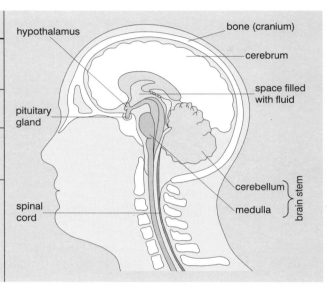

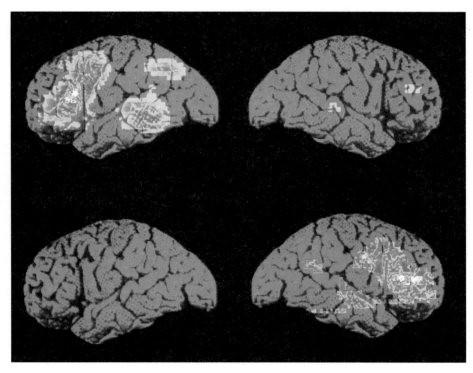

Brain scans of two different people thinking about verbs to go with a list of nouns. The right-handed person (top) shows a lot of activity in the left half of their brain. The left-handed person (bottom) shows the same activity in the right half of their brain.

Check yourself

QUESTIONS

Q1
a) What is a stimulus?
b) What is the difference between a receptor and an effector?

Q2
a) Which two parts of the eye focus light?
b) In dim light we do not see colour very well. Which receptors are mainly being used?
c) Why are **two** sets of muscles needed in the iris?

Q3
a) What are the two parts of the CNS?
b) What is the difference between a nerve and a neurone?
c) Describe three features that neurones have in common with other cells.
d) Describe two features that neurones have that make them different from other cells.

Q4
a) Which of the following are reflexes?
A coughing when something 'goes down the wrong way'
B changing channels on the TV when a programme ends
C jumping up when you sit on something sharp
D always buying the same brand of soft drink
b) Why are reflexes important?

Q5
In the brain what is the difference between the cerebrum and the cerebellum?

REMEMBER! Cover the answers if you want to.

ANSWERS

A1
a) A change, either in the surroundings or inside the body, that is detected by the body.
b) Receptors detect (sense) stimuli. The sense organs are receptors. Effectors are the parts of the body that respond. Effectors are usually muscles but can also be glands.

A2
a) The cornea and the lens.
b) Rods.
c) When the radial muscles contract the pupil gets larger and at the same time the circular muscles relax. To then make the pupil smaller it is not enough for the radials to relax, the circular muscles must also contract.

A3
a) Brain and spinal cord.
b) Neurones are nerve cells. A nerve in the body is usually a bundle of neurones wrapped in connective tissue.
c) Membrane, cytoplasm, nucleus.
d) Cell body, nerve fibres (axons or dendrons) nerve endings or dendrites. Any two.

TUTORIAL

T1
a) *For example if you were crossing a road and noticed a bus approaching, the stimuli would have been the sight and sound of the bus.*
b) *Some candidates get confused between effectors and the 'effects' that they produce.*

T2
a) *Many people are surprised that the cornea actually bends light. The lens only makes fine adjustments necessary for accurate focusing.*
b) *Remember Cones are for Colour and rods are not.*
c) *This is an example of antagonistic muscle action where different muscles work against each other to produce movement in different directions.*

T3
a) *Many reflexes do not need the brain in order to function and the spinal cord is therefore not simply a nerve pathway to and from the brain but also processes some information itself.*
b) *In exam questions you may pick up extra marks for using the correct terms.*
c) *Although neurones are specialised they are still cells and so share these features with other cells.*
d) *If you are asked for a certain number of answers in an exam only answer up to that number as any more will not be marked.*

ANSWERS

A4 a) A and C.
b) Since they do not have to be thought about, reflexes happen very quickly. This means if there is some kind of danger you can respond very quickly to reduce or prevent any harm.

A5 The cerebrum controls conscious processes like thought, speech and memory. The cerebellum controls movement and balance.

TUTORIAL

T4 a) *The others do involve some kind of response, but reflexes happen **automatically** without thought.*
b) *It is possible to override reflexes by thinking about it, so you could hold on to something hot even though that is not usually a very wise thing to do.*

T5 *More 'intelligent' animals, such as ourselves, have larger cerebrums.*

THE ENDOCRINE SYSTEM

Hormones are chemical messengers. They are made in **endocrine glands**. Endocrine glands do not have ducts (tubes) to carry away these substance, instead the hormones are **secreted** directly into the blood to be carried around the body. (There are other types of glands, called exocrine glands, such as salivary or sweat glands, that do have ducts.) Most hormones affect several parts of the body, others only affect one part of the body which is termed the **target organ**. Compared with changes brought about by the nervous system, changes caused by hormones are usually slower and longer-lived.

Adrenal glands

Hormone: **adrenaline**. This is released in times of excitement, anger, fright or stress and prepares the body for 'flight or fight' – the crucial moments when an animal must instantly decide whether to attack or run for its life. Adrenaline:

- increases heart rate
- increases depth of breathing and breathing rate
- increases sweating
- makes hair stands on end (this makes an animal look larger but only gives goose bumps in humans)
- releases glucose from liver and muscles
- makes pupils dilate
- makes skin go pale as blood is redirected to muscles.

Pituitary gland

Hormones: **growth hormone** and many others. Growth hormone encourages mental and physical development in children. The pituitary gland also controls many other glands.

Thyroid gland

Hormone: **thyroxine**. Encourages mental and physical development in children.

Pancreas gland

Hormone: **insulin**. (Note that the pancreas also secretes digestive enzymes through the pancreatic duct into the duodenum.) Insulin controls **glucose** levels in the blood. It is important that the blood glucose level remains as steady as possible. If it rises or falls too much you can become very ill.

KEY POINTS

Controlling the body via the endocrine system – slow response, lasting effects.

- **glands produce hormones**
- **hormones signal to the body**
- **medical uses of hormones to control the body**

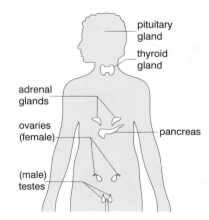

The endocrine system.

pituitary gland

thyroid gland

adrenal glands

ovaries (female)

(male) testes

pancreas

After a meal, the level of glucose in the blood tends to rise. This causes the pancreas to release insulin which travels in the blood to the liver. Here it causes any excess glucose to be converted to another carbohydrate, **glycogen**, which is stored in the liver.

Between meals, glucose in the blood is constantly being used up, so the level of glucose in the blood falls. When a low level of glucose is detected, the pancreas stops secreting insulin and secretes the hormone **glucagon** instead. Glucagon converts some of the stored glycogen back into glucose which is released into the blood raising the blood glucose level back to normal.

Testes (males only)

Hormone: **testosterone** (male sex hormone). Causes secondary sexual characteristics in boys:

- growth spurt
- hair grows on face and body
- penis, testes and scrotum grow and develop
- sperm is produced
- voice breaks
- body becomes broader and more muscular
- sexual 'drive' develops.

Ovaries (females only)

Hormones: **progesterone** and **oestrogen** (female sex hormones). Cause secondary sexual characteristics in girls:

- growth spurt
- breasts develop
- vagina, oviducts and uterus develop
- menstrual cycle (periods) start
- hips widen
- pubic hair and hair under the arms grows
- sexual 'drive' develops.

They also control the changes that occur during the menstrual cycle (see chart on the next page):

- oestrogen encourages the repair of the uterus lining after bleeding
- progesterone maintains the lining
- oestrogen and progesterone control **ovulation** (egg release).

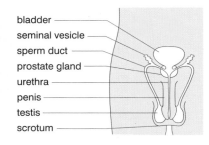

bladder
seminal vesicle
sperm duct
prostate gland
urethra
penis
testis
scrotum

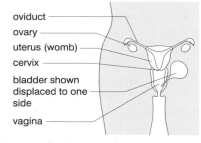

oviduct
ovary
uterus (womb)
cervix
bladder shown displaced to one side
vagina

Male and female reproductive systems

THE MENSTRUAL CYCLE

About every 28 days an egg is released from either one or the other of a woman's ovaries. (The egg develops from one of the thousands of **follicles** present in the ovaries.) The egg travels down the oviduct to the uterus (womb) where, if it has been fertilised, it can implant and grow into a baby. (To prepare for this the lining of the uterus has become thickened – controlled by the release of progesterone from the **corpus luteum**, the remains of the follicle left behind in the ovary). If the egg has not been fertilised then the lining breaks down and is released (menstruation). Oestrogen from the ovary will encourage the uterus lining to grow again for the next released egg. If the egg has been fertilised then progesterone continues to be released from the corpus luteum which maintains the uterus lining during pregnancy and also prevents further ovulation.

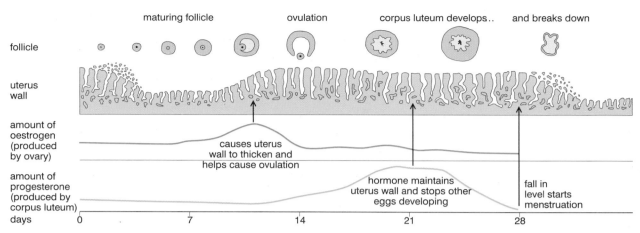

maturing follicle · ovulation · corpus luteum develops.. · and breaks down

follicle

uterus wall

amount of oestrogen (produced by ovary)

causes uterus wall to thicken and helps cause ovulation

amount of progesterone (produced by corpus luteum)

hormone maintains uterus wall and stops other eggs developing

fall in level starts menstruation

days | 0 | 7 | 14 | 21 | 28

The menstrual cycle.

MEDICAL USES

Hormones can be used to treat various medical conditions, for example in those cases caused by the hormone not being made naturally in the correct quantities, as well as having other medical uses.

Insulin

There are many types of **diabetes**, but in 'diabetes mellitus' the body is unable to make enough of the hormone insulin. One of insulin's main functions is to control the level of glucose in the blood. The excess glucose in a diabetic's blood cannot be stored and is excreted in urine instead. Other symptoms of diabetes include: thirst, weakness, weight loss, coma.

Some people with diabetes control their diet and activity. For example: they make sure they do not go a long time without a meal; eat snacks in between meals; make sure that they eat some high-glucose food before any energetic activity.

Insulin injected just before a meal is another way of controlling the problem. Insulin can now be extracted from animals' blood or produced by genetically engineered bacteria.

Progesterone and oestrogen

These are used in **fertility drugs** which encourage women's ovaries to release more eggs. A side effect is that lots of eggs may be released possibly resulting in multiple pregnancies.

Contraceptive pills work by mimicking the levels of these hormones during pregnancy. As eggs are not released during pregnancy a woman taking the pill should not get pregnant.

Growth hormone

Injections of growth hormone can be given during childhood to ensure that growth is normal.

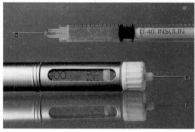

Insulin injection devices. An ordinary hypodermic needle and syringe (above) and a 'pen' device that meters the required dose.

Check yourself

QUESTIONS

Q1 Signals can be sent around the body by the nervous system and the endocrine system. How are they different?

Q2
a) How do the changes that adrenaline causes help the body react to a stressful situation (fight or flight)?
b) After it has been released, adrenaline is broken down fairly quickly. Why is this important?

Q3 Explain the differences between glucose, glycogen and glucagon.

Q4 Why do people with diabetes try to make sure they are not late eating their meals or, that if they are, they eat a snack in between meals?

REMEMBER! Cover the answers if you want to.

ANSWERS

A1 In the nervous system, signals are sent very quickly as electro-chemical impulses along neurones and the effects of the signal usually last a short time. In the endocrine system, signals are sent more slowly as hormones through the blood and the effects usually last longer.

A2
a) Fighting and running away both need energy. This is released by respiration which uses oxygen to release energy from glucose. So breathing increases to take in more oxygen, and glucose is released from stores in the liver and muscles. The blood is redirected and heart rate increased to carry the oxygen and glucose quickly to the muscles that need it. Sweating starts because the body will need cooling. Hairs stand up and pupils dilate to make an animal look larger and more fierce.
b) Otherwise all the effects caused by adrenaline would persist.

A3 Glucose and glycogen are both carbohydrates, but glucose is made of small molecules that dissolve in the blood whereas glycogen is made of larger molecules (lots of glucose molecules joined together) that are insoluble and do not travel through the blood. Glucagon is a hormone that converts glycogen back into glucose (i.e. does the opposite of insulin).

A4 As they lack the hormones to maintain the correct amount of glucose in the blood, they have to do it themselves by making sure there is a steady supply of glucose into the blood from the digestive system.

TUTORIAL

T1 *Both of these system are involved in co-ordination (responding to stimuli). Both are needed because of the different nature of their effects.*

T2
a) *Although an exam question may be apparently about one particular topic, to gain full marks you might have to include ideas from other topics. In this example ideas about respiration are relevant. This is likely to happen in questions that require extended answers.*
b) *Where hormones are involved, constant fine adjustments need to be made, for example in the amounts of insulin required in the blood. If hormones are not being constantly broken down and at the same time secreted this balance could not be maintained.*

T3 *It is very common for candidates to confuse these especially glycogen and glucagon.*

T4 *Many diabetics are able to control their diabetes by diet alone without the need for injections of insulin.*

HOMEOSTASIS

For our cells to stay alive and work properly they need the conditions around them, such as their temperature and the amount of water and other substances, to stay within acceptable limits. How the body keeps inside these limits is called **homeostasis**.

TEMPERATURE CONTROL

Inside your body the temperature is about 37 °C regardless of the weather or how hot or cold you may feel on the outside. This **core temperature** may naturally vary a little, but it is never very different unless you are ill. Heat is constantly being released by respiration and other chemical reactions in the body, and is transferred to the surroundings outside the body. To maintain a constant body temperature these two processes have to balance. If the core temperature rises above or falls below 37 °C various changes happen, mostly in the skin, to restore normal temperature. The core temperature is monitored by the hypothalamus gland – a part of the brain that monitors the temperature of the blood passing through it.

Thermography shows skin and body temperature as different colours – from black (cool) to red, yellow and white (warm).

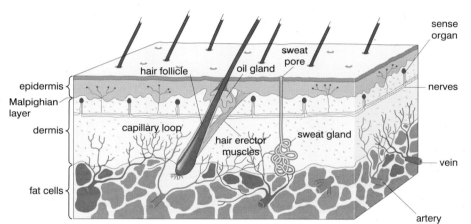

Labels: sense organ, sweat pore, hair follicle, oil gland, epidermis, Malpighian layer, dermis, capillary loop, hair erector muscles, sweat gland, nerves, vein, fat cells, artery

Section through skin.

Too cold?	Too hot?
Vasoconstriction: blood capillaries in the skin become narrower so they carry less blood close to the surface. Heat is kept inside the body.	**Vasodilation**: blood capillaries in the skin widen so they carry more blood close to the surface. Heat is transferred from the blood to the skin by **conduction**, then to the environment by **radiation**.
Sweating is reduced.	Sweating: sweat is released onto the skin surface and as it **evaporates** heat is taken away.
Hair erection: muscles make the hairs stand up trapping a layer of air as insulation (air is a poor conductor of heat). This is more beneficial in animals but still occurs in humans (goose bumps).	Hairs lay flat so less air is trapped and more heat is transferred from the skin.
Shivering: muscle action releases extra heat from the increased respiration.	No shivering.
A layer of fat under the skin also acts as insulation.	

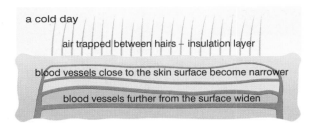

a cold day

air trapped between hairs – insulation layer

blood vessels close to the skin surface become narrower

blood vessels further from the surface widen

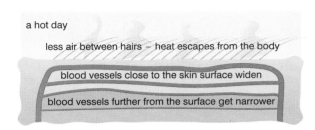

a hot day

less air between hairs – heat escapes from the body

blood vessels close to the skin surface widen

blood vessels further from the surface get narrower

These walruses are blue in the cool water (look at the one at the back). When they are too hot they go pink because vasodilation leads to more blood circulating near the surface of the skin.

WATER BALANCE

Our bodies are about two-thirds water. The average person loses and gains about three litres of water a day under normal conditions. Water is constantly being lost from the body:

- in the air we breathe out
- in sweat
- in **urine**
- in faeces.

Water is gained by the body:

- in food and drink
- from respiration and other chemical reactions.

The water lost cannot be more or less than the water gained. So on a hot day, when the body sweats more, less water is lost as urine producing a smaller amount of more concentrated (usually darker) urine than normal.

The amount of water that is lost as urine is controlled by the **kidneys**. The kidneys' main jobs are:

- to regulate the amounts of water and salt in the body by controlling the amounts in the blood
- to remove waste products such as **urea** from the blood. Urea is formed in the liver from the breakdown of proteins in the body. This is an example of **excretion**.

The water balance for an average person.

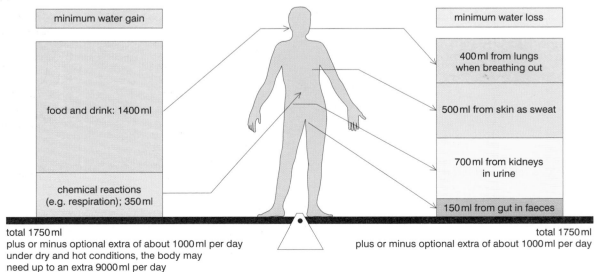

minimum water gain

food and drink: 1400 ml

chemical reactions (e.g. respiration); 350 ml

total 1750 ml
plus or minus optional extra of about 1000 ml per day under dry and hot conditions, the body may need up to an extra 9000 ml per day

minimum water loss

400 ml from lungs when breathing out

500 ml from skin as sweat

700 ml from kidneys in urine

150 ml from gut in faeces

total 1750 ml
plus or minus optional extra of about 1000 ml per day

As blood flows around the body it passes through the kidneys which remove urea, excess water and salt from the blood.

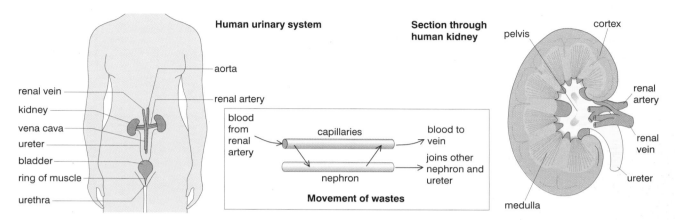

(1) Blood enters the kidneys through the **renal arteries** which then divide to form many tiny capillaries. (2) A lot of the blood plasma (water, urea and other substances) is forced under great pressure into tiny tubules called **nephrons**. (3) Further along each nephron, the more useful contents and some water is reabsorbed into the blood leaving urea, excess water and salt – urine. (4) The capillaries join up to form the **renal veins** which carry blood away from the kidneys. (5) The nephrons join up to form a **ureter** which carries the urine to the **bladder** where it is stored until you go to the toilet when the urine leaves the body through the **urethra**.

Like temperature, the water content of the blood is monitored by the hypothalamus. If there is too little water in the blood, if you have been sweating a lot for example, then the hypothalamus is stimulated and sends a hormone to the kidneys making them reabsorb more water back into the blood so less is lost in urine. If your blood water level is low, if you have been drinking a lot for example, the hypothalamus is much less stimulated so less hormone is released, less water is reabsorbed and more water is lost as urine.

Monitoring water level.

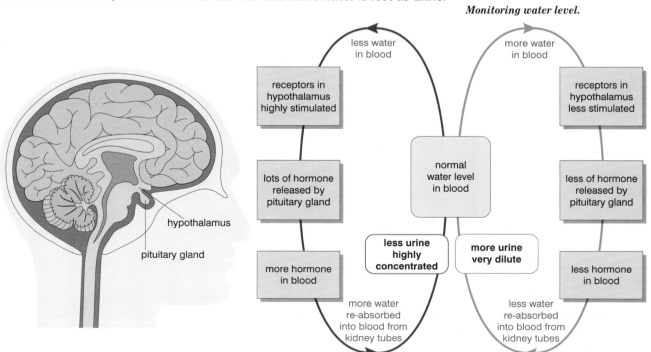

NEGATIVE FEEDBACK

The body maintains a constant internal environment by monitoring changes and reacting to reduce these changes. This process is an example of **negative feedback**. It is 'negative' because the body attempts to *reduce* changes that occur. We have already looked at some examples such as:

- if the amount of glucose in the blood increases after a meal the body releases insulin to reduce the amount
- if the body temperature falls then shivering and vasoconstriction will increase the temperature
- if you drink a lot, increasing the amount of water in the blood, then the kidneys will remove more water in urine.

There are many other cases of negative feedback, for example when you exercise you also breathe faster. Muscles respiring more than usual use up oxygen: faster breathing increases the amount of oxygen in the blood. These two processes counter each other producing a steady oxygen level. The faster breathing also removes the extra carbon dioxide that has been produced by the muscles respiring. In fact it is mainly the carbon dioxide levels in the blood that the brain monitors to control the rate of breathing. This is because high levels of carbon dioxide in the blood are toxic and have to be reduced quickly.

Check yourself

QUESTIONS

Q1 If you took the temperature under your tongue standing outside on a hot summer's day and again on a cold winter's day, which would be higher?

Q2 Why might your skin look red after you have run a race?

Q3 Why do you feel cold when you first get out of a bath but feel much warmer when you are dry?

Q4 Why do birds look more fluffy on a cold day in winter?

Q5 A farmer who has been working outside all day on a hot summer's day produces a much smaller amount of urine then normal. It is also a much darker yellow colour than normal. Explain these points.

Q6 Why are the lungs organs of excretion?

Q7 How would the blood in the renal arteries be different to the blood in the renal veins?

Q8 What is the difference between a ureter and a urethra?

Q9 What is the difference between homeostasis and negative feedback?

ANSWERS

A1 Neither, they would both be the same, about 37 °C.

A2 Your body has to remove the extra heat released from respiration in muscles and so vasodilation occurs causing more blood to flow near the skin surface so you look redder.

A3 When water evaporates it requires energy in the form of heat to change from a liquid to a gas. This heat comes from the skin so it cools down. When the skin is dry, evaporation stops and so there is no more cooling.

A4 Just as animals fluff up hair or fur, birds fluff up their feathers to trap a layer of insulating air.

A5 The farmer will have been sweating to stay cool and so will have lost water this way. Unless he/she has drunk a lot, his/her kidneys will have to reduce how much water is lost as urine. This is why there is less. However there will still be just as much urea so the urine is more concentrated and darker in colour.

A6 They remove carbon dioxide and water from the body.

A7 The veins would have a lower concentration of salt and urea. Plus, as with any other organ, the oxygen level and blood pressure would be lower.

A8 Ureters carry urine from the kidneys to the bladder. The urethra carries urine from the bladder to the outside.

A9 Homeostasis means keeping conditions inside an organism within certain limits. Negative feedback is the process of opposing changes that happen. The body achieves homeostasis using negative feedback processes.

TUTORIAL

T1 *Don't be confused. The skin temperature on the two days may be different and you would certainly 'feel' hot or cold but your actual core temperatures would be the same unless you were ill.*

T2 *Many candidates lose marks in exams because they describe blood vessels 'moving up to the skin surface'. The capillaries themselves do not move although there is more blood flowing closer to the surface.*

T3 *Many exam candidates show by their answers that they mistakenly think that sweat is cold and so cools for that reason. Obviously sweat is the same temperature as you are and it is the evaporation that causes the cooling. If the sweat cannot evaporate, for example if the air is very humid, then we are not cooled and feel very uncomfortable.*

T4 *In an exam do not describe the fur or feathers as trapping a layer of 'warm air'. The important point is that the air is an insulator and so there is little heat transfer by conduction and as the air is trapped convection is also reduced. You can find out more about the different ways that heat is transferred in chapter 17.*

T5 *The body has to lose extra water by sweating to maintain body temperature. Humans cannot reduce how much is lost in breathing as the lining of the alveoli in the lungs is always moist, nor can the amount lost in faeces be reduced greatly. The only way therefore to cut down on losses is to reduce urine output.*

T6 *Excretion is the removal of substances that have been produced in the body. Carbon dioxide is a waste product of respiration as is some of the water.*

T7 *Although most other substances in the blood, e.g. glucose, are also initially filtered from the blood most are then absorbed back before the blood leaves the kidneys. Also don't forget that the arteries are carrying blood away from the heart into the kidneys, and the veins the reverse.*

T8 *These are easy to confuse.*

T9 *These are two ideas that many candidates find difficult. You may find it easier to explain them to yourself and to the examiner by using examples. There are many examples in this section.*

GOOD HEALTH

One main cause of disease is the presence of foreign organisms – **micro-organisms**, or **microbes**. Most of the time our bodies are able to prevent microbes, such as **viruses** or **bacteria**, from getting in or from spreading.

Barriers to infection	How they work
Skin	The skin provides a barrier that micro-organisms cannot penetrate unless through a wound or a natural opening.
Mucus and ciliated cells in lining of respiratory tract	The lining of the nasal passages, and the trachea, bronchi and bronchioles in the lungs are covered with a slimy mucus which traps air-borne micro-organisms as well as dirt. Tiny hair-like cilia use a waving motion to move the mucus upwards where it usually ends up going down the oesophagus.
Acid in the stomach	Stomach acid kills micro-organisms present in food or drink.
Blood clots	At the site of a wound, blood **platelets** break open triggering off a series of chemical changes that result in the formation of a network of threads at the wound. This network traps red blood cells forming a clot. The clot prevents blood escaping and provides a barrier to infection.

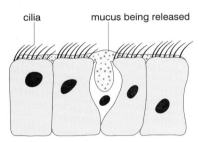

cilia   mucus being released

Cells lining the air passages in the trachea, bronchi and bronchioles.

THE IMMUNE SYSTEM

If micro-organisms do manage to enter the body they can cause damage by either damaging the cells around them or by releasing toxins (poisons) that make us ill. In either case they are attacked by **white blood cells** which are part of our **immune system**.

There are two main types of white blood cells:

- **phagocytes** which, because of their flexible shape, can engulf and then digest micro-organisms. This is called **phagocytosis**.

- **lymphocytes** which produce chemicals called **antibodies**. Antibodies destroy the micro-organisms in various ways, for example by making them clump together so that they are more easily engulfed by phagocytes, by killing the micro-organisms themselves or by destroying the toxins they release.

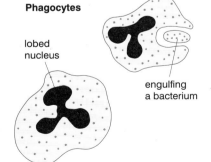

Phagocytes

lobed nucleus

engulfing a bacterium

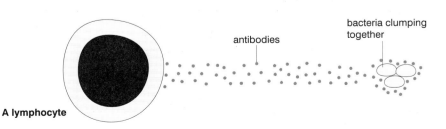

antibodies

bacteria clumping together

A lymphocyte

On the surface of all cells are chemicals called **antigens**. Each different type of micro-organism has its own **specific** antigens which can only be attacked by the right type of antibody. When micro-organisms first invade the body, the white blood cells start to make the antibody to destroy the micro-organisms. This may take a while and this is usually when you show the symptoms of the disease. When the white blood cells have made enough antibodies to attack the micro-organisms you (hopefully) recover from the disease. Now if you are ever invaded by the same type of micro-organisms, white blood cells making the right type of antibody can quickly destroy the micro-organisms before the symptoms of the disease appear because the white blood cells 'remember' the antigen. You are then said to be **immune** to the disease.

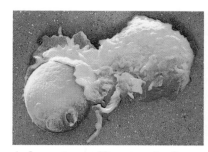

A phagocyte engulfing a yeast cell (yellow).

The reason that organ transplants are ideally taken from a close relative is that the organs are more likely to have antigens similar to the patient's own cells, and are therefore less likely to be attacked by the patient's immune system. When the immune system attacks a transplanted organ, the organ is said to be **rejected**. If the donor is not a close relative, immuno-suppressant drugs have to be used to stop the transplant being rejected.

You can be artificially **immunised** against a disease by having a 'jab'. In this case you are injected with the antigen in the form of dead or harmless forms of the micro-organism or a harmless form of the toxin they release. The harmless foreign antigens stimulate your white blood cells to produce antibodies against them in the normal way. This means that your white cells will 'remember' how to do this in the future and you will be protected if you ever catch the real disease-causing micro-organisms. Sometimes this immunity will last for the rest of your life but in other cases you may need 'booster' injections.

DRUGS

Another health problem is posed by various **drugs**, both illegal and otherwise, that people may take. Drugs in this case are defined as substances that affect the way our bodies behave. The effects may be useful but many can be dangerous.

Different drugs can affect us in different ways:

- **stimulants** (e.g. caffeine, nicotine, ecstasy, cocaine, amphetamines) increase nervous activity – they can speed up your heart rate, keep you awake or help depression.

- **depressants** (e.g. alcohol, painkillers, tranquillisers, solvents, heroin) decrease nervous activity – they can slow your heart rate or reaction times, deaden pain or relax you. They do this by restricting the passage of nervous impulses across synapses.

- **hallucinogens** (e.g. LSD, marijuana) affect the way we perceive things so people see or hear things that are not there or perceive things more vividly.

Some drugs can be **addictive** which means that if you stop taking the drug you will show various **withdrawal** symptoms which may include cravings for the drug, nausea and sickness. Another danger with drugs is that you build up a **tolerance** to them. This means that your body 'gets used' to them and you have to take larger amounts for the drug to have the same effect.

Tobacco

Contains **nicotine** which is a stimulant, increases blood pressure and is addictive. It can also lead to the formation of blood clots, increasing the chances of heart disease.

Contains **tar** which irritates the lining of the air passages in the lungs making them inflamed causing **bronchitis**. It can also cause the lining cells to multiply leading to lung cancer. It also damages the cilia lining the air passages and causes extra mucus to be made which trickles down into the lungs as the cilia can no longer remove it. Bacteria can breed in the mucus so you are more likely to get chest infections and smokers' cough as you try to get rid of the mucus. The alveoli are also damaged so it is more difficult to absorb oxygen into the blood. This condition is known as **emphysema**.

Tobacco smoke contains carbon monoxide which stops red blood cells from carrying oxygen by combining with the haemoglobin in the cells. This could also seriously affect the development of the fetus in the womb of a pregnant smoker.

Alcohol

Long term use can cause **liver damage** (because the liver has the job of breaking down the alcohol), heart disease, brain damage and other dangerous conditions. Alcohol is a depressant. Short term effects include: making you more relaxed – which is why in small amounts it can be pleasant; slower reactions and impaired judgement – which is why drinking and driving is so dangerous. Larger amounts can cause lack of co-ordination, slurred speech, unconsciousness and even death.

Solvents

These are depressants that slow down brain activity. The effects last for a much shorter time than alcohol but can include dizziness, loss of co-ordination and sometimes unconsciousness. The solvent fumes are toxic and can kill.

Threat of heart disease

Both drinking large amounts of alcohol and smoking, as well as a diet containing a lot of fatty foods, little exercise and stress, can increase the chance of **coronary heart disease** – the main cause of death in Britain today. This disease affects the **coronary arteries** which are the blood vessels that supply the heart muscle itself. Fatty substances such as **cholesterol** can build up on the inside of the vessels, narrowing them and restricting the blood flow. If a blood clot gets stuck in an artery the blood flow may be stopped altogether so oxygen and food cannot be supplied to part of the heart. If this stops the heart beating then the result is a **heart attack**.

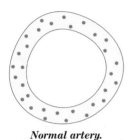

Normal artery.

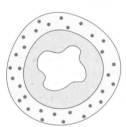

Artery affected by build up of cholesterol.

Check yourself

QUESTIONS

Q1 How can micro-organisms enter the body?

Q2
a) Most of the time there are very few white blood cells in the blood compared to red blood cells. Explain why.
b) When would you expect the numbers of white blood cells to increase?

Q3 If you have had measles once why are you unlikely to catch it again?

Q4 If you need a kidney transplant why is the ideal donor an identical twin?

Q5 Why is it dangerous to drive after drinking alcohol?

Q6 Why might a drink of coffee help to keep you awake at night?

ANSWERS

A1 Through wounds in the skin or natural openings – mouth, nose, eyes, anus, genitals.

A2
a) Red blood cells are needed in large quantities to carry oxygen whereas white blood cells are fighting infection and most of the time we are not recovering from diseases.
b) When there is an infection to fight off.

A3 Your white blood cells remember how to make large quantities of the measles antibody quickly. If the measles virus ever entered your body again you would fight it off before it had a chance to multiply and cause the symptoms of measles.

A4 Only an identical twin would have exactly the same antigens on their cells and the white blood cells would not recognise them as being foreign cells.

A5 As reaction times are slower and judgement impaired you are much more likely to have an accident.

A6 Coffee contains caffeine which is a stimulant.

TUTORIAL

T1 *Through the mouth and nose are most common because micro-organisms are present in the air, food and drink that we take into our bodies.*

T2 *Even when we are making lots of white blood cells to fight an infection there are many times more red blood cells.*

T3 *You only 'have a disease' when the micro-organism has had a chance to multiply and spread.*

T4 *In exam questions, try to use the correct technical words, like 'antigen', if it is appropriate. You might correctly explain a point but there could be other marks specifically for using the correct scientific words.*

T5 *This does not just apply if you are 'drunk'. Drinking and driving is dangerous because the alcohol affects you to some extent even after one drink. In exams a common shortcoming is to say 'because you may have an accident' which is not wrong but might not get full marks. Always try to go beyond common-sense type answers and give a scientific reason.*

T6 *The effects may not last very long but in the short term caffeine will increase nervous activity so you are more alert.*

EXAMINATION CHECKLIST

The facts and ideas that you should understand by studying this topic.

To be successful at Foundation GCSE Tier you should be able to:

- recall which receptors gather information about touch, taste, smell, light, sound and balance
- locate the main parts of the eye and recall their functions
- name and locate the main parts of the nervous system: brain, spinal cord, peripheral nerves
- recall the structure of a motor neurone: cell body, axon, sheath
- describe the paths taken by nervous impulses, specifically the path taken by spinal reflexes
- recall that reflexes are fast, automatic and protective responses to a stimulus
- name and locate the major endocrine glands: pituitary, adrenal glands, pancreas, ovaries and testes
- recall the functions of insulin, adrenaline, and the male and female sex hormones
- recall that hormones travel in the blood
- recall that diabetes is caused by a failure of the pancreas to produce insulin and that symptoms include excessive fluctuation of blood sugar level
- explain how the body works to maintain steady levels of temperature, water, oxygen and carbon dioxide and that this is essential to life
- recall that maintaining a constant internal environment is called homeostasis and that it involves balancing bodily inputs and outputs
- locate and recall the functions of: the kidney, bladder, ureter and urethra
- recall that carbon dioxide from respiration is removed from the body by the lungs
- explain how the body is defended against disease organisms: the skin, mucous membranes (breathing system), stomach acid
- recall the functions of white blood cells in the immune system
- recall that immunity to disease organisms arises from prior infection
- recall the physical effects and problems associated with common drugs: alcohol, tobacco, solvents, caffeine
- explain the terms 'addiction' and 'withdrawal symptoms'.

In addition, to be successful at Higher GCSE Tier you should be able to:

- describe how muscles in the iris control pupil size in response to light
- describe how neurones are adapted to their functions
- explain the value of the pathways that make the brain aware of reflexes
- recall that neurones are connected by synapses

- recall the functions of oestrogen and progesterone in the menstrual cycle
- explain how insulin regulates blood sugar levels
- explain how adrenaline prepares the body for 'flight or fight'
- explain how female fertility can be affected by the use of female sex hormones
- explain why the dosage of insulin for diabetics depends upon diet and activity
- explain how homeostasis is achieved by negative feedback mechanisms including the control of temperature, blood water and carbon dioxide level
- explain the control of heat transfer to the environment by sweating, vasodilation and vasoconstriction
- recall that each disease organism has its own antigens requiring specific antibodies
- explain how immunisation is a means of giving the antigen without the disease
- explain the action of depressants and stimulants upon the synapses of the nervous system
- recall that as addicts become used to drugs they need larger doses (tolerance).

Key Words

Tick each word when you are sure of its meaning

antibody	immunisation	reflex
antigen	immunity	relay neurone
axon	insulin	sensory neurone
cilia	lymphocyte	stimulant
depressant	motor neurone	stimulus
effector	negative feedback	synapse
glucagon	phagocyte	tolerance
glycogen	platelet	urea
homeostasis	receptor	vasoconstriction

EXAM PRACTICE

Sample Student's Answers & Examiner's Comments

a) *If you are provided with information in a question, it will be there to help you to answer the question. If it tells you to use the information, you may not gain full marks if you do not use it.*
In this case you have to name two places 'on the diagram' so if the candidate had for example written 'mouth', which is a place where bacteria could enter, they would not have gained a mark.

b) **i)** *An acceptable answer. 'Immune system' would be better.*
ii) *The candidate gains one mark for the idea that antibodies can be made quickly. One mark would also have been given for stating that the antibodies would recognise or attack the bacteria and another mark (up to a maximum total of two) would have been given for explaining that some antibodies could remain from the previous infection.*

c) **i)** *The candidate has not read the question carefully. The correct answer would have been that less sweat would be produced (one mark) so there would be less evaporation cooling the skin (one mark).*
ii) *The question has been misread. The correct answer would have been that the hairs stand up (one mark) trapping a layer of air as insulation (one mark).*
iii) *One mark has been gained for explaining that less blood is flowing through the skin because the capillaries contract. The second mark would have been awarded for explaining that this means less heat loss through the skin.*

d) *The candidate should have given three differences to gain full marks. There were two other answers worth one mark each: that nerve signals are electrical and hormonal signals are chemical, and that nerve signals tend to go to one part of the body but hormones travel through the whole body.*

● *This candidate has gained 7 marks out of a possible 14. A grade C candidate should have gained at least 9 marks.*

56

1 Your skin plays a vital role in protecting you from infection and in controlling your temperature. The diagram shows a section through human skin.

a) Name **two** places on the diagram where bacteria might gain entry.
   pore ✓
   hair follicle ✓ (2)

b) (i) Name the body defence system which recognises that bacteria are not part of the body and should be destroyed.
   White blood cells ✓ (1)

   (ii) Explain why the body can react more quickly to infection if it has been previously infected by the same type of bacteria.
   The white blood cells remember how to make the right kind of antibodies ✓ (2)

c) Some of the structures shown help to control body temperature. For each one listed below, explain how it responds to cold conditions, and how this response helps to keep the body warm.
   (i) sweat glands sweat evaporates cooling the skin ✗

   (ii) hair
   Muscles pull the hairs flat so they don't trap a layer of insulating air ✗

   (iii) blood capillaries ✓
   Blood capillaries get smaller so they carry less blood so the skin is paler and cooler (6)

d) 'Signals' to control responses in the skin may be sent either by the nervous system or by the hormone system. Describe the main differences in the way these two systems work.
   The nervous system sends signals along nerves but the hormone system sends signals by hormones in the blood. ✓
   Signals travel faster ✓ along nerves. 7/14 (3)

Midland Examination Group

Questions to Answer

Answers to questions 2 – 6 can be found on page 359.

2 This question is about the nervous system. Look at the diagram of a motor neurone.

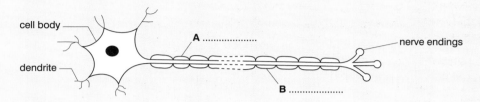

a) Label parts A and B. (2)

b) Describe how one neurone (nerve cell) connects with another neurone. (1)

c) South American Indians use a nerve poison to kill animals. The poison affect the places where neurones connect. It stops signals passing from one neurone to another. Use this information to suggest **one** way the poison would affect the animals before killing them. Explain your answer. (2)

Midland Examining Group

3 The graph below shows the relative levels of two hormones, oestrogen and hormone **Q**, during a woman's menstrual cycle.

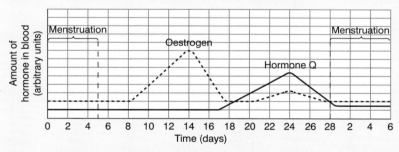

a) When is the level of oestrogen equal to the level of hormone Q? (2)

b) What effect does the build up of oestrogen have from day 9 to day 14? (2)

c) (i) Name hormone Q. (1)

(ii) In which part of the body is hormone Q produced? (1)

(iii) Which event is caused by a fall in the level of hormone Q? (1)

(iv) What would be indicated if the high level of hormone Q continued beyond day 24? (1)

London Examinations

4 The concentration of glucose in the blood is not constant. It varies between 0.6 and 1.4 g per litre depending on conditions. After a meal it rises and after going without food for a time it becomes lower. 0.6 g per litre is the lowest level that allows brain and nerve cells to function properly.

After a meal the level of glucose in the blood increases. Glucose is removed from the blood for respiration by the body cells.

Describe how the concentration of glucose in the blood is controlled in humans. In your description refer to the parts played by the liver, the pancreas and hormones.

(5)

Midland Examining Group

5 Bronchitis is an infection of the bronchial tubes which lead to the lungs.

a)

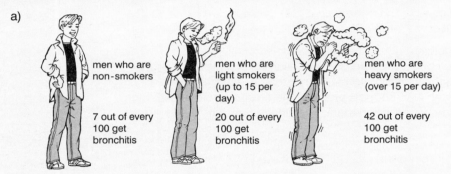

men who are non-smokers

men who are light smokers (up to 15 per day)

men who are heavy smokers (over 15 per day)

7 out of every 100 get bronchitis

20 out of every 100 get bronchitis

42 out of every 100 get bronchitis

What general pattern is shown by this information?

(1)

b) The diagrams show the effects of cigarette smoke on small hairs called cilia in the bronchial tubes.

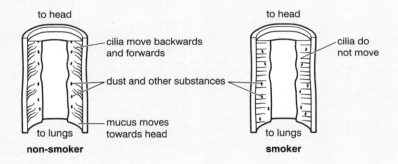

to head

cilia move backwards and forwards

dust and other substances

mucus moves towards head

to lungs

non-smoker

to head

cilia do not move

to lungs

smoker

(i) What effect does cigarette smoke have on the movement of mucus?

(1)

(ii) Suggest why smokers tend to cough more than non-smokers.

(2)

(iii) Smokers are sometimes short of breath. Suggest how smoking affects the lungs to cause this.

(2)

Midland Examining Group

6 Prostaglandins are produced when the body is cut, burned or becomes infected. They are released into the blood. They raise the temperature of the affected area and nerve endings are stimulated causing irritation.

a) Explain why prostaglandins can be described as acting like hormones. (3)

b) Aspirin, codeine and paracetamol are all pain relievers.

Aspirin stops production of prostaglandins. It also reduces the blood's ability to form clots.

Codeine blocks the nerve impulses inside the brain and central nervous system. It can cause drowsiness and may be habit forming.

Paracetamol stops the production of prostaglandins in the brain instead of at the site of injury. Too much paracetamol can cause liver damage.

(i) Aspirin is given to people who have suffered a heart attack. Explain why this might reduce the chances of another heart attack. (2)

Information on the active ingredients of four brands of pain relievers is shown in the table.

Brand \ Ingredient	aspirin	codeine	paracetamol
Veganin	✓	✓	✓
Solpadeine		✓	✓
Panadol			✓
Disprin	✓		

✓ = present

Arthritis is a long term condition which causes pain in joints.

(ii) Use the information to suggest which brand sufferers should take. Explain your choice. (2)

Midland Examining Group

CHAPTER 4

PLANTS

Plants are like animals in that they show the same characteristics that all living things do (see chapter 1). Like animals their bodies are also made up of different organs that carry out different roles.

Flower – needed for reproduction, seeds are formed here

Leaf – for photosynthesis to make food

Buds – growing points on the stem, some are flower buds

Stem – for support, also contains transport system

Root – for water and mineral salt uptake, also anchors the plant in the soil

Anatomy of a plant.

PHOTOSYNTHESIS

One fundamental way in which plants are different from animals is how they get their food. Plants need and use the same types of foods as animals (carbohydrates, proteins and fats) but while animals have to eat other things to get their food, plants *make it themselves*. The way they do this is called **photosynthesis**. The other ways that plants are different from animals, such as having leaves and roots, or being green, are all linked with photosynthesis.

In photosynthesis, plants take carbon dioxide from the air and water from the soil, and use the energy from sunlight to convert them into food. The first food they make is **glucose** but that can later be changed into other food types. **Oxygen** is also produced in photosynthesis, and although some is used inside the plant for respiration (releasing energy from food) most is not needed and is given out as a waste product (although it is obviously vital for other living things). The sunlight is absorbed by the green pigment **chlorophyll**. The process of photosynthesis can be summarised in a word equation:

$$\text{carbon dioxide} + \text{water} \xrightarrow[\text{light}]{\text{chlorophyll}} \text{glucose} + \text{oxygen}$$

It can also be written as a balanced chemical equation:

$$6CO_2 + 6H_2O \xrightarrow[\text{light}]{\text{chlorophyll}} C_6H_{12}O_6 + 6O_2$$

Much of the glucose is converted in to other substances such as **starch**. Recall from what you know on animal digestion (see chapter 2), that starch molecules are made of lots of glucose molecules joined together. Unlike glucose, starch is insoluble and so can be stored in the leaf without affecting water movement into and out of cells by **osmosis**. Some glucose is converted to **sucrose** (a type of sugar consisting of two glucose molecules joined together) which is still soluble, but not as reactive as glucose, so it is easily carried around the plant in solution.

The raw materials and products in a plant.

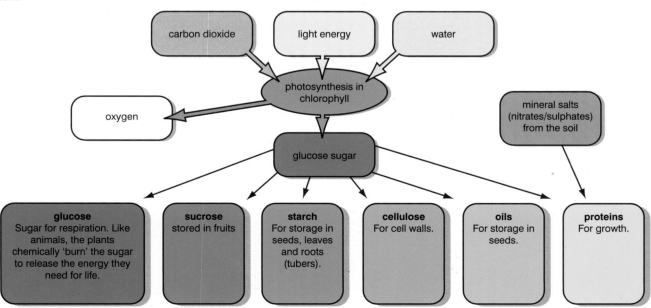

Photosynthesis takes place mainly in the leaves although it can occur in any cells that contain green chlorophyll. The leaves are adapted to make them very efficient at photosynthesis.

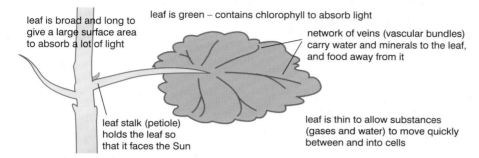

leaf is broad and long to give a large surface area to absorb a lot of light

leaf is green – contains chlorophyll to absorb light

network of veins (vascular bundles) carry water and minerals to the leaf, and food away from it

leaf stalk (petiole) holds the leaf so that it faces the Sun

leaf is thin to allow substances (gases and water) to move quickly between and into cells

Just by looking at a leaf you can see some of the ways that it is adapted to be efficient at photosynthesis in that it:

- is **broad**, so as much light as possible can be absorbed
- is **thin**, so it is easy for carbon dioxide to diffuse in to reach the cells in the centre of the leaf
- contains green **chlorophyll**, in the **chloroplasts**, which absorbs the light energy
- contains **veins** to bring up water from the roots and carry food products to other parts of the plant
- has a stalk, or **petiole**, that holds the leaf so it can absorb as much light as possible.

If you look inside a leaf using a microscope there are many other features that also help:

- a transparent **epidermis** to allow light to penetrate inside the leaf
- many **palisade** cells, in which most of the photosynthesis takes place, tightly packed together in the top half of the leaf so as many as possible collect sunlight
- chloroplasts containing chlorophyll concentrated in the top half of the leaf to absorb as much sunlight as possible

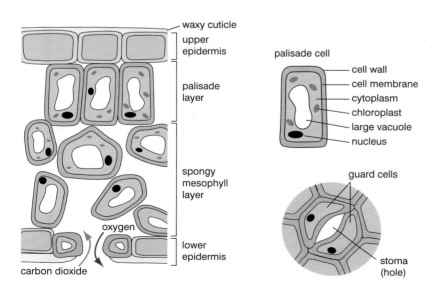

waxy cuticle

upper epidermis

palisade layer

spongy mesophyll layer

oxygen

lower epidermis

carbon dioxide

palisade cell

cell wall

cell membrane

cytoplasm

chloroplast

large vacuole

nucleus

guard cells

stoma (hole)

- air spaces in the **spongy mesophyll** layer to allow the movement of gases (carbon dioxide and oxygen) through the leaf to and from each cell
- a large internal surface area to volume ratio to allow the efficient absorption of carbon dioxide and removal of oxygen by the photosynthesising cells
- many pores or **stomata** (singular: stoma) which allow the movement of gases into and out of the leaf.

PHOTOSYNTHESIS AND RESPIRATION

Many people forget that once plants have made food using sunlight, they will at some stage need to release the energy in the food. They do this in the same way that humans and other animals do – **respiration**. Plants are *always* respiring. During the day, plants respire 'slower' than they photosynthesize so we only detect carbon dioxide entering and oxygen leaving the plant. During the night, photosynthesis stops and only then can we detect oxygen entering and carbon dioxide leaving during respiration. There are two times in the day, at dawn and dusk, when the rates of photosynthesis and respiration are the same and no gases enter or leave the plant because any oxygen produced by photosynthesis is immediately used up in respiration and vice versa for carbon dioxide. These occasions are known as **compensation points**.

Limiting factors

If a plant is given more light, carbon dioxide, water or a higher temperature, then it may be able to photosynthesise at a faster rate. However the rate of photosynthesis will eventually reach a maximum because there is not enough of one of the other factors needed. In other words there is some **limiting factor**. For example, a farmer might pump extra carbon dioxide into a greenhouse and the rate of photosynthesis might increase so the crop will grow faster. However it might be that the light levels are not sufficient to allow the plants to use the carbon dioxide as quickly as it is supplied. In this case the light intensity would be the limiting factor.

The following graphs show how increasing the levels of light and carbon dioxide, two of the factors necessary for photosynthesis, will increase the rate of photosynthesis until the rate is halted by some other limiting factor.

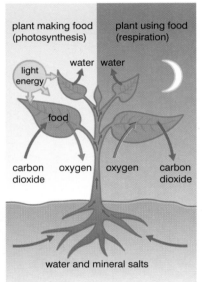

Photosynthesis is limited by a number of factors.

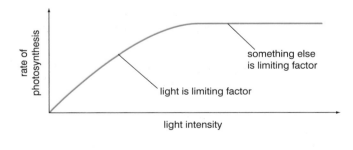

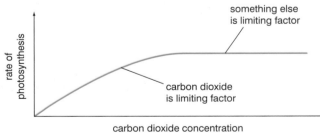

If the limiting factor in the first graph was the amount of carbon dioxide and the plants were then given more carbon dioxide, the graph would look like this:

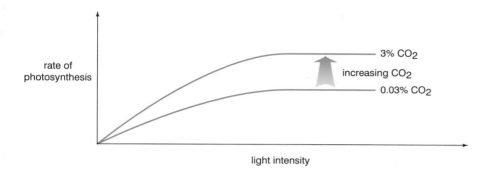

Temperature is also a limiting factor. Temperature affects the enzymes that control the different chemical reactions that happen during photosynthesis. Compare the shape of the graph to the right with the graph showing the effect of temperature on enzyme activity on page 26.

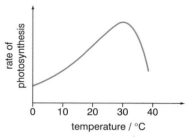

Check yourself

QUESTIONS

Q1 Animals and plants both need food. How are they different in the way that they get that food?

Q2 a) When does a plant photosynthesise?
b) When does a plant respire?

Q3 How does a plant get the raw materials it needs for photosynthesis?

Q4 Carbohydrates, fats and proteins are all found in plants and all contain the element carbon. Where do plants get the carbon from?

Q5 Why isn't the food that is made in leaves transported around the plant as starch?

Q6 Why is the top surface of leaves usually a darker green than the lower surface?

Q7 Why are leaves usually broad and thin?

Q8 Which cells in a leaf can photosynthesise?

Q9 Why do farmers who grow crops in greenhouses often add extra carbon dioxide to the air inside?

REMEMBER! Cover the answers if you want to.

ANSWERS

A1 Animals get their food by eating other things. Plants make their food by photosynthesis.

TUTORIAL

T1 *Many people incorrectly think that plants get their food from the soil. This is not helped by fertilisers sometimes being called plant 'foods'. Fertilisers (minerals) are needed for growth but they are not the plant's food. See the later sections of this chapter.*

ANSWERS

A2
a)   When there is light.
b)   All the time

A3
Carbon dioxide from the air enters the leaves through the stomata. Water moves from the soil into the roots and up the stem to the leaves.

A4
From the carbon dioxide they take in.

A5
Starch is not soluble.

A6
There is more chlorophyll near the top surface.

A7
Being broad allows as much light as possible to be absorbed. Being thin allows carbon dioxide to quickly reach the photosynthesising cells.

A8
The palisade cells and the spongy mesophyll cells to a lesser extent.

A9
To increase the rate of photosynthesis so the plants grow faster and larger.

TUTORIAL

T2
a)   *Light is necessary for photosynthesis. Even if there is enough light, a plant may be unable to photosynthesise if other factors are not present, e.g. if it was too cold.*
b)   *It is very common for candidates to say that photosynthesis happens in the day and respiration happens at night time. This is wrong. Respiration happens all the time but during the day it is masked by photosynthesis in that any carbon dioxide made is not given out but immediately used up in photosynthesis.*

T3
As plants cannot move around, their raw materials have to be readily available. Note that sunlight is not a raw material. It is the energy source that drives the chemical reaction.

T4
Look at the chemical equation for photosynthesis on page 60. There are in fact many steps in the change from the raw materials to the products shown but you can clearly see that the only place that carbon comes from is the carbon dioxide (CO_2).

T5
Starch is a useful storage material but needs to be broken down into smaller molecules to be able to dissolve.

T6
This is so it is available to absorb as much sunlight as possible.

T7
There are exceptions to this general principle. See page 67 about cactus spines and pine needles.

T8
The epidermal cells are transparent because they contain no chlorophyll and so do not absorb any light themselves.

T9
Remember that there is only about 0.03% carbon dioxide in the air normally and under warm sunny conditions when plants are well watered it will be this

TRANSPORT IN PLANTS

In humans and many other animals, substances are transported around the body in the blood through blood vessels. In plants, water and dissolved substances are also transported through a series of tubes or vessels. There are two types of transport vessel, called **xylem** and **phloem**.

Xylem vessels are long tubes made of the hollow remains of cells that are now dead. They carry water and dissolved minerals up from the roots, through the stem, to the leaves. They also give **support** to the plant.

Phloem vessels are living cells. They carry dissolved food materials, mainly sucrose, from the leaves to other parts of the plant such as growing roots or shoots, or storage areas such as fruit. This movement of food materials is called **translocation**.

In roots the xylem and phloem vessels are usually grouped together separately but in the stem and leaves they are found together as **vascular bundles** or 'veins'.

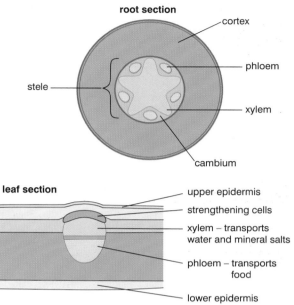

root section

- cortex
- phloem
- stele
- xylem
- cambium

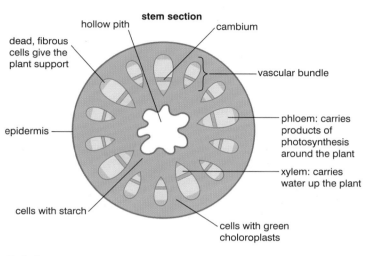

stem section

- hollow pith
- cambium
- dead, fibrous cells give the plant support
- vascular bundle
- epidermis
- phloem: carries products of photosynthesis around the plant
- xylem: carries water up the plant
- cells with starch
- cells with green choloroplasts

leaf section

- upper epidermis
- strengthening cells
- xylem – transports water and mineral salts
- phloem – transports food
- lower epidermis

Gaining water

Roots are covered in tiny **root hair cells** which increase the surface area for absorption. Water enters by osmosis because the solution inside the cells is more concentrated than the water in the soil.

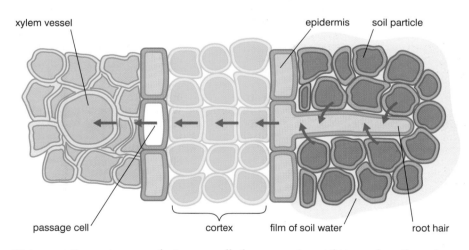

- xylem vessel
- epidermis
- soil particle
- passage cell
- cortex
- film of soil water
- root hair

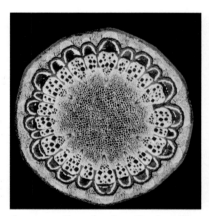

A cross-section of a clematis stem. You can clearly see the ring of yellow and brown vascular bundles.

Water continues to move between cells by osmosis until it reaches the xylem vessels which carry it up to the leaves. Some water will also travel between cells as shown in the diagram.

Losing water

In the leaves, water moves out of the xylem and enters the leaf cells by osmosis (as they contain many dissolved substances). Water **evaporates** from the surface of the cells inside the leaf and then **diffuses** out through the open stomata. The evaporation of water causes more water to rise up the xylem from the roots rather like a drink flows up a straw when you suck at the top. This movement of water from the leaves is known as **transpiration**. The flow of water through the plant from the roots to the leaves is known as the **transpiration stream**.

very little water lost
through waxy cuticle

palisade
layer

water passes from
xylem cells to other cells
by osmosis

water seeps from cells
and evaporates into air
spaces of the spongy
mesophyll layer

water diffuses out of the
leaf, through the
stomatal pores

guard cell

How water leaves a plant.

Transpiration is useful to plants in several ways:

- It brings up water and **minerals** from the soil.
- As the water evaporates it cools the plant. (You can find out more about the cooling effect of evaporation in chapter 17.)
- The water-filled xylem vessels help to hold the plant upright so that the leaves 'catch' the Sun.

Transpiration happens faster in conditions that encourage evaporation, that is when it is:

- warm
- windy
- dry
- very sunny (because this is when the stomata are most open)
- and when the plant has a good water supply.

WATER BALANCE

Sometimes plants may lose more water by transpiration than they take in from the roots. When this happens the plant will **wilt**. This is because in a healthy plant the cells are full of water and the cytoplasm presses hard against the inelastic cell wall making the cell rigid. The cell is described as being **turgid** or having **turgor**. Most plants do not have wood to hold them up, they are only supported upright because of cell turgor. If cells lose water, the vacuole shrinks and the cytoplasm stops pressing against the cell wall. The cell has lost its rigidity and become **flaccid**. The plant will start to droop. If this continues the cytoplasm may shrink so much that it starts to come away from the cell wall. This process of shrinking away from the cell wall is known as **plasmolysis**.

Water plays a key role in the structure of plant cells.

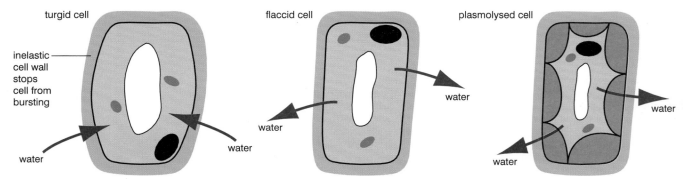

turgid cell

inelastic
cell wall
stops
cell from
bursting

water

water

flaccid cell

water

water

plasmolysed cell

water

water

There is a conflict. Plants need to have stomata open to allow the movement of the gases involved in photosynthesis. However this means that they will also lose water through the stomata. Losing too much water can be a real problem. Plants have many ways of reducing water losses. This particularly applies to plants living in places where water is not readily available, such as dry deserts or cold regions where the water in the soil is frozen.
Ways of reducing water losses include:

- a waxy cuticle – so water can not evaporate from epidermal cells
- leaves covered with hairs – this creates a thin layer of still, moist air close to the leaf surface which reduces evaporation
- stomata located mostly on the underside of leaves – where it is cooler so reducing evaporation
- reducing leaves to spines, e.g. cactus spines or pine needles – so less surface area is exposed
- stomata sunken below the surface of the leaf – so they are less exposed
- closing the stomata.

Guard cells

Around each **stoma** there are two **guard cells**. During daylight hours they absorb water by osmosis which makes them swell. The inner walls are thickened and can not stretch, unlike the outer ones. This means that as the cells swell they form a sausage shape and the gap between them opens up. During darkness the guard cells lose water by osmosis and the stomata close. Some other plants have a regular rhythm of opening and closing that is not so simply 'open during the light, closed during the dark'.

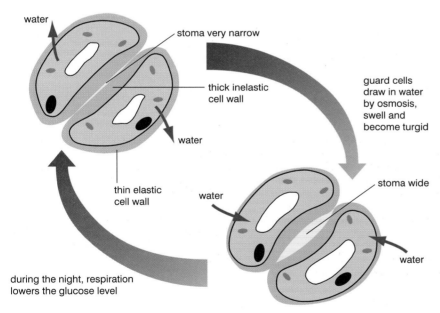

The cell walls in guard cells are designed to open the stoma when the cell swells.

Check yourself

QUESTIONS

Q1 What are the differences between xylem and phloem?

Q2 Why does transpiration happen faster on a hot, sunny day than on a cool, dull day?

Q3 What is the difference between a turgid cell and a plasmolysed cell?

Q4 Some grasses that grow on sand dunes have the leaf blades rolled into a hollow cylinder with the stomata on the inside. Suggest why the stomata are on the inside.

Q5 What is the difference between transpiration and translocation?

REMEMBER! Cover the answers if you want to.

ANSWERS

A1 Xylem is made of dead cells. It carries water and minerals from the roots, up the stem, to the leaves. Phloem is made of living cells. It carries dissolved food substances from the leaves to other parts of the plant.

A2 On a sunny day the guard cells will fully open the stomata allowing water vapour to pass out more easily. When it is hot, water evaporates more quickly into the air spaces inside the leaf which causes the water vapour to diffuse more quickly out of the leaf.

A3 In a turgid cell the cytoplasm presses hard against the cell wall because the cell contains as much water as it can. This gives the cell rigidity. In a plasmolysed cell so much water has been lost that the cytoplasm shrinks and comes away from the cell wall. A plasmolysed cell is flexible and can't keep a fixed shape.

A4 This is to cut down on water losses by transpiration. The space inside the cylinder will become humid because the water vapour is not quickly removed by moving air. This reduces the concentration gradient so reducing water losses. This will be very important for the plants as water will not be readily available to the roots.

A5 Transpiration is the loss of water vapour from the leaves when it evaporates into the air. Translocation is the movement of materials through a plant, for example sugars being moved from the leaves through the phloem to other parts of the plant.

TUTORIAL

T1 *Make sure you can also identify the xylem and phloem in diagrams of sections of roots, stems and leaves.*

T2 *Don't forget that if the plant is losing more water than it is taking in it will start to wilt. In this case the stomata will start to close so reducing transpiration.*

T3 *In a healthy plant none of the cells should be plasmolysed. Remember that when cells become plasmolysed, although the cytoplasm shrinks and the membrane with it, the cell wall does not.*

T4 *Most plants have some way to ensure that they do not lose more water than they can replace.*

T5 *Make sure before an exam that you can explain the meaning of keywords such as these. Use the keyword lists at the end of chapters to help check yourself.*

MINERALS

To grow properly, plants need minerals which they get from the soil. There are many different minerals but some are more important than others as they may be needed in larger amounts.

Mineral	Element	Use in plant	Problems caused by deficiency
Nitrates	Nitrogen	To make proteins which are needed to make new cells	• poor growth • pale or yellow leaves
Phosphates	Phosphorus	Involved in respiration	• poor growth of roots and stems • small leaves • low fruit yield
Potassium salts	Potassium	To control salt balance in cells	• mottled leaves • low fruit yield • low disease resistance
Magnesium salts	Magnesium	The chlorophyll molecule contains magnesium	• yellow patches between leaf veins
Sulphates	Sulphur	To make proteins	• poor growth • pale or yellow leaves
Iron salts	Iron	To help in the production of chlorophyll	• yellowing of younger leaves

Absorption

Minerals may only be present in the soil in small amounts and so may need to be absorbed by the root hairs against a concentration gradient. This means that they have to be absorbed by **active transport** (see chapter 1). They enter the roots in solution (dissolved in water) and are carried in the transpiration stream through the xylem up to the leaves.

Although minerals are constantly being taken from the soil by plants they are restored when animal and plant materials decay. You will find out more about this and artificial fertilisers in chapter 5.

Check yourself

QUESTIONS

Q1 a) Plants make proteins from carbohydrates. Which extra elements do they need to do this?
b) Why do plants need proteins?

Q2 Suggest why a lack of magnesium can cause yellow leaves.

Q3 Root hair cells contain many mitochondria. Suggest why.

ANSWERS

A1
a) Nitrogen and sulphur.
b) Protein is used to make new cells. For example, cell membranes contain protein.

A2 Magnesium is found in chlorophyll, the green pigment that gives plants their colour. Without it they cannot therefore be green.

A3 Minerals are absorbed by active transport, this means that energy is required which is provided by the mitochondria.

TUTORIAL

T1
a) *Plants do not take in and use the pure element. There is a lot of nitrogen in the air (78%) but in its pure form it is very unreactive and the plants cannot use it.*
b) *This applies to plants as well as to animals such as humans.*

T2 *Chlorophyll is actually chemically similar to haemoglobin in the blood, but containing magnesium instead of iron.*

T3 *Remember that although water and minerals both enter through the roots, water enters by osmosis which is a 'passive' process – it does not require energy.*

PLANT HORMONES

Many of the ways that plants grow and develop are controlled by chemicals called **plant hormones** or **plant growth regulators**. Like animal hormones they are made in one part of the organism and travel to other parts where they have their effects. They are different to animal hormones in that they are not made in glands and obviously do not travel through blood but rather they diffuse through the plant.

The things they control include:

- growth of roots and shoots
- growth of buds
- flowering time
- fruit formation and ripening
- germination
- leaf fall
- healing of wounds.

There are many commercial uses of hormones such as:

- Selective weedkillers that kill the weeds in a lawn without harming the grass. The hormones make the weeds quickly overgrow, which usually results in death through a number of causes (e.g. an insupportably weighty structure, constricted veins).

- Rooting powder which is applied to cuttings to make them grow roots.

- Keeping potatoes and cereals dormant so that they do not germinate during transport or storage.

- Delaying the ripening of soft fruit and vegetables so that they are not damaged during transport.

- Making crops such as apples ripen at the same time to make picking them less time-consuming.

TROPISMS

Tropisms are directional growth responses to stimuli. Examples include shoots growing towards the light (known as **phototropism**) or roots always growing downwards (known as **geotropism**). They are controlled by a hormone called **auxin**. Auxin is made in the tips of shoots and roots and diffuses away from the tip before it affects growth. One effect of auxin is to inhibit the growth of side shoots. This is why if a gardener wants a plant to stop growing taller and become more bushy he or she will pinch off the shoot tip so removing a source of auxin.

The growth of shoots towards light can be explained by the behaviour of auxin. Auxin moves away from the light side of a plant to the dark side. Here it encourages growth by increasing cell elongation and, as this means that the dark side grows more than the light side, the shoot bends towards the light source.

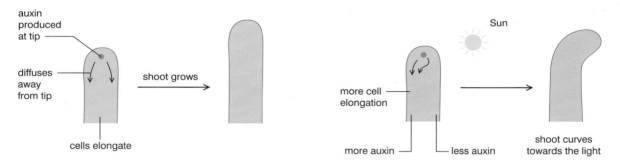

In roots, gravity causes the auxin to collect on the lower side. Here it *stops* the cells elongating which causes the root to bend downwards.

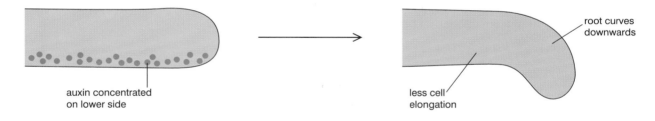

Seedlings growing towards a light source.

71

Check yourself

QUESTIONS

Q1 How are plant hormones different to animal hormones?

Q2 Astronauts have investigated root geotropism in space. They germinated seeds to see in which direction the roots grew. What do you think happened?

Q3 Which of the following are controlled by plant hormones?
A flowering,
B pollination,
C germination,
D leaf fall.

Q4 Why would a hormone that delays bud opening be useful to flower growers?

Q5 Once a plant shoot has changed direction so it is growing towards the light why does it stop bending?

REMEMBER! Cover the answers if you want to.

ANSWERS

A1 Plant hormones are not made in glands. Nor are they carried in the blood.

A2 The roots were not affected by gravity and so grew in any direction. They would however grow away from a directional light source.

A3 A, C and D.

A4 It would allow the growers to transport flowers before the buds have opened so reducing damage and increasing the longevity of the flowers.

A5 Auxin is no longer concentrated more on one side than the other so one side does not grow more than the other.

TUTORIAL

T1 *It is because of these differences that many people refer to them as growth regulators instead. Refer back to chapter 3 if you need to revise about animal hormones.*

T2 *It is important that roots normally grow down to ensure anchorage for the seedling and good absorption of water and minerals.*

T3 *Plant hormones affect growth and development. Pollination is simply the transfer of pollen between flowers.*

T4 *Other hormones can speed up flowering so all the flowers arrive at the shops when they have just opened.*

T5 *Remember that growth in tropisms is by cell elongation not the formation of new cells by cell division.*

EXAMINATION CHECKLIST

The facts and ideas that you should understand by studying this topic.

To be successful at Foundation GCSE Tier you should be able to:

- name the functions of: stem, flower, root, leaf and leaf veins
- recall that plants make their own food in the form of glucose and starch by a process called photosynthesis
- recall the equation: carbon dioxide + water (+ light energy) (+ chlorophyll) → glucose + oxygen
- recall which food substances in plants are used for providing energy, growth and storage products
- recall that food is made as glucose, is transported as soluble sugar and is stored as insoluble starch
- explain how photosynthesis can be increased by providing more CO_2, light and a higher temperature
- recall how leaves are adapted for efficient photosynthesis
- recall the arrangement and function of xylem and phloem in a stem and leaf
- recall that by increasing surface area, root hairs increase the ability of roots to take up water
- explain the loss of water from leaves in terms of the diffusion of water molecules
- explain how the structure of a leaf is adapted to reduce water loss
- explain the function of transpiration and recall the path of water through a plant
- recall that fertilisers contain minerals needed for plant growth and are absorbed by the roots
- recall that plant hormones (e.g. auxins) are chemicals that control growth and development, they move through the plant in solution and are involved in the response to light and gravity
- relate the action of plant hormones to their commercial uses.

In addition, to be successful at Higher GCSE Tier you should be able to:

- recall the equation: $6CO_2 + 6H_2O$ (+ chlorophyll) (+ light energy) → $C_6H_{12}O_6 + 6O_2$
- explain the relative effects of limiting factors on photosynthetic rates
- explain how turgor pressure supports plants
- explain how the cellular structure of a leaf is well adapted for efficient photosynthesis
- explain how plants are supported by the turgor pressure within cells
- explain how water loss from leaves is a consequence of the way in which the leaf is adapted for efficient photosynthesis
- explain how the cellular structure of the leaf is adapted to reduce water loss
- explain how stomatal apertures depend upon light intensity and availability of water

- recall that plants require nitrates to make proteins, and magnesium to make chlorophyll
- recall that minerals are taken up into roots by active transport
- recall that auxin brings about shoot growth in terms of cell elongation
- explain phototropism experiments in terms of auxin action.

Key Words

Tick each word when you are sure of its meaning

auxin	palisade	transpiration
chlorophyll	photosynthesis	tropism
chloroplast	plasmolysis	turgor
flaccid	root hair cells	vascular bundle
limiting factor	stoma	
osmosis	translocation	

EXAM PRACTICE

Sample Student's Answers & Examiner's Comments

1 This question is about plant hormones.

a) Plant hormones can speed up or slow down plant growth. Use this information to help you answer these questions.

(i) Why are some plant hormones used as rooting powder?

to make roots grow on cuttings ✓ .... (1)

(ii) Some plants are used as weedkillers. Explain as fully as you can how they work.

they make the weeds grow ✓ *so fast they are weak* ✓ *and can't support themselves and die.* ....

.... (2)

(iii) Some hormone weedkillers are selective. They kill weeds but not grass. How do they do this?

they do not poison grass ✗ ....

.... (1)

b) Look at the diagrams. They show a plant growing in a room with one window.

← light   ← light

at the start   a few weeks later

The plant shoot grows towards the light. This growth involves the plant hormone called auxin. Explain how auxin causes the shoot to grow towards the light.

Auxin makes the dark side ✓ *grow more so the shoot bends towards the light.* ....

.... (2)

4/6

Midland Examining Group

a) **i)** *To make or help the roots grow is the correct answer and so this candidate gains the mark. Note that there are some plants that can be planted as cuttings and will grow very well without rooting powder (such as pelagoniums, also known as 'geraniums') but many will not without the powder.*
ii) *A common mistake would be to say that weedkillers simply kill weeds. Remember that strangely they actually make them grow! However the growth is much faster than normal and they become weak and die. The candidate has correctly described this and gains the two marks which are for firstly stating that the growth would be more rapid and secondly for describing the harmful effect of this on the weeds.*
iii) *The candidate has basically repeated information already given in the question, that the weedkillers kill the weeds, but has not explained why. Therefore no marks can be awarded. A mark would have been gained for explaining that weeds respond to the hormone more than grass does.*

b) *There are two marks available. One mark is for explaining that the auxin moves from the light side to the dark side. The candidate may have known this but as it has not been stated the mark can not be given. Remember you can only be given marks for what you show that you know or understand. The second mark is for explaining that there is more growth on the dark side for which a mark has been awarded. A better answer would be to explain that the growth is because of cell elongation on the dark side. There are only two marks for this question so that further information has not been necessary to get full marks. The final part of the candidate's answer, that the shoot bends towards the light, is repeating information from the question and so does not qualify for credit.*

● *This candidate has gained 4 marks out of a possible 6. This is a grade C answer.*

● *A grade A or B candidate would be expected to get full marks.*

Questions to Answer

Answers to questions 2 – 4 can be found on page 360.

2 A green plant was placed in a dark cupboard. After 24 hours, some of the leaves were tested for starch. No starch was found in any leaf. The same plant was then placed in sunlight. One leaf, Q, was treated as shown in the diagram below. After a further 24 hours, leaves P and Q on the plant were tested for starch.

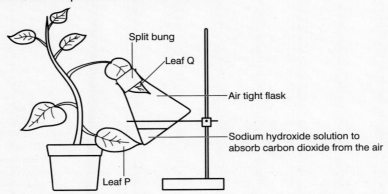

Split bung

Leaf Q

Air tight flask

Sodium hydroxide solution to absorb carbon dioxide from the air

Leaf P

a) Why was there no starch in the leaves after the plant had been kept in the dark cupboard for 24 hours? (2)

b) (i) Name the substance in green leaves which helps plants to make glucose. (1)

(ii) Name the gas produced by green leaves when they make glucose for starch production. (1)

c) The diagrams below show the results of starch tests on discs taken from leaves **P** and **Q**. The diagrams also show the parts of the leaves from which the discs were taken.

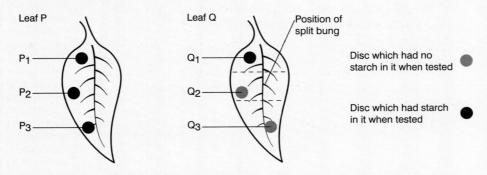

Leaf P

P_1

P_2

P_3

Leaf Q

Position of split bung

Q_1

Q_2

Q_3

Disc which had no starch in it when tested

Disc which had starch in it when tested

Explain the reasons for the differences between discs **P3** and **Q3** and between discs **P2** and **Q2**. (3)

London Examinations

3 Many people like to keep pot plants inside their houses.
These plants need watering regularly.

a) Draw arrows on the diagram to show the path taken by the
water from the time it enters the plant to the time it leaves as
water vapour.

(1)

b) (i) Name the process by which the water enters the plant
from the soil. (1)

(ii) Explain how this process works. (2)

c) (i) Explain briefly how a plant loses water and how it
controls the rate of loss. (2)

(ii) Suggest **two** reasons why it is important for a plant to
be able to control its rate of water loss. (2)

Midland Examining Group

4

Farmers often put fertilisers on their crops.

a) What effect do the fertilisers have on the crops? (1)

b) Nitrates are found in fertilisers. Which **two** other minerals are
also important in fertilisers? Choose **two** answers from:

aluminium   fluoride   lead   phosphate   potassium (1)

c) The minerals in fertilisers enter plants through the roots.
The roots cells use a lot of energy taking in minerals.
Explain why. (2)

d) Why do plants need nitrates? (2)

Midland Examining Group

Ecology is the study of how living things affect and are affected by other living things as well as by other factors in the environment.

Some ecological terms:

- **ecosystem** – a community and its environment, e.g. a pond
- **environment** – often used to refer to only the physical features of an area (but can include living creatures)
- **community** – the different species in an area, e.g. the grassland community
- **species** – one type of organism, e.g. a robin. (Only members of the same species can breed successfully with each other.)
- **population** – the numbers of a particular species in an area or ecosystem
- **niche** – an organism's way of life or part it plays in an ecosystem
- **habitat** – where an organism lives.

COMPETITION AND POPULATION

KEY POINTS

Competition for resources
- **Relationship of predator–prey population**
- **Exponential increase in human population**

Animals and plants are always trying to survive and reproduce. However there is always a 'struggle' for survival for various reasons. Animals struggle:

- for food
- for water
- for protection against the weather
- against being eaten by predators
- against disease
- against accidents.

Plants similarly may not survive through:

- a lack of water
- a lack of light
- a lack of minerals in the soil
- weather
- disease
- being eaten.

Animals and plants generally produce many young but their population sizes usually do not vary significantly because most of the young do not survive to adulthood. One reason for this is that they are having to **compete** for the resources they need, such as food, as there is not enough to go around. To help them survive, animals and plants are **adapted** to the environments in which they live. You will find out more about adaptations in chapter 6.

PREDATION

One of the factors affecting population size, of animals or plants, is the number of animals trying to eat them. The numbers of **predators** and **prey** are intimately connected. A famous example is that of snowshoe hares and their predators, lynxes, in northern Canada. Both animals were hunted for their fur and the fur company kept records of how many animals were caught, so we can estimate the sizes of the populations over nearly a hundred years. This is such a vivid example because when a predator's prey reduces in number, the predator usually eat something else instead. In this case the lynx did not.

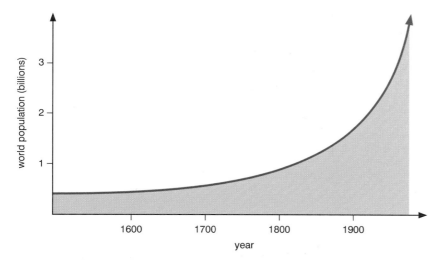

a good year for hares...

...followed by a good year for the lynx...

...followed by a poor year for the hares...

...followed by a poor year for the lynx...

lynx (eat hares)

snowshoe hare

Predator follows prey! Use this phrase to remember how the population graphs relate to each other.

HUMAN POPULATION

Unlike that of most other animals and plants the human population is very rapidly increasing in size. You can see from the graph that it is getting steeper and steeper. This means that not only is the population increasing but the *rate* of increase is also increasing. This is described as an **exponential** increase.

This rapid increase has been caused by:

- an increase in food production (through farming techniques)
- advances in medicine such as immunisations and antibiotics
- improved living conditions.

Check yourself

QUESTIONS

Q1 What is the difference between a habitat and an ecosystem?

Q2 In dense woodland there may not be many small plants growing on the ground. Suggest why.

Q3 In this country, foxes eat rabbits. However a graph of the two populations would not show the same oscillating (up and down) pattern as the lynxes and snowshoe hares on page 79. Why not?

Q4 In some summers there are much greater numbers of ladybirds than usual. (i) How would this affect the numbers of greenfly, their prey? (ii) Why don't the high numbers of ladybirds persist year after year?

REMEMBER! Cover the answers if you want to.

ANSWERS

A1 An organism's habitat is the place where it lives. The ecosystem includes this but also includes the other living things and other factors such as climate.

A2 This is because the taller trees will be taking most of the light and with their roots taking most of the water and minerals from the soil.

A3 If the numbers of rabbits went down, foxes would just concentrate their diet on other things – many live in towns scavenging in rubbish. If foxes died out, there are other things, including humans, that would still kill rabbits.

A4 (i) The greenfly numbers would go down. (ii) There would have to be enough food to feed them and they would have already eaten most of their prey. Also birds that feed on ladybirds would help to bring their numbers back down.

TUTORIAL

T1 *If you use scientific words in an exam make sure you use them properly. If you don't you could be saying something very different from what you intend. If you are not sure about a word try to think of another way of giving your answer.*

T2 *This is an example of **competition** in which the trees are out-competing the smaller plants for these important resources.*

T3 *There are very few other large animals in northern Canada which is why the example of lynxes and snowshoe hares is unusual in showing so clearly the way that predators and prey affect each other.*

T4 *Numbers of most species alter a little each year but they are usually kept more or less constant because of different factors like these.*

FOOD CHAINS

Food chains show how living things get their food. They also show how they get their **energy**. This is why they are sometimes written to include the **Sun**. Here is an example of a food chain:

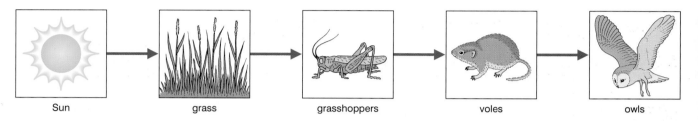

| Sun | grass | grasshoppers | voles | owls |

You may see food chains written going up or down a page, or even from right to left. All are correct as long as the arrows always point *towards* the living thing that is taking in the food or energy. The different parts of a food chain get their food in different ways and are given different names:

- **Producers** – e.g. grass. These are the green plants. They make their own food in **photosynthesis** using energy from the Sun. (You can read more about photosynthesis in chapter 4.)

- **Primary consumers** – e.g. grasshoppers. These are animals that eat plants or parts of plants such as fruit. They are also called **herbivores**.

- **Secondary consumers** – e.g. voles. These are animals that eat other animals. They may be called **carnivores** or predators.

- **Tertiary consumers** – e.g. owls. These are animals that eat some secondary consumers. They are also called **carnivores** or predators.

The animals hunted and eaten by predators are called prey. Animals that eat both plants and animals are called **omnivores**. If you eat meat and vegetables *you* are an omnivore.

The different stages of a food chain are sometimes called **trophic levels**. (Trophic means 'feeding'.) So for example producers make up the first trophic level, primary consumers make up the second trophic level and so on. Most food chains are not very long. They usually end with a secondary consumer or a tertiary consumer. Occasionally there might be an animal that feeds on tertiary consumers. This would be called a **quaternary consumer**. You will read about why most food chains are short on the next two pages.

> **KEY POINTS**
>
> **Energy is a resource**
> - **producers and consumers**
> - **disturbing the food web**
> - **pyramids of number**
> - **pyramids of biomass**
> - **energy 'loss' at each trophic level**

FOOD WEBS

Food webs are different food chains joined together. It would be unusual to find a food chain that was not part of a larger food web. Here is part of a food web:

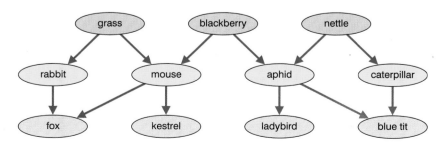

Food webs show how different animals feed, and they can also help us see what might happen if the food web is disturbed in some way. For example, what might happen if there was a disease that killed many, if not all, of the rabbits? Looking at the food web you can see that the foxes would have to eat more of the other animals they eat. This could mean that there would be *fewer* mice as a result. On the other hand because there would now be more grass for the mice to eat, so it could be that there would be *more* mice as a result. It is not possible to say for sure what would happen because there would also be other living things involved.

PYRAMIDS

There are other ways of describing the relationship between organisms in an ecosystem. Two ways you need to know about are **pyramids of numbers** and **pyramids of biomass**.

Pyramids of numbers show the relative numbers of each type of living thing in a food chain or web by trophic level. Sometimes pyramids of numbers can be 'inverted' if, for example, the producers are much larger than the consumers.

foxes
rabbits
grass

A pyramid of number.

blue tits
caterpillars
oak tree

An 'inverted' pyramid of numbers.

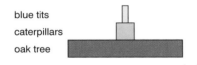

blue tits
caterpillars
oak tree

A pyramid of biomass.

Pyramids of biomass show the mass of living material at each stage in the chain or web, so this shows what you would have if you could weigh all the producers together, then all the primary consumers and so on. Pyramids of biomass are *never* inverted.

Pyramids either show organisms on a single food chain or trophic levels. In this second case, each stage of the pyramid may be labelled with all the living things at a particular trophic level, for example producers or primary consumers.

ENERGY FLOW

The arrows in food chains and webs show the transfer of energy. Not all the energy that enters an animal or plant is available to the next trophic level. Only energy that has resulted in an organism's growth will be available to the animal that eats it. For example think about cows feeding on grass in a field:

Sun → grass → cow

Only a small proportion of the energy in the sunlight falling on a field is used by the grass in photosynthesis. This is because:

- some may be absorbed or reflected by clouds or dust in the air
- some may be absorbed or reflected by trees or other plants
- some may miss the grass and hit the ground
- some will be reflected by the grass (grass reflects the green part of the spectrum and absorbs the red and blue wavelengths)
- some will enter the grass but will pass through the leaves.

The remainder of the energy can be used in photosynthesis. Some of the food made will be released in respiration to provide the grass's energy requirements, but some will be used for growth and will be available for a future consumer. Not even all this will be available to cows:

- some grass may be eaten by other animals (the farmer would probably call these pests)
- the cows will not eat all of the grass (e.g. the roots remain).

Of the energy available in the grass **ingested** (eaten) only a small proportion is used for growth by the cow. The diagram below shows what happens to the energy in an animal's food.

Energy is lost from the animal in two main ways:

- **Egestion** – the removal from the body, in faeces, of material that may contain energy but can not be digested.
- **Respiration** – the release of energy from food necessary for all the processes that go on in living things, e.g. movement, growth, repair, heat to keep warm. Ultimately most of this energy is lost as heat. (There is more about respiration in chapter 2.)

It is the fact that energy is lost from the chain at each stage that is responsible for the pyramid shape that you almost always get with a pyramid of biomass.

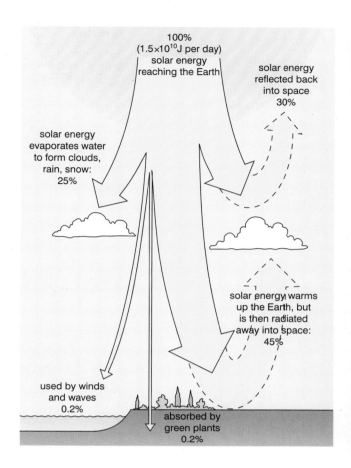

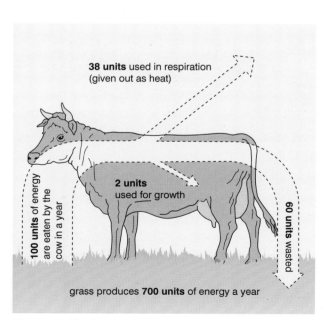

The energy flow in a young cow.

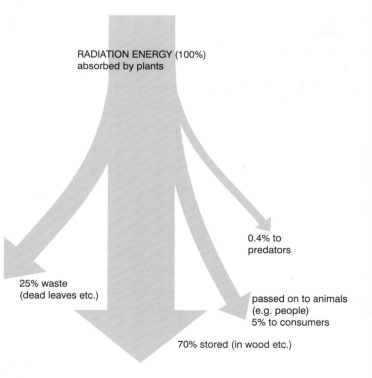

Farming

For a farmer raising plant or animal crops, a greater yield will be achieved if energy losses from the food chain can be reduced. There are different ways to do this. For a plant crop:

- remove plants that are competing for light, e.g. weeds
- remove other organisms that might damage the plant crop, i.e. pests
- ensure the plants have sufficient water and minerals to be able to photosynthesise and grow efficiently.

For animals:

- keep them indoors so they do not have to use as much energy to generate heat
- keep them penned up to reduce their movement so heat losses from respiration are reduced.

This is sometimes known as **battery farming** and some people disagree with it on moral grounds.

Check yourself

QUESTIONS

Q1 In Africa, zebras feed on grass. The zebras are eaten by lions.
a) Write this out as a food chain.
b) Label the producer, primary consumer and secondary consumer.

Q2 What do producers do?

Q3 Look at this food web from a pond and use it to answer the questions.

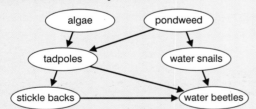

a) Which animal is both a predator and a prey animal?
b) From the web, (i) name one herbivore, (ii) name one carnivore.
c) Which animal is feeding at both the third and the fourth trophic levels?
d) If a disease killed most of the tadpoles what effect would this have on the water snails? Explain your answer.

Q4 Look at this food chain in a garden.

rose bushes → aphids → ladybirds

a) Draw and label a pyramid of numbers for this food chain.
b) Draw and label a pyramid of biomass for this food chain.

Q5 Even though a person eats throughout a week they may not weigh any more at the end. Why not?

Q6 Which is more energy efficient, to grow crops that we eat ourselves or to use those crops to feed animals which we then eat?

Q7 Why are there usually not more than five stages in a food chain?

ANSWERS

A1 Sun → grass (producer) → zebras (primary consumer) → lions (secondary consumer)

A2 Produce (make) food (glucose) from carbon dioxide and water using energy from the Sun in photosynthesis.

A3
a) Stickleback.
b) (i) Tadpole or water snail,
(ii) Stickleback or water beetle.
c) Water beetle.
d) Either the number of water snails would go up because there would be more pondweed to feed on, or the number could go down because the water beetles would need to eat more of them.

A4
a)
lady birds
aphids
rose bushes

b)
lady birds
aphids
rose bushes

A5 Some of the food is passed out, undigested, in faeces (egestion). Food that has been digested and absorbed will have been used mainly in respiration and the waste products will have been excreted.

A6 It is more energy efficient to eat plant crops rather than to raise animals. This is because we are eating at a lower trophic level. At each transfer between trophic levels energy is transferred away from a food chain.

A7 Energy is lost at each stage of a food chain so there is a limit as to how long a food chain can be.

TUTORIAL

T1
a) Do not forget that the arrows always point **towards** the thing that is taking in the food or energy. Including the Sun is not always necessary but if in doubt include it.
b) Producers are always the green plants. The primary consumers are the herbivores. The secondary consumers are the carnivores.

T2 'Making food' is a correct but simplistic answer. In an exam question play safe and give more details, especially if there is more than one mark for the answer.

T3
a) In food chains there are often animals that are both predator and prey.
b) Look carefully at the direction of the arrows in the web so you know which animals are feeding on which.
c) A food web is several different linked food chains. An animal that is at one trophic level in one chain can be at a different level in another chain.
d) It is not often possible to predict with certainty how a food web will respond to a change, however you can make reasoned guesses looking at the information you have. In an examination, the marks would not usually be for the prediction alone (which could be guessed), but rather for the explanation (which shows understanding).

T4 In a food chain where the producers are much larger than their consumers the pyramid of numbers is inverted. However the pyramid of biomass is always a normal pyramid shape.

T5 In other words very little of our food is used for growth and this applies to other organisms as well. This is why energy transfer along food chains is not very efficient.

T6 **Ecologically** therefore it is much more efficient to be vegetarian.

T7 This is why pyramids of biomass are always a pyramid shape.

NATURAL RECYCLING

Chapter 4 explains about the minerals that plants need from the soil. These minerals are mostly released from the decayed remains of animals and plants and their waste. This is one example of natural recycling. There is only a limited amount (on Earth) of the elements that living things need and use. Four of the most important elements in living things are **carbon** (C), **hydrogen** (H), **oxygen** (O) and **nitrogen** (N).

Important substances such as carbohydrates, fats and proteins are made up of carbon, hydrogen and oxygen. Proteins also contain nitrogen. The only way that animals and plants can continue to take in and use substances containing these elements are if the substances are constantly cycled around the ecosystem for reuse. We will look at two important examples of this biological recycling: the **Carbon Cycle** and the **Nitrogen Cycle**.

The Carbon Cycle

Plants take in carbon dioxide because they need the carbon (and oxygen) to use in photosynthesis to make carbohydrates and then other substances like protein.

When animals eat plants some of the carbon-containing compounds will be used to grow and some will be used to release energy in respiration.

As a waste product of respiration animals breathe out carbon as carbon dioxide which is then available for plants to use. (Don't forget that plants also respire producing carbon dioxide.)

Carbon dioxide is also released when animal and plant remains decay (decomposition) and when wood, peat or fossil fuels are burnt (combustion).

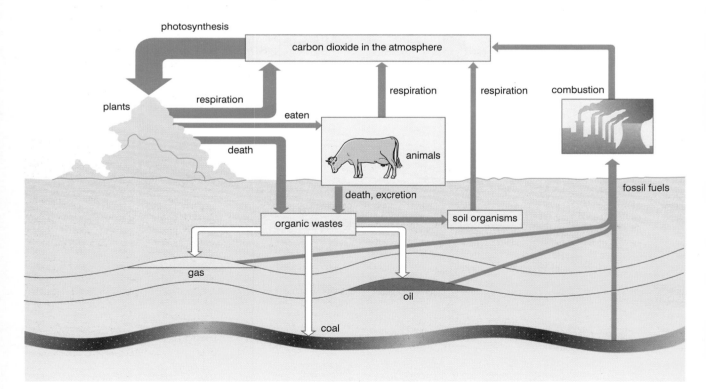

The Nitrogen Cycle

Living things need nitrogen to make proteins which are needed, for example, to make new cells for growth. The air is 78% nitrogen gas (N_2), but nitrogen gas is very unreactive and cannot be used by plants or animals. Instead plants use nitrogen in the form of **nitrates** (NO_3^- ions). The process of getting nitrogen into this useful form is called **nitrogen fixation**.

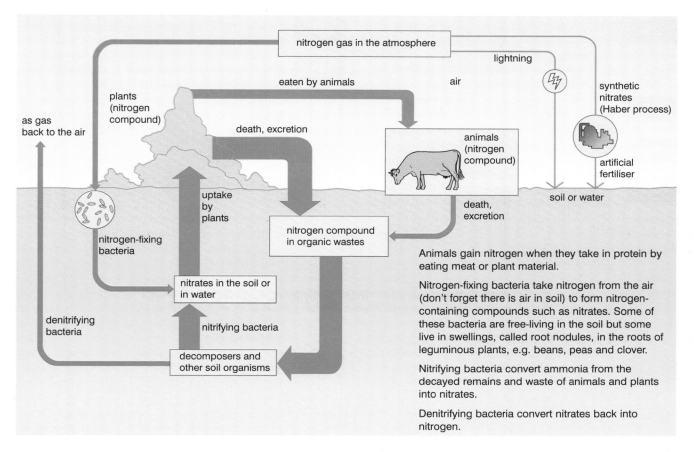

Animals gain nitrogen when they take in protein by eating meat or plant material.

Nitrogen-fixing bacteria take nitrogen from the air (don't forget there is air in soil) to form nitrogen-containing compounds such as nitrates. Some of these bacteria are free-living in the soil but some live in swellings, called root nodules, in the roots of leguminous plants, e.g. beans, peas and clover.

Nitrifying bacteria convert ammonia from the decayed remains and waste of animals and plants into nitrates.

Denitrifying bacteria convert nitrates back into nitrogen.

Decomposers

The decay of dead animal and plant remains is an important part of biological recycling. Decay happens because of the action of various types of micro-organisms, namely **bacteria** and **fungi**. These are known as **decomposers**. Conditions that allow these to thrive are conditions that will help decay:

● warmth (but not too hot to kill the micro-organisms)

● moisture

● presence of oxygen.

There are other organisms such as worms or woodlice that are known as **detritivores** that feed on and break down dead remains (known as detritus) exposing a greater surface area for the decomposers to act upon. Materials that can decompose are known as **biodegradable**.

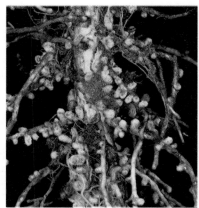

These root nodules contain nitrogen-fixing bacteria.

Check yourself

QUESTIONS

Q1 a) In the carbon cycle, which main process removes carbon dioxide from the atmosphere?

b) How is carbon dioxide put back in the atmosphere?

Q2 Why do living things need carbon?

Q3 What is nitrogen fixation?

Q4 Nitrogen-fixing bacteria cannot live in waterlogged soil but denitrifying bacteria can, which is why this kind of soil is low in nitrates. Some plants that live in these conditions are carnivorous, trapping and digesting insects. Suggest why you often find carnivorous plants in bogs.

Q5 What is the difference between nitrifying and denitrifying bacteria?

Q6 Which of the following are biodegradable: glass, paper, wood, pottery, steel?

REMEMBER! Cover the answers if you want to.

ANSWERS

A1 a) Photosynthesis.
b) Respiration and combustion.

A2 Carbon is one of the elements that important biological substances such as carbohydrates, proteins and fats, are made from.

A3 Changing nitrogen gas into more reactive forms such as soluble nitrates.

A4 This is the only way they can get the nitrogen they need.

A5 Nitrifying bacteria produce nitrates from decayed remains. Denitrifying bacteria convert nitrates and other nitrogen compounds into nitrogen gas.

A6 Paper and wood.

TUTORIAL

T1 a) *Make sure you remember the details of photosynthesis from chapter 4.*

b) *Until the last few hundred years combustion played quite a small part in this. It is now playing a bigger role and this is the reason for the build up of carbon dioxide in the atmosphere.*

T2 *This is why life on Earth is sometimes described as being 'carbon based'.*

T3 *Remember although most of the air around us is nitrogen few organisms can use it in this form.*

T4 *The nitrogen is made available by digesting the protein in the insects' bodies.*

T5 *Denitrifying bacteria therefore have the opposite effect to nitrogen-fixing bacteria.*

T6 *Steel may eventually corrode but this is not decomposition which is breakdown caused by microbes.*

HUMAN INFLUENCES ON THE ENVIRONMENT

The effects of humans on the environment are increasing for several reasons:

- An increasing population means that more resources are needed and used (e.g. land, raw materials for industry, sources of energy and food).

- Technological advances and an overall increase in the standard of living also mean an increased use in resources.

- Following on from this there is an increase in waste production – **pollution**.

These problems are caused particularly by the developed countries of the world. There has recently been an increase in people's awareness of these problems and of the consequences for the future if they persist:

- The use of some resources like minerals and fossil fuels cannot continue forever as they are **finite**. This means there is only a limited amount of these resources and once they have been used they cannot be replaced. In the future we will have to recycle these materials or find replacements that are renewable.

- Pollution caused by putting waste into the environment is causing changes that may be irreversible.

There are various ways of tackling these problems such as:

- Reducing the amount of raw materials we use (e.g. by using less packaging on products).

- Recycling more materials so reducing the need for, and energy cost of, extracting more raw materials.

- Being more energy efficient (e.g. by reducing heat losses from buildings or turning off lights when not needed).

- Using renewable energy sources that will not run out and cause little pollution (e.g. solar, wind or wave power).

- Using sustainable or renewable raw materials that can be grown again (such as wood).

There are many examples of the ways humans affect the environment. We will look at some of them.

KEY POINTS

Human use of resources
- **Agrochemicals**
- **Greenhouse gases**
- **Production of acid rain**
- **CFCs and ozone depletion**
- **Household waste**
- **Habitat destruction**

FARMING

Many farmers add different types of chemicals (known as **agrochemicals**) to their crops to increase crop production. **Insecticides** are chemicals which kill insects that damage the crops. They can cause unwanted effects by killing animals that are not pests. Some of them may also get into food chains. Insecticides such as DDT, which is now banned in many countries, have caused problems because they enter the food chain and as they cannot be broken down or excreted they remain in animals' bodies. Such substances are described as being **persistent**. This means that predators can contain much higher levels of the substance than the animals below them in a food chain. This is the reason why many birds of prey in Britain, like sparrowhawks and peregrine falcons, suffered in the 1950s and 1960s. Most modern insecticides are not persistent and breakdown naturally in the environment after a short while.

DDT affected many food chains. In this one, the figures give the relative concentration of DDT at each stage

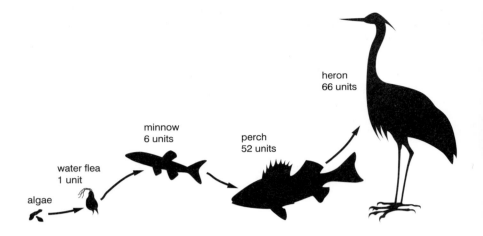

This scatter graph shows the relative thickness of a sample of sparrowhawk eggshells over a period of years. The drop in thickness was traced to the introduction of DDT.

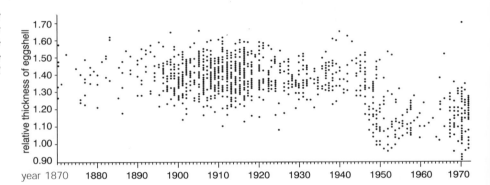

An alternative to using chemical insecticides is to introduce another organism that will kill the pests. This is known as **biological control**. This has advantages in that it does not cause pollution and pests cannot become resistant to it (as they can to chemical insecticides). However you have to be very careful that you do not get unwanted effects such as the introduced predators also attacking harmless or even useful animals.

Fungicides are chemicals that kill the fungi responsible for many plant diseases. Insecticides and fungicides together are often known as **pesticides**.

Herbicides, or weedkillers, are chemicals that kill plants that would otherwise grow among the crop and compete with them for light, water and minerals. These other plants might also be the home of pests. By removing such plants, however, you may be removing sources of food for other animals. Also any pests that normally feed on the crop will have plenty of food and may rapidly increase in numbers and become a bigger problem.

Fertilisers are added to improve the growth of crops. They may be either **organic** (natural) fertilisers such as compost or manure, or **inorganic** (artificial) fertilisers which are mined or manufactured to contain the necessary minerals.

Artificial fertilisers come in powdered or liquid form so they are easy to spread, but they can cause pollution. When it rains, the water will seep down through the soil and find its way into rivers and lakes. The soluble fertiliser dissolves in the water and is carried away – it is said to be **leached** from the soil. In the rivers and lakes the fertiliser causes excessive plant growth, especially of algae. Algae is a small plant that, when overgrown, makes the water murky and blocks much of the light. This causes plants under the surface to die and decay. The bacteria that cause the decay use up the oxygen in the water causing fish and other water animals to die. This whole process is called **eutrophication**.

Eutrophication can be reduced by using less fertiliser. This may actually help the farmer as there is a point beyond which adding more fertiliser will not help, and may even hinder, crop growth. Other ways of maintaining the soil fertility include:

- Using organic fertilisers which release minerals into the soil at a slower rate. This also helps to improve soil structure making it easier to plough for example.

- Using crop rotation which includes growing crops that contain nitrogen-fixing bacteria in their root nodules. This restores soil fertility when the crops or their roots are ploughed back into the soil.

This river is blocked with algae because of excess nitrates washed in from nearby fields.

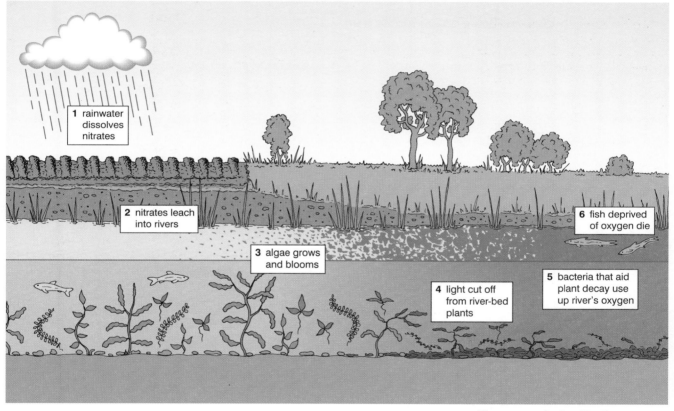

1 rainwater dissolves nitrates

2 nitrates leach into rivers

3 algae grows and blooms

4 light cut off from river-bed plants

5 bacteria that aid plant decay use up river's oxygen

6 fish deprived of oxygen die

The process of eutrophication.

GREENHOUSE EFFECT

Greenhouse gases include carbon dioxide, methane and CFCs. The levels of some greenhouse gases in the atmosphere are increasing due to the burning of fossil fuels, pollution from farm animals, and the use of aerosol and refrigerator gases. These gases allow short-wave radiation from the Sun to reach the Earth and warm the ground. The warm Earth gives off heat as long-wave radiation – much of which is stopped from escaping from the Earth by the greenhouse gases. This is known as the **greenhouse effect** and is responsible for keeping the Earth warmer than it otherwise would be.

The greenhouse effect is normal and important for life on Earth, however it is thought that the increasing levels of greenhouse gases is stopping even more heat escaping and causing the Earth to slowly warm up. This is known as **global warming** and if it continues this may lead to climatic changes and a rising sea level as polar ice melts. The temperature of the Earth is gradually increasing, but it is not known for certain if the greenhouse effect is responsible.

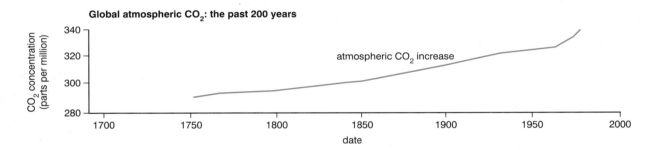

Global atmospheric CO_2: the past 200 years

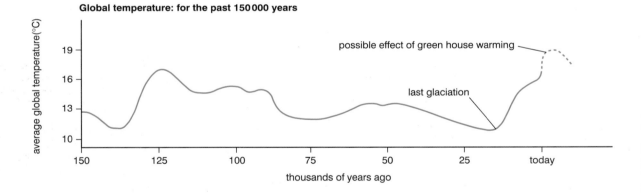

Global temperature: for the past 150 000 years

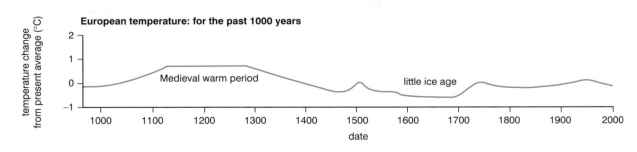

European temperature: for the past 1000 years

It may be that the observed rise in recent global temperatures is part of a natural cycle – there have been Ice Ages and intermediate warm periods before. Many people are concerned that it is not part of a cycle and it is important to act now to reduce emissions of these greenhouse gases by using less fossil fuels and turning to renewable energy sources.

ACID RAIN

Burning fossil fuels gives off many gases including sulphur dioxide and various nitrogen oxides. Sulphur dioxide combines with water to form sulphuric acid. Nitrogen oxide combines with water to form nitric acid. These substances can make the rain acidic. Acid rain harms plants that take in the water as well as the animals that live in the affected rivers and lakes. Acid rain also washes ions such as calcium and magnesium out of the soil, so depleting the minerals available to plants. It also washes aluminium out of the soil and into rivers and lakes where it is poisonous to fish.

It is possible to reduce the emissions of the gases causing acid rain but it is expensive and part of the problem is that the acid rain usually falls a long way from the places where the gases were given off. Emission of the gases from cars can be stopped using catalytic converters.

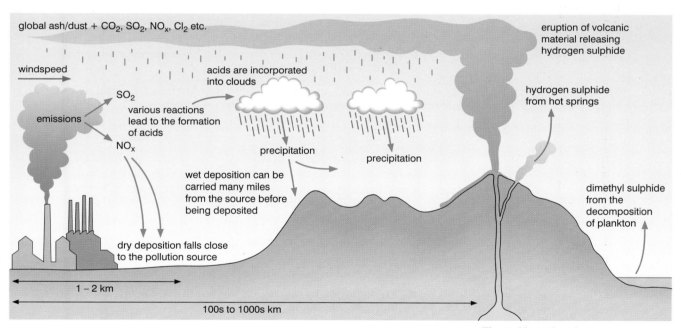

The problem of acid rain.

DEPLETION OF THE OZONE LAYER

High in the atmosphere there is a layer of gas called the **ozone layer**. Ozone is a form of oxygen in which the molecules contain three oxygen atoms giving it the formula O_3. Ozone is continually being broken down and reformed as shown below:

ozone formation

UV light

$O_2 \longrightarrow O + O$
$O_2 + O \rightarrow O_3$

ozone breakdown

$O_3 \rightarrow O_2 + O$
$O_3 + O \rightarrow 2O_2$

The ozone layer is important as it reduces the amount of **ultraviolet radiation** (or UV) that reaches the Earth's surface. This is important as UV radiation can damage cells causing skin cancer. Normally the rates at which

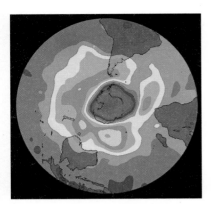

You can see the ozone 'hole' in this image drawn from data gathered by NASA satellites.

the ozone forms and breaks down are about the same. Certain gases, called CFCs, which are used as aerosol propellants and in refrigerator cooling systems, speed up how quickly the ozone breaks down without affecting how quickly it reforms. This means that more UV radiation will be reaching the Earth's surface so for example there will be an increase in skin cancer if people are exposed to a lot of sunlight.

We can reduce the damage by finding alternatives to CFCs, but many of these seem to contribute to other problems such as the greenhouse effect. Even if all CFC emissions stopped today it will still take around a hundred years for all the CFCs *presently* in the atmosphere to break down.

HOUSEHOLD WASTE

Houses produce a lot of waste including sewage and throw-away rubbish. Sewage has to be treated, by removing disease organisms and the nutrients that cause eutrophication, before it can be discharged into the sea. If household rubbish is burnt it can cause acid gas pollution and acid rain. Rubbish tips create their own problems:

- they are unsightly and can smell
- they can encourage rats and other pests
- tips covered with soil may build up explosive methane gas produced by rotting material
- covered over tips are not suitable for building on because the ground settles.

We can reduce the amount of material in our dustbins by recycling or reusing materials and not buying highly packaged materials.

HABITAT DESTRUCTION

Many natural habitats are being destroyed to create land for farming or building on. The rainforests are being cut down for their wood and to create farming or grazing land. Many species of animals and plants are only found in wetland areas which are being drained to 'reclaim' the land. This habitat destruction causes problems such as:

- reducing species diversity – many drugs and medicines were first discovered in plant extracts
- exposing soil to rain which washes the soil away to block rivers and cause flooding
- altering the climate – less water is transpired into the atmosphere.

Check yourself

Q1 What is the difference between 'finite' resources and 'renewable' resources?

Q2 Why are some pesticides described as 'persistent'?

Q3 What is the difference between the greenhouse effect and global warming?

Q4 What is eutrophication?

Q5 Acid rain can be reduced by passing waste gases through lime to remove sulphur dioxide. Why don't all factories automatically do this?

Q6 Does cutting down rainforests contribute to the greenhouse effect? Explain your answer.

REMEMBER! Cover the answers if you want to.

ANSWERS

A1 Finite resources are those that there is only a certain, limited, amount of. When that is used up then there is no more. Examples are coal, oil, aluminium and other metal ores. Renewable resources are those that can be replaced and will not run out. Examples are solar power, wind power, wood.

A2 They do not easily break down and so remain in the bodies of organisms that have taken them in.

A3 The greenhouse effect describes how some of the gases in the atmosphere restrict heat escaping from the Earth. Global warming is an increase in the Earth's temperature that may well be caused by an *increased* greenhouse effect.

A4 Eutrophication starts with the increased growth of algae and other water plants caused by an increase in available nitrates. The resulting blockage of light causes underwater plants to die. Bacteria in the water that cause the dead plants to decay use up the oxygen in the water. This deprives fish and other animals of oxygen and kills them.

A5 This would increase their costs which would have to passed on to their customers who may not want to pay more.

TUTORIAL

T1 *It is a common mistake to think that fossil fuels are renewable. They are not classed as such because they take millions of years to form.*

T2 *This is the reason they accumulate in animals and are passed along food chains.*

T3 *Remember that it is the **increased** greenhouse effect that people are concerned about. A greenhouse effect is not only normal but essential for life on Earth.*

T4 *It is common for an exam question asking about eutrophication to carry several marks, perhaps 4. Therefore you would be expected to give a full description. Depending on how many marks are available you might get full marks even though some details have been missed out, but play safe and give as much detail as you can. Look at the figure on page 91 again.*

T5 *An added problem is that, unlike pollution into rivers, the problems caused by acid rain take place a long way*

ANSWERS

A6 It can but *only* if the land is not used for growing trees again. As plants grow and increase in size they take in carbon dioxide from the air for photosynthesis to make the materials they need to grow. (Sometimes this is described by saying that carbon dioxide is 'locked up' in the trees.) This would reduce the amount of carbon dioxide in the atmosphere. However a forest does not continue getting bigger and bigger forever. In a mature forest, the death of some plants is balanced by the growth of new ones. When dead plants decay, carbon dioxide is released into the atmosphere which means that in a mature forest there is no *overall* uptake of carbon dioxide.

from the source of the pollution.

T6 ## TUTORIAL

Many people get confused by this idea which is not helped by such phrases as describing rainforests as 'the lungs of the world'. In mature forests just as much carbon dioxide is released as is taken in. If trees are cleared to make room for farming land then this would increase the greenhouse effect because less carbon dioxide is now 'locked up' as there is less plant material.

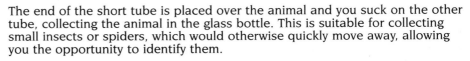

SAMPLING TECHNIQUES

You may have already carried out an ecological survey of an area such as your school grounds. This involves finding out what organisms live there, where they live, and the sizes of their populations. To do this different equipment and techniques may be used.

Pooter

The end of the short tube is placed over the animal and you suck on the other tube, collecting the animal in the glass bottle. This is suitable for collecting small insects or spiders, which would otherwise quickly move away, allowing you the opportunity to identify them.

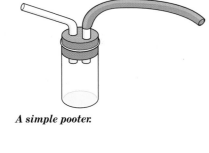

A simple pooter.

Pitfall trap

The trap can be left overnight. It is useful for collecting insects such as beetles that are more active at night and are too heavy to climb out. A lid or drainage holes will prevent water collecting and drowning the animals.

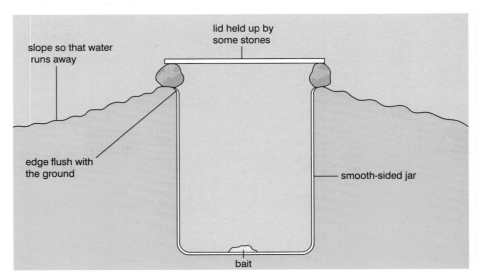

slope so that water runs away

lid held up by some stones

edge flush with the ground

smooth-sided jar

bait

The standard way of using a pitfall trap.

Tullgren funnel

Soil contains many small animals. One way of extracting these is to put a soil sample inside the funnel. The heat from the lamp makes the animals go deeper until they fall through the sieve inside the funnel and are collected in a pot underneath. This can be filled with ethanol if you wish to preserve the animals.

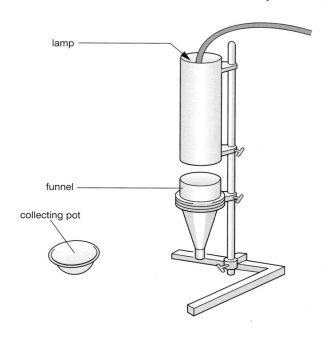

lamp

funnel

collecting pot

The apparatus used to make a Tullgen funnel.

Sweep net

A sweep net can be waved or brushed over grass or other vegetation to collect any insects there.

Beating tray

The tray is held under a bush or branch which is then shaken to dislodge any small animals which can then be caught using a pooter.

Quadrat

Quadrats can be used to carry out a plant survey, for example to compare the plant life in two different areas. In this case the quadrat would be randomly placed a number of times in each area. Each time, the plants within the quadrat are identified and counted. The results can be pooled to find the average for the two areas so that they can be compared. If the area of the region being investigated and the area of the quadrat are both known, then an estimate can be made of the total numbers of each species. Using quadrats in this way is called **sampling** because only a sample of the region is inspected. Inspecting every plant throughout the region would take too long.

Quadrats work best if:

- they are placed randomly
- a reasonable number of samples is taken, so any 'odd' results will be 'evened out' when averages are taken
- the plants being investigated are reasonably evenly spread throughout the area.

Quadrats can also be placed at regular intervals along a line to see how the habitat changes. This is called a **transect**. Quadrats could also be used to investigate the numbers or distribution of animals in a habitat as long as they are evenly distributed and not constantly moving around the area.

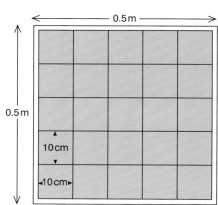

0.5 m

0.5 m

10 cm

10 cm

Check yourself

QUESTIONS

Q1 What equipment would be best to collect:
(i) tiny insects living in the soil, (ii) large ground beetles that come out at night, (iii) small insects living in a thick bush, (iv) ants.

Q2 A pupil wanted to estimate the number of dandelions in a field. She used a large quadrat that was 1 m² and the field was 200 m². She used the quadrat 10 times and counted a total of 25 dandelions. Estimate the total number of dandelions in the field.

Q3 Why is it important that quadrats are placed randomly?

REMEMBER! Cover the answers if you want to.

ANSWERS

A1 (i) Tullgren funnel, (ii) pitfall trap, (iii) beating tray and then a pooter, (iv) pooter.

A2 The average number of dandelions in 1 quadrat (1 m²) was $25 \div 10 = 2.5$. The field is 200 m² so the total number of dandelions is $2.5 \times 200 = 500$

A3 If the pupil in the previous question had only placed her quadrat where there were dandelions and had deliberately missed out those areas without any she would have calculated a much higher estimate.

TUTORIAL

T1 *Check with your teacher whether your science exam will ask you questions about these.*

T2 *You might have to do some maths so don't forget your calculator in your exam.*

T3 *It is for the same reason, to get a more reliable average and total estimate, that you also need to take a reasonably large number of measurements before you work out your average.*

EXAMINATION CHECKLIST

The facts and ideas that you should understand by studying this topic.

To be successful at Foundation GCSE Tier you should be able to:

- describe how the use of collecting and counting methods allows quantitative estimates of population size and distribution to be made
- explain how competition influences the population size of animals and plants
- explain how a predator population affects the prey population and vice versa
- construct food chains and pyramids of numbers from given information
- recall that energy enters food chains when plants absorb sunlight
- recall how farmers can produce more food if they use herbicides and pesticides, but that these practices can cause harm to the environment and health
- explain how fertilisers can leach into rivers and cause excessive plant growth
- recall the key factors in the process of decay
- recognise materials that can decay and be recycled
- recall how materials are recycled in a biological community
- recall that the human population is increasing and that this is causing an increase in pollution and the use of finite resources.

In addition, to be successful at Higher GCSE Tier you should be able to:

- explain the limitations of counting and collecting methods
- describe how the populations of predators and their prey regulate one another
- construct and interpret food webs and pyramids of biomass from given information
- recall that energy from the Sun flows through food chains by photosynthesis and feeding
- explain how energy is transferred to less useful forms at each stage in the food chain
- explain the shape of pyramids of biomass in terms of the efficiency of energy transfers
- explain how intensive food production and battery farming improves the efficiency of energy transfer
- explain how the excessive use of nitrogenous fertilisers, and other organic pollution containing nitrates, can lead to eutrophication
- describe the cycling of carbon and nitrogen by the activity of microbes and other living things
- recall that as the human population increases exponentially, there is a related increase in the use of resources and the production of pollution
- recall that the developed countries of the world, with a small proportion of the world's population, have the greatest impact on the use of resources and production of pollution.

Key Words

Tick each word when you are sure of its meaning.

biological control
carbon cycle
community
competition
consumer
denitrification
ecosystem
eutrophication
food chain
greenhouse effect
habitat
nitrogen-fixation
persistent
population
predator
producer
pyramid of biomass
quadrat
sampling
trophic level

EXAM PRACTICE

Sample Student's Answers & Examiner's Comments

a) The arrows all point in the correct direction, gaining one mark. The connections are all correct but an extra arrow has been added showing the foxes eating the owls so instead of gaining two marks for the connections only one is gained.

b) **i)** This is a satisfactory explanation. Although you could argue that the greenfly do not 'eat the tree', the explanation for why the moth population would increase is clear and one mark would be awarded.

ii) This is a satisfactory explanation and would also gain one mark.

c) The candidate has incorrectly drawn the pyramid. It would have been correct if a pyramid of numbers had been asked for. Remember that a pyramid of **biomass** always gets smaller at each trophic level.

● This candidate has gained 4 marks out of a possible 6. This is a grade C answer. A grade A candidate would be expected to score full marks.

---

1 This question is about a food web in an oak woodland.

Foxes and tawny owls eat woodmice.

The tawny owls also eat blue tits.

The blue tits feed on moths and greenfly.

Woodmice, moths and greenfly feed on the oak trees.

a) Use the information given to complete the food web described.

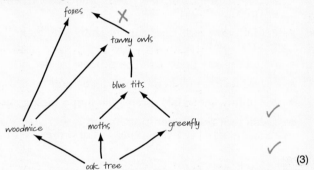

(3)

One year there are not as many greenfly as usual. This could make the moth population increase or it could make it decrease.

b) (i) Explain why this could make the moth population **increase**.

The moths would have more food as the green flies are not eating the tree so their numbers would go up ✓ (1)

(ii) Explain why this could make the moth population **decrease**.

Bluetits would have no greenflies to eat so they would have to eat more moths so their numbers would go down ✓ (1)

c) Draw a pyramid of biomass for the oak trees, woodmice and tawny owls. Label the parts of your diagram.

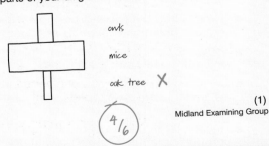

(1)

4/6

Midland Examining Group

Questions to Answer

Answers to questions 2 – 4 can be found on page 361.

2

hedge

Some students were surveying dandelions in a field. They noticed that the heights of the dandelions were all different. One student suggested that the differences in height could have been caused by the different conditions in the field.

a) (i) Was he correct? Give reasons for your answer. (2)

(ii) Explain how you would test to see if his answer was correct. (2)

b) The hedge was cut down and removed. What would happen to the height of the dandelions after some time? Explain your answer. (2)

Northern Examination and Assessment Board

3 The figure below represents the nitrogen cycle.

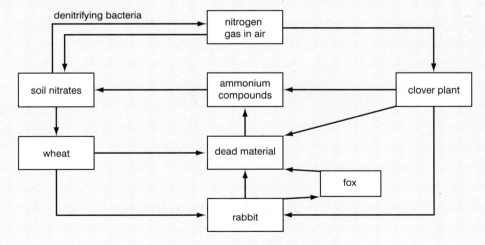

a) Bacteria play an important part in the nitrogen cycle.

(i) Denitrifying bacteria are anaerobic. Explain why plants growing in waterlogged soil may show signs of nitrogen deficiency. (3)

(ii) Why might a nitrogen deficient soil limit the growth of the plants? (2)

b) Explain why a good crop of wheat would require the addition of a fertiliser containing nitrogen compounds, while a crop of clover grown in the same field would not need the addition of nitrate fertiliser. (4)

c) Why are artificial fertilisers normally applied directly to crops in the **spring** but manure is added to the soil in **autumn**? (4)

Midland Examining Group

4 a) Give two disadvantages of using chemical pesticides. (2)

b) Biological control is another method of reducing the population of some pests. The graph below shows the changing population of a pest organism and a biological control agent over a 42 month period.

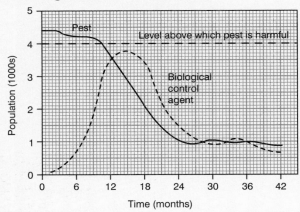

Explain what is happening between:

(i) 0 and 6 months (2)

(ii) 9 and 15 months (2)

(iii) 27 and 42 months. (2)

(iv) Explain why the system of control shown on the graph was considered successful. (2)

London Examinations

No two people are the same. Similarly, no two oak trees will be exactly the same in every way – they will be different heights, have different trunk widths and different numbers of leaves. This is known as variation.

VARIATION

There are two types of variation:

- **Discontinuous variation** – where a characteristic can have one of a certain number of specific alternatives. Examples would be gender where you are either male or female, or blood groups where you are either A, B, AB or O. This is sometimes referred to as **discrete** variation.

- **Continuous variation** – where a characteristic can have any value in a range. Examples would be body weight or length of hair.

There are two causes of variation:

- **Environmental** causes, for example your diet, the climate you live in, or accidents. Your surroundings, the way you have been brought up or your lifestyle influences characteristics.

- **Genetic** causes where the characteristic is controlled by your **genes**. Genes are inherited from your parents. Examples of characteristics in humans purely influenced by genes are eye colour and gender.

Many characteristics are influenced by environment *and* genes. It might run in your family that family members tend to be tall, but unless you eat correctly when you are growing you will not become tall even though genetically you have the tendency to be tall. There are other examples that are more controversial, such as human intelligence, where it is unclear whether the environment or genes is more influential.

GENES

Genes are instructions that direct the processes going on inside cells. If they affect the way cells grow and work then they can affect features of your body such as the shape of your face or the colour of your eyes. All the information needed to make a fertilised egg grow into an adult is contained in its genes. Inside virtually every cell in the body is a **nucleus**. The nucleus contains long threads called **chromosomes**. These threads are usually spread throughout the nucleus, but when the cell splits they gather into X-shaped bundles that can be seen through a microscope. The chromosomes are made of a chemical called deoxyribonucleic acid (**DNA**).

DNA is a very long molecule and along its length are a series of chemicals called **bases**. There are four different types of base in DNA: thymine (T), adenine (A), guanine (G) and cytosine (C). The order in which they occur is a genetic code. The code spells out instructions that control how the cell works. Each length of DNA that spells out a different instruction is known as a **gene**. Each chromosome contains thousands of genes. All the genes in a particular living thing are known as its **genome**.

The way that genes actually work is by telling cells how to make different proteins. Many proteins are enzymes that control the chemical processes inside cells. When you talk about a gene for blue eyes, what you are actually talking about is a gene that ensures that a person has blue pigment in their eyes. Most of our features that are controlled genetically are in fact affected by several genes.

KEY POINTS

Genes:

- **cause some variation**
- **are lengths of DNA**
- **are copied in mitosis**
- **are copied and shuffled in meiosis.**

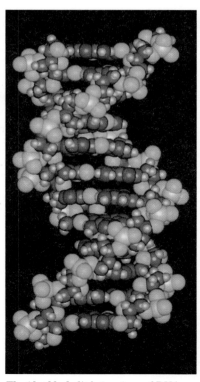

The 'double-helix' structure of DNA was discovered by James Watson and Francis Crick working in Cambridge in 1953. Rosalind Franklin, working in London, contributed important work on DNA's structure.

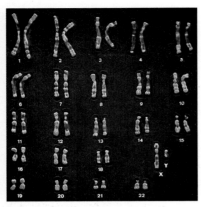

The 23 pairs of chromosomes in a human cell, including the x and tiny y.

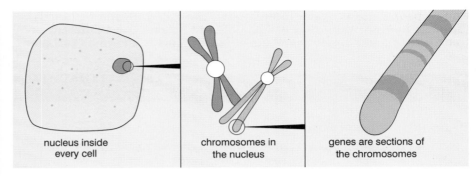

nucleus inside every cell

chromosomes in the nucleus

genes are sections of the chromosomes

Sometimes genes can be altered so that their message becomes meaningless or a different instruction. This is known as a **mutation**. Mutations can be caused by:

- radiation
- certain chemicals, called mutagens
- spontaneous change.

If mutations occur in sex cells then they may be passed on to offspring.

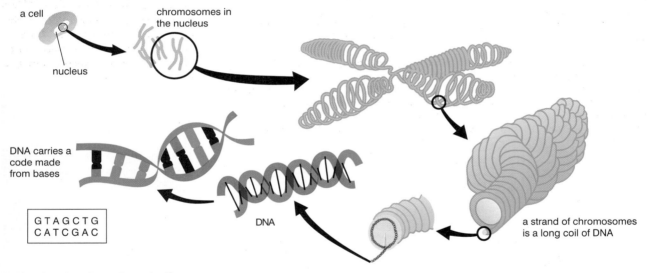

a cell

nucleus

chromosomes in the nucleus

DNA carries a code made from bases

GTAGCTG
CATCGAC

DNA

a strand of chromosomes is a long coil of DNA

DNA is found in the nucleus of cells.

CELL DIVISION

Cells grow by splitting in two. This is called **cell division** and is part of normal body growth. It is also the way that single-celled organisms reproduce and is the only type of cell division involved during **asexual reproduction** (reproduction that does not involve sex cells).

Before a cell splits, its chromosomes duplicate themselves. The new cells formed, sometimes called the **daughter cells**, contain chromosomes identical with the original cell. This type of cell division, in which the new cells are genetically identical to the original, is known as **mitosis**. Cells or organisms that are genetically identical to each other are known as **clones**.

Stage 1. The chromosomes get fatter and become visible.

Stage 2. Each chromosome makes an exact copy of itself

Stage 3. The chromosomes line up in the middle of the cell

Stage 4. The colour chromosomes part and move to opposite halves of the cell.

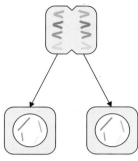

Stage 5. The cell splits in two.

A spider plant forms new plants at the end of stalks ('runners'). These new plants eventually grow independently of the parent plant. This is an example of asexual reproduction.

There is another type of reproduction in which the new individuals have been formed by two special cells, called **sex cells** or **gametes**, usually one from each parent. In humans the gametes are sperm cells and egg cells. In flowering plants the gametes are the pollen cells and the female 'egg' cells or ovules. This type of reproduction is called **sexual reproduction**. The offspring produced by sexual reproduction are genetically different to their parents. The gametes contain half the normal number of chromosomes so that when they join together (at **fertilisation**) the normal number is restored. For example, human cells normally contain 23 pairs of chromosomes: 46 in total. Human egg and sperm cells therefore contain only 23 chromosomes, so when they join together the new cell formed has the normal 46. (The number of chromosomes depends on an organism's species.) A different type of cell division called **meiosis** ensures that the new cells only have the half number of chromosomes.

Stage 1. A normal cell in the sex organs. It has the usual number of chromosomes, which fatten and become visible.

Stage 2. Each chromosome makes a copy of itself

Stage 3. Remember that each chromosome has its own 'partner'. They carry similar 'blueprints' – one from the original father and one from the mother. These partner chromosomes get together and line up across the middle of the cell.

Stage 4. The chromosome pairs swap genes, split up and move to opposite halves of the cell.

Stage 6. In each new cell, the chromosomes part and move to opposite halves of their cells.

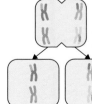

Stage 5. The cell splits into two new cells. These special cells have half the usual number of chromosomes.

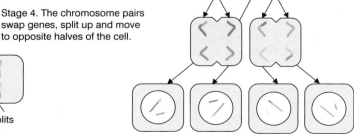

Stage 7. The two new cells split once more to make four new cells. Each has half the usual number of chromosomes.

The process of meiosis.

Every gamete formed has one of each pair of chromosomes. However which one of each chromosome pair ends up in a particular gamete is purely random. In humans this means that an individual can produce gametes with billions of different combinations of chromosomes. Add this together with the fact that when chromosome pairs are lying next to each other they swap lengths of DNA (this is called **crossing over**) altering the combination of genes on a chromosome, and it is not surprising that even with the same parents, we can look very different to our brothers and sisters.

Check yourself

QUESTIONS

Q1 Which of the following are examples of (a) continuous and (b) discontinuous variation? Hair colour, blood group, foot size, gender, hair length.

Q2 Bob and Dave are identical twins. Dave emigrates to Australia where he works outside on building sites. Dave stays in England and works in an office. Suggest how they might look (a) similar and (b) different. Explain your answers.

Q3 Put the following in order of size starting with the smallest: cell nucleus, chromosome, gene, cell, chemical base.

Q4 Most human body cells contain 46 chromosomes. Why is it important that the sex cells contain only 23?

Q5 Write down the differences between mitosis and meiosis.

REMEMBER! Cover the answers if you want to.

ANSWERS

A1
a) Hair colour, hair length, foot size.
b) Gender, blood group.

A2
a) Eye colour, height, shape of face (there are other examples). These are genetic features and as they are identical twins they will have the same genes.
b) Skin colour, accent, how muscular they are (there are other examples). These features are affected by the environment. They are doing different jobs in different climates, so their environments are different.

A3 Chemical base, gene, chromosome, cell nucleus, cell.

A4 When an egg and sperm join together at fertilisation the full number is restored. As the 23 chromosomes in each gamete are one of each pair of chromosomes, it means that the offspring receive one copy of each chromosome from each parent.

TUTORIAL

T1 *If the features fall into distinct groups they are discontinuous, if not, they are continuous.*

T2 *Some features like eye colour are purely genetically controlled and so will be identical. Some features like accent are purely environmentally controlled and so would probably differ. Many features, like skin colour, are controlled both by genes and the environment (i.e. tanning) so they would also show differences.*

T3 *Look back at the diagram on p. 104 if you're not sure.*

T4 *Other organisms may contain different numbers of chromosomes but the gametes will always contain half of this number. The full number of chromosomes is known as the diploid number. The half number found in the gametes is known as the haploid number.*

INHERITANCE

Sometimes features, such as the colour of your eyes, are passed on, or **inherited**, from your parents, but other features may not be passed on. Sometimes features may appear to miss a generation, e.g. your grandmother may have ginger hair and so do you, but neither of your parents do. You may have the same parents as your brother or sister but you may have inherited a different combination of features from them. This section is about how features are inherited and you should be able to explain these examples by the end of the section.

Lets look at one example of how features may be passed on. Leopards may occasionally have a cub that grows up being completely black in colour instead of having the usual spotted pattern. It would be known as a black panther but is still the same species as the ordinary leopard.

Just as in humans, leopard chromosomes occur in *pairs*. One pair will carry a gene for fur colour. There will be a copy of the gene on each chromosome so there are *two* copies of the gene in a normal body cell. Both copies of the gene may be identical but sometimes they may be different, one being for a spotted coat and the other for a black coat. Different versions of the same gene like this are called **alleles**. Different combinations of alleles will produce different colour fur:

spotted coat allele + spotted coat allele = spotted coat
spotted coat allele + black coat allele = spotted coat
black coat allele + black coat allele = black coat

The black coat only appears when *both* of the alleles are for the black coat. As long as there is at least one allele for a spotted coat present, the coat will be spotted. The allele for a spotted coat overrides the allele for a black coat. This is quite common. Alleles like the one for the spotted coat are described as **dominant**. Alleles like the one for the black coat are described as **recessive**.

Leopard cubs receive their genes from their parents, half from each one. Remember that eggs and sperm cells only contain half the normal number of chromosomes as normal body cells. This is because they only contain one of each chromosome pair. When an egg and sperm join together at fertilisation the new cell formed, or **zygote**, which will develop into the new individual, now has two copies, or alleles, of each gene. An individual's combination of genes is called their **genotype**. An individual's combination of physical features is called their **phenotype**. Your genotype influences your phenotype.

We can show the influence of the genotype in a **genetic diagram**, using the example of the leopards. The letter **S** stands for the allele for a spotted coat, which is dominant. The lower case letter **s** will stand for the recessive allele for the black coat. Two spotted parents who have a black cub must each be carrying an S and an s:

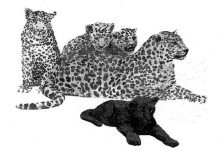

> **KEY POINTS**
>
> **Alleles:**
> - **are types of the same gene**
> - **can be recessive or dominant**

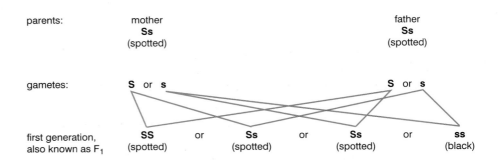

This is an example of a **monohybrid cross**. 'Mono' means one, and there is only one characteristic, colour, being looked at. A 'hybrid' is produced when two different types breed, or cross. Exam questions will often ask you to explain a cross 'using a genetic diagram'. This means that you are expected to set it out either like this or as another type of genetic diagram known as a **Punnett Square**. The same example as above is shown in the Punnett Square below.

		male gametes	
		S	s
female gametes	S	SS (spotted)	Ss (spotted)
	s	Ss (spotted)	ss (black)

Having two identical alleles, e.g. SS or ss, is described as being **homozygous**.
Having two alleles that are different, e.g. Ss, is described as being **heterozygous**.

You can see from both diagrams that when two heterozygous parents, both carrying two different alleles, are crossed, there is a **3:1 ratio** among the offspring of those showing the feature controlled by the dominant allele to those showing the feature controlled by the recessive allele. This is because of the random way that gametes combine, i.e. a sperm cell carrying an **S** allele does not have more chance of fertilising an egg than a sperm cell carrying an **s** allele. The 3:1 ratio refers to the *probabilities* of particular combinations of alleles. So there is a 1 in 4 chance of a cub being black for example. If there is a large number of offspring you would expect something near the 3:1 ratio of spotted to black cubs. However, because it does refer only to probabilities you would not be too surprised if in a litter of four, two of the cubs were black or if none were black.

How genes are inherited was first worked out in the nineteenth century by an Austrian monk, Gregor Mendel, who did lots of breeding experiments using pea plants.

INHERITED DISEASES

There are certain diseases or disorders that can be inherited.
Examples include:

- cystic fibrosis – in which the lungs become clogged up with mucus
- haemophilia – in which blood does not clot as normal
- diabetes – in which insulin is not made in the pancreas (see chapter 3)
- sickle-cell anaemia – in which the red blood cells are misshapen and do not carry oxygen properly.

Most inherited diseases are caused by faulty genes. For example, one form of diabetes is caused by a fault in the gene carrying the instructions telling the pancreas cells how to make insulin. Most of these faulty alleles are recessive which means that you have to have two copies of the faulty allele to show the disorder. Many more people will be **carriers** having one normal allele as well as the faulty version. The diagram below shows how two carrier parents could have a child with the disorder.

		male gametes	
		D	**d**
female gametes	**D**	**DD** (makes insulin)	**Dd** (makes insulin)
	d	**Dd** (makes insulin)	**dd** (has diabetes)

SEX DETERMINATION

Whether a baby is a boy or a girl is also genetic and is decided by one pair of chromosomes called the **sex chromosomes**. There are two sex chromosomes, the **X chromosome** and the **Y chromosome**. Boys have one of each and girls have two X chromosomes. This means that egg cells always contain one X chromosome but sperm cells have an equal chance of containing either an X chromosome or a Y chromosome. This means that a baby has an equal chance of being either a boy or a girl. This is shown in the diagram below.

		male gametes	
		X	**Y**
female gametes	**X**	**XX** (girl)	**XY** (boy)
	X	**XX** (girl)	**XY** (boy)

Sex-linked inheritance

Some genetic disorders such as red–green colour blindness are more common in men than women. The recessive allele causing the disorder is found on the X chromosome, on a part for which there is no equivalent on the Y chromosome as it is smaller (see photograph on page 104). This means that males only need to have *one* copy of the recessive allele to show the disorder while females would need two copies.

Check yourself

QUESTIONS

Q1 What is the difference between the following genetic terms:
(i) allele and gene, (ii) genotype and phenotype, (iii) homozygous and heterozygous, (iv) dominant and recessive?

Q2 In humans, the gene that allows you to roll your tongue is a dominant allele. Use a genetic diagram to show how two 'roller' parents could have a child who is a 'non-roller'.

Q3 In leopards the allele for being spotted, S, is dominant to the allele for being black, s. How could you tell whether a spotted leopard was Ss or SS, given that you have black leopards that you know are ss?

Q4 John and Mary have had a little girl who has cystic fibrosis. Cystic fibrosis is caused by a recessive allele. What is the probability of their next child also having cystic fibrosis? Neither John nor Mary have the condition.

Q5 A couple have had two baby girls. What is the chance of their next baby being a boy?

REMEMBER! Cover the answers if you want to.

ANSWERS

A1 (i) Alleles are alternative forms of the same gene, e.g. the gene that controls fur colour in leopards has two alleles S (spotted) and s (black). (ii) Genotype refers to the combinations of alleles, e.g. Ss, whereas the phenotype refers to the characteristics shown, e.g. spotted coat. (iii) Homozygous is having two identical alleles, e.g. SS or ss, but heterozygous is having two different alleles, e.g. Ss. (iv) The dominant allele is the one that is shown in the phenotype by a heterozygous individual, e.g. an Ss leopard is spotted therefore the S (spotted) allele is dominant and the s allele is recessive.

TUTORIAL

T1 *You must make sure you are clear about these terms. Do not be put off by them. Just learn them and make sure you use them correctly.*

ANSWERS

A2 Let R be the dominant allele for tongue rolling and r the recessive allele for not being able to roll your tongue.

parents: **Rr**   **Rr**

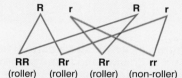

gametes: **R**   **r**   **R**   **r**

F_1 generation: **RR**   **Rr**   **Rr**   **rr**
                (roller) (roller) (roller) (non-roller)

A3 Cross the spotted leopard with a known ss leopard. If there are any black cubs then they must be ss and have received one of the s alleles from the spotted parent, which would therefore be Ss. If there are no black cubs the spotted parent is likely to be an SS. (The random nature of fertilisation means that an Ss leopard may still produce no ss cubs. This is unlikely with a large litter).

A4 As neither has the condition both must be heterozygous carriers of the recessive allele if their child has two copies of the recessive allele. The probability of a child being homozygous recessive, i.e. having cystic fibrosis, is 1 in 4.

A5 It does not matter how many children they have had. With each baby there is still a 1 in 2 (or 50:50) probability of it being a boy.

TUTORIAL

T2 *If not being able to roll your tongue is a recessive condition then the 'non-roller' child must be rr. This means each parent must have at least one r allele. As the parents can roll their tongues therefore they must both be Rr.*

T3 *Crossing with a homozygous recessive is known as a test cross or back cross. It is used because the homozygous recessive is the only genotype that can be identified by its phenotype alone.*

T4 *Look back at the monohybrid cross shown on page 108. You can see that the offspring showing the recessive condition are in the ratio of 1:3 to those showing the dominant condition. Another way of saying this is that there is a 1 in 4 probability of a child being homozygous recessive and showing the recessive condition. The fact that the couple have one child is not a consideration (see next question).*

T5 *Some people find this difficult to understand. Imagine tossing a coin. It is a 50:50 chance whether it lands heads or tails. It does not matter how it landed the last time it was thrown.*

KEY POINTS

Desirable genes can be maintained by:
- **selective breeding**
- **plant cuttings**
- **tissue culture**
- **genetic engineering.**

APPLICATIONS

Understanding about inheritance and genes has allowed people to change the genetic makeup of living things and therefore the features they show. We will look at three areas:

- selective breeding
- cloning
- genetic engineering.

SELECTIVE BREEDING

Selective breeding (also known as **artificial selection**) has been used by farmers for hundreds of years to change and improve the features of the animals or plants they grow. Examples would be producing:

- apples that are bigger and more tasty
- sheep that produce more or better quality wool
- cows that produce more milk or meat
- wheat that can be planted earlier so giving an extra crop
- plants that are more resistant to disease.

In every case the principle is the same. You select two individuals with the features that you want and breed them together. From the next generation you again select the best offspring and use these for breeding. This will be repeated over many generations. So for example if you wanted to produce dogs that had short tails, you would first select a short-tailed male and a short-tailed female from the ones you had. From their offspring you would select the ones with the shortest tails to use for breeding. After repeating this over many generations you would find that you almost always produced short-tailed dogs. In other words they would be 'breeding true'. Simply selecting for different features such as size, shape and so on has produced all the breeds of dogs we have today which are all descended from domesticated wolves thousands of years ago.

The power of selection. Selective breeding has produced an enormous variety of dog breeds from wild wolf stock (right).

CLONING

If you have an animal or plant with features that you want, there is a possibility that its offspring may not show those features. However there are ways of reproducing the organism to produce new ones that are genetically identical to the original. A simple way of doing this with plants is by taking **cuttings**. A cutting is a shoot or branch that is removed from the original and planted in soil to grow on its own. Some plants may do this easily, others may need rooting-powder hormones (see chapter 4) to encourage them to grow roots.

A more modern version of taking cuttings is **tissue culture**. This procedure is used to grow large numbers of plants quickly as only a tiny part of the original is needed to grow a new plant. However special procedures have to be followed:

1. Cut many small pieces, called **explants**, from the chosen plant.
2. Sterilise the pieces by washing them in mild bleach to kill any microbes.
3. In sterile conditions transfer the explants on to a jelly-like growth medium that contains nutrients as well as plant hormones to encourage growth.
4. The explants should develop roots and shoots and leaves.
5. When the plants are large enough they can be transferred to other growth media and eventually to a normal growth medium like compost.

A small cutting from a geranium that has been encouraged to grow fresh roots.

These tiny sundew plants have all grown from tiny cuttings of a single plant.

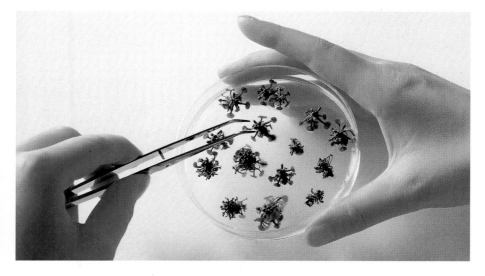

Another way of cloning has been recently developed that allows clones of animals to be produced. This involves making ordinary body cells grow into new animals. Dolly the sheep was the first animal clone grown from the cell of a fully grown organism – in Dolly's case a cell from the udder of a sheep.

GENETIC ENGINEERING

Selective breeding works by trying to bring together desirable combinations of genes. **Genetic engineering** is a more recent development that allows this to be done quickly and directly. It involves taking genes from one organism and inserting them into cells of another. For example, genes have been inserted into crop plants to make them more resistant to attack by pests. The gene that carries the instructions for making the human hormone insulin has been inserted into bacteria which now produce insulin. This is where much of the insulin to treat diabetes comes from. The gene was originally from a human cell so the insulin is identical to that made normally in the pancreas.

There are several steps involved in genetic engineering:

1. Find the part of the DNA that contains the gene you are looking for.
2. Remove the gene using special enzymes that cut the DNA either side of the gene.
3. Insert the gene into the cells of the organism you want to change.

Step 3 may be done directly but it often involves using other organisms, such as bacteria or viruses, to act as **vectors**. This means the gene is first inserted in them and they transfer the gene to the host when they infect it. Inserting the gene into bacteria means that many copies can be made of the gene because bacteria reproduce asexually producing new cells that are genetically identical to the original.

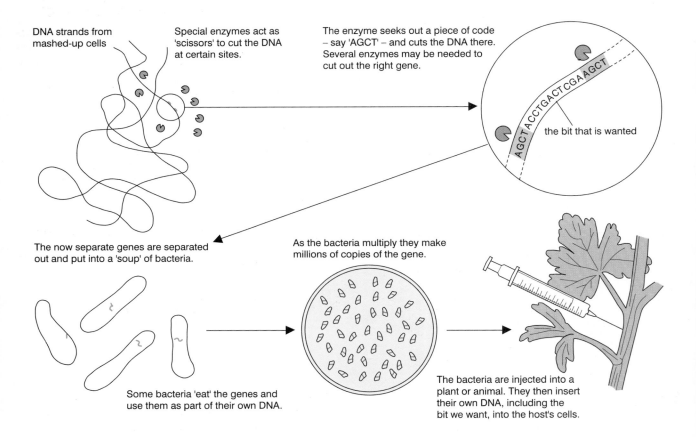

DNA strands from mashed-up cells

Special enzymes act as 'scissors' to cut the DNA at certain sites.

The enzyme seeks out a piece of code – say 'AGCT' – and cuts the DNA there. Several enzymes may be needed to cut out the right gene.

AGCTACCTGACTCGAAGCT

the bit that is wanted

The now separate genes are separated out and put into a 'soup' of bacteria.

As the bacteria multiply they make millions of copies of the gene.

Some bacteria 'eat' the genes and use them as part of their own DNA.

The bacteria are injected into a plant or animal. They then insert their own DNA, including the bit we want, into the host's cells.

Although genetic engineering promises many benefits for the future, such as curing genetic diseases or changing crops to make them disease resistant, there are possible dangers. For example the wrong genes could accidentally be transferred or viral infection could transfer inserted genes to other unintended organisms. Imagine the problems if instead of crops becoming resistant to disease, weeds became immediately resistant to herbicides or harmful bacteria became immediately resistant to medicines like antibiotics.

Check yourself

QUESTIONS

Q1 Suggest useful features that a farmer might want in the following: (i) cows, (ii) tomatoes.

Q2 A farmer wants to produce sheep with finer wool. How could this be done by selective breeding?

Q3 What are the advantages and disadvantages of producing new roses by tissue culture (cloning)?

Q4 What is the advantage of tissue culture over taking cuttings?

Q5 Both genetic engineering and selective breeding are ways of producing new combinations of genes in an organism. What are the advantages of genetic engineering?

REMEMBER! Cover the answers if you want to.

ANSWERS

A1 (i) Higher milk production or more muscular bodies so more meat is produced.
(ii) Larger size, better taste, brighter coloured fruit, increased resistance to disease. (All good answers.)

A2 Pick sheep with finer wool than the others. Breed these together. From the offspring pick those with the finest wool and breed these. Continue this over many generations.

A3 Advantages: Many roses with identical flowers can be produced relatively quickly. This would be useful if the roses on this particular plant sold well. Disadvantages: There is no variation, so if one rose was attacked by a plant disease then all the rest would be susceptible too.

A4 Tissue culture allows many new plants to be grown directly from the original whereas with cuttings, the piece removed has to be much bigger and so fewer can be taken.

A5 New combinations of genes can be produced quickly and reliably with genetic engineering. Genetic engineering allows genes to be taken from one species and given to another, which could not be done with selective breeding as different species cannot breed together.

TUTORIAL

T1 These are all features that are important to a farmer as they would affect the quality of the produce and therefore the profits. They could be improved by selective breeding or genetic engineering

T2 The principle of selective breeding is the same whatever the example.

T3 The key point about cloning is that the new plants are genetically identical. This may or may not be an advantage.

T4 The disadvantage of tissue culture is that sterile (or aseptic) conditions have to be maintained and so it is more involved and costly in this respect.

T5 The second answer is probably the more important. For example a gene for resistance to a particular disease could be taken from one species and given to another.

EVOLUTION

Techniques like selective breeding can change the features of animals and plants. Natural processes also allow animals and plants to change over millions of years. This change is called **evolution**. Strictly speaking we cannot *prove* that evolution has happened as we can't go back in time and watch. However there is a lot of evidence that evolution has happened.

Fossils show that very different animals and plants existed in the past. In some cases we can see steady progressive changes over time. In other cases there is no evidence of a particular organism's ancestors or descendants – the likelihood of a fossil forming in the first place is very rare. An animal or plant has to die and its body lie undisturbed, become covered in silt and eventually turn to rock for a fossil to form. Even if fossils form they may not be discovered or they may be destroyed by erosion. It is also likely that evolution happened more quickly or slowly in different cases. If it happened relatively quickly there would be less chance of finding fossils of the intermediate stages. Even when fossils are found they usually only provide a record of the hard tissues like bone.

Affinities are similarities that exist between different species. For example animals as different as humans, dolphins and rats have much the same number and basic arrangement of bones in their skeletons. The easiest explanation is that they are similar because they are *related*, in other words they have evolved from a *common ancestor*.

Ideas about evolution have been around for a very long time, but it was only after 1859, when Charles Darwin published his famous book O*n the Origin of* S*pecies*, that the ideas became widely discussed. (The scientist Alfred Wallace also came up with the same ideas as Darwin.) Many people were angered by Darwin's ideas as they believed they went against the Christian Churches' teachings that God had created the world and all the living things in it. Many newspaper cartoons and articles ridiculed his ideas. Some of these were based on the mistaken idea that Darwin had said that humans had evolved from apes like those present today such as chimpanzees.

The idea of evolution is not that the animals and plants we see around us evolved from each other, but rather that some have similar features because they have evolved from some common ancestor. The diagram below illustrates how humans and apes may have evolved from a common ancestor. Humans and chimpanzees share more similarities than the other types of apes which means that they probably share a more recent ancestor than do the others. Today there are people who still do not believe in the idea of evolution.

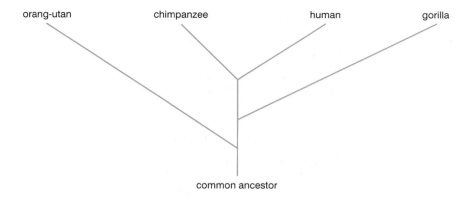

NATURAL SELECTION

Darwin suggested a theory of how evolution could have happened. He called his theory **natural selection**. Darwin partly based his ideas on his observations of the different wildlife he found on a lengthy sea voyage. He also used his knowledge of selective breeding. Natural selection is the same basic idea except it is 'nature' rather than the breeder that determines which animals or plants are able to reproduce and pass on their features.

These are the stages in Darwin's argument:

1. Animals and plants produce more offspring than will ever survive (see chapter 5). This means that if some individuals are slightly better adapted they will have a better chance of surviving than the others. For example a zebra that could run just that bit faster than the others would be less likely to be caught by lions, or a cactus with slightly sharper spines would be less likely to be eaten. This idea is sometimes known as the 'survival of the fittest'.

2. Those animals and plants that are more likely to survive are more likely to reproduce. If those features that helped them survive are genetically controlled they may be passed onto their offspring. Offspring who inherit those good features are in turn more likely to survive and reproduce than those offspring that do not inherit the best features.

3. As this process continues over many generations, organisms with the best features will make up a larger proportion of the population and those with features that are not as helpful will gradually disappear. For example if in every generation it is the fastest zebras that survive, then over many generations zebras will get faster and faster.

Natural selection explains why animals and plants are adapted to their surroundings. If the environment changes though, those organisms that do not evolve and adapt may become extinct.

The polar bear is well adapted to its environment.
- *Eyes on front of head for good judgement of size and distance.*
- *Sharp teeth for killing prey and tearing through flesh.*
- *White fur is good camouflage to hide from prey.*
- *Thick fur to trap heat to keep warm.*
- *Lots of fat to act as a food store and to keep warm.*
- *Strong legs for walking and swimming long distances.*
- *Sharp claws to grip ice.*
- *Hair on the soles of the feet provides a grip on icy surfaces.*

Check yourself

QUESTIONS

Q1
(a) Why do we have a clearer idea of how vertebrate animals like mammals, birds and reptiles evolved than of how invertebrates like worms and slugs evolved?

(b) Why are there many gaps in the fossil record even for animals with skeletons?

Q2 The long necks of giraffes are an adaptation allowing them to feed on the high branches of trees where many other animals cannot reach. Use the theory of natural selection to explain how giraffes could have evolved long necks from shorter-necked ancestors.

REMEMBER! Cover the answers if you want to.

ANSWERS

A1
a) Vertebrates have hard body parts, e.g. bones and teeth, which can leave fossil remains, whereas many invertebrates have only soft body parts which rarely leave fossils.

b) There are many reasons such as: the animals did not die and their bodies lay undisturbed in the right environmental conditions to form fossils; fossils were formed but have since been destroyed, e.g. by erosion; or they have not been found yet.

A2 Among the shorter-necked ancestors there would have been *variation* in height. A giraffe that was only slightly taller than others would have been able to reach and feed on some leaves that others would not be able to. They would have had a slight *advantage* in the *competition* for food and so would be slightly better fed and slightly more healthy than shorter ones. They would therefore probably on average live a little longer and be able to successfully rear slightly more offspring. If they were able to pass on the longer neck to their offspring because it was a *genetic* feature, it would mean both that there would be more of their offspring than of the shorter-necked giraffes, and that the overall average height of the giraffe population would have increased. If this continued for many, many generations it would lead to the giraffes we see today.

TUTORIAL

T1
a) *This does not apply to all invertebrates. There are many remains of the shells of marine invertebrates. Even so, these leave little evidence of what the soft tissues were like.*

b) *Although the fossil record of evolution is incomplete and may not give detailed information about how all living things evolved it is still a very important source of evidence.*

T2 *In an exam, questions about natural selection will often require extended answers and may carry four or more marks. To get full marks you will therefore need to give a detailed response. Do not be put off if the example used is unfamiliar. This is to test that you understand the **process** which will be the same in each case. Note that natural selection favours features that give an advantage but will not always result in evolutionary change. In the example of the giraffe, being taller than they are today would pose disadvantages, e.g. the difficulty of pumping blood to the brain or problems of stability, which means that they will not just continue to get taller. Some animals alive today are known as 'living fossils' as they appear similar to ancient fossil remains. If an organism's environment has not changed much and it is well adapted to it, then natural selection would work to keep the animal's features the same.*

EXAMINATION CHECKLIST

The facts and ideas that you should understand by studying this topic.

To be successful at Foundation GCSE Tier you should be able to:

- list observed variations within a population
- recall that variation can be caused by genetic and environmental factors
- recall that genetic variation can be caused by mutations and fertilisation
- recall that chromosomes are held in the nucleus and are strings of genes that instruct each cell
- recall that most body cells have the same number of chromosomes but that gametes have half that number
- recall that organisms grow by cells dividing
- recall that in sexual reproduction gametes join to produce a new individual and that at fertilisation genetic material from both parents combines to produce a unique individual
- recall that in asexual reproduction cell division produces new individuals
- recall that it is the X and Y chromosomes that determine gender
- describe the process of selective breeding
- know that animals and plants have evolved over long periods of time and that the incomplete fossil record provides evidence for this
- recall that animals and plants that are better adapted to their environment are more likely to survive.

In addition, to be successful at Higher GCSE Tier you should be able to:

- recall that genes are made of a chemical called DNA
- recall that mutations are changes to the DNA's genetic code
- recall that in mitosis the chromosomes are copied to produce genetically identical cells; the chromosomes duplicate, separate and head to opposite poles of the cell
- explain why in meiosis the chromosome number is halved and each cell is different
- explain how the random events of fertilisation produce unique individuals
- explain the advantages and disadvantages associated with the commercial use of cloned plants
- describe the process of cloning by tissue culture
- explain a monohybrid cross involving dominant and recessive alleles using genetic diagrams
- explain sex inheritance and the production of equal numbers of male and female offspring using genetic diagrams
- use genetic diagrams to predict the probabilities of inherited diseases passing to the next generation
- recall the principles of genetic engineering: selection of characteristics, isolation, replication and insertion

- explain the potential advantages and risks of genetic engineering and selective breeding
- explain the main steps in Darwin's theory of natural selection leading to the evolution or extinction of organisms
- explain why the theory of evolution by natural selection met with a hostile response.

Key Words

Tick each word when you are sure of its meaning.

allele	gene	natural selection
asexual reproduction	genotype	phenotype
chromosome	heterozygous	recessive
continuous	homozygous	sexual reproduction
discontinuous	meiosis	variation
dominant	mitosis	zygote
evolution	monohybrid cross	
gamete	mutation	

EXAM PRACTICE

Sample Student's Answers & Examiner's Comments

1

Red-green colour blindness is a sex-linked inherited characteristic. The allele (gene) n for colour-blindness is recessive to the allele N for normal vision. The allele n is carried only on the X chromosome.

Different combinations of alleles produce different characteristics.

a) Describe the characteristics (phenotypes) produced by these combinations.

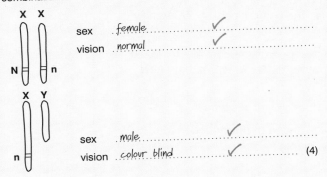

sex   female   ✓
vision   normal   ✓

sex   male   ✓
vision   colour blind   ✓   (4)

b) In a family, Arthur could not understand why he was red-green colour blind when his brother Colin and his parents had normal vision. Use the symbols shown above to explain how this was possible.

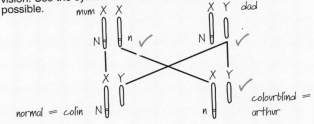

mum X X   X Y dad

normal = colin   colourblind = arthur

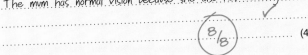

You need only one n gene to be colourblind if you are male.
The mum has normal vision because she also has an N gene.   ✓

8/8   (4)

Midland Examining Group

a) *Full marks have been awarded. The candidate correctly identified XX as female and XY as male. As the allele N for normal vision is dominant, the female, who is Nn, has normal vision. Normally you would have to have two copies of a recessive allele to show the recessive condition but in this case the male only needs one copy of the n allele to be red–green colour blind. This is because there is no corresponding part of the Y chromosome to carry a dominant allele. Therefore it is much more likely that more males will show the recessive condition than females. This is why red–green colour blindness is described as a sex-linked condition.*

b) *All the necessary marking points have been identified and the candidate has gained full marks. One mark each has been gained for identifying the genotypes of the parents. As Arthur is red–green colour blind he must have one copy of the n allele. His father has normal vision and as males only carry one allele his father must have one N allele. Therefore the n must have come from his mother, but she too has normal vision so she must have the genotype Nn. The third mark is for showing Arthur's genotype and phenotype and the fourth mark is for showing how Arthur had inherited his genotype.*

Parts of the diagram have been missed out, but it seems clear that the candidate understands what is going on, so marks have not been lost for these omissions. Nevertheless it would be wise in an exam to set out genetic diagrams in full. The chromosomes can be shown as in the answer but there are acceptable alternatives such as: $X^N X^n$ for the mother, $X^N Y$ for the father and $X^n Y$ for Arthur. Full marks could also have been gained for a labelled Punnett Square instead.

● *This candidate has gained 8 marks out of a possible 8. This is the mark that a grade A or B candidate might be expected to get.*

Questions to Answer

Answers to questions 2 – 5 can be found on pages 362–363.

2 A grower found some small strawberries with a nice taste, growing
on a strawberry plant.

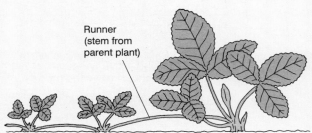

Runner
(stem from
parent plant)

a) How could the grower develop plants with strawberries
which were larger but had the same nice taste? Carefully
describe each step. (4)

b) Once the grower had developed his new plants, he could
use runners to produce more plants which had the new
large and tasty strawberries

   (i) What type of reproduction is this called? (1)

   (ii) Why would he use this type of reproduction to
produce more new plants? (1)

Northern Examination and Assessment Board

3 Tall high-yielding tomato plants are deliberately crossed with short, disease-resistant tomato plants.

The genotype of the tall plant is HH and that of the short plant is hh. The dominant allele is H.

a) Write the genotype of the F1 generation plants. (1)

b) What proportion of the F1 generation plants would be tall? (1)

c) The F1 generation plants are allowed to self pollinate. The seeds from this cross are grown.
State and explain the probable proportions of tall and short plants in the offspring. (2)

Midland Examining Group

4 The drawing shows the Galapagos Cormorant. It feeds by swimming under the surface of the sea to catch fish.

a) Use information from the drawing to describe two ways in which the Galapagos Cormorant is adapted for swimming. Explain how each adaptation helps the bird to swim. (4)

b) The theory of evolution states that birds evolved from reptiles. The reptiles from which birds evolved are now extinct. The drawing shows an extinct reptile.

(i) What evidence is there that reptiles like the one in the drawing once lived on Earth? (1)

(ii) Give one piece of evidence that all species of living things have evolved from simple life forms. (1)

Northern Examination and Assessment Board

5 Read the following extract from an article in 'The Independent'. Then answer the questions below

---

Thanks To The Test Tube Banana

Specially bred resistant varieties may save African crops from disease

A banana is a fruit, but it has no seeds and if there are no seeds, how do the plants reproduce? At one level the answer is easy – centuries of selective breeding have resulted in varieties with plenty of tasty flesh but few bitter inedible seeds, and propagation is carried out by means of root corms.

Most bananas we eat are thus actually 'clones' of a few successful plants. Banana clones are genetically identical to their parents, so growers can be completely sure their fruits will be big and tasty.

Genetic variability of these cloned plants is extremely low. Resistance to new diseases, therefore, is almost nil.

Bananas and plantains are being ravaged by a new fungal disease called Black Sigatoka. Some uncultivated varieties of these plants are resistant to the disease.

---

a) Explain, as fully as you can, why 'genetic variability of these cloned plants is extremely low' compared with natural populations. (4)

b) Explain, as fully as you can, how scientists might use genetic engineering to produce varieties of banana resistant to Black Sigatoka disease. (4)

Northern Examination and Assessment Board

Using chemical symbols to write formulae and equations is an important short-hand for a scientist. Languages change from country to country but the chemical symbols are the same throughout the world. To be successful in chemistry at GCSE and beyond you need to become familiar with this very specialised form of short-hand.

CHEMICAL FORMULAE

All substances are made up from simple building blocks called **elements**. These are arranged in the periodic table. Each element has a unique **chemical symbol**, containing one or two letters. Those discovered a long time ago often have symbols which don't seem to match their name. For example, sodium has the chemical symbol Na. This is derived from the Latin name for sodium – 'nadium'. You won't need to start learning Latin but you will have to learn some of the common symbols!

When elements chemically combine together they form **compounds**. Just as an element can be represented by chemical symbol so a compound can be represented by a **chemical formula**. There are thousands upon thousands of chemical compounds and most of these have a unique chemical formula. Remembering all these formulae would be extremely difficult but there is a way of working them out.

SIMPLE COMPOUNDS

Many compounds contain just two elements. For example, when magnesium burns in oxygen a white ash of magnesium oxide is formed. What is the chemical formula of magnesium oxide?

1. Write down the name of the compound.
2. Write down the chemical symbols for the elements in the compound.
3. Use the periodic table to find the 'combining power' of each element. Write the combining power of each element under its symbol.
4. If the numbers can be cancelled down do so.
5. Swap over the combining powers. Write them after the symbol, slightly below the line (as a 'subscript').
6. If any of the numbers are one there is no need to write them.

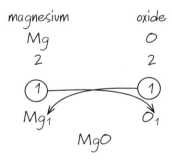

Magnesium oxide has the chemical formula you would have probably guessed. Are all formulae the same? What about calcium chloride?

1. Write down the name of the compound.
2. Write down the chemical symbols for the elements in the compound.
3. Use the periodic table to find the 'combining power' of each element. Write the combining power of each element under its symbol.
4. If the numbers can be cancelled down do so.
5. Swap over the combining powers. Write them after the symbol, slightly below the line (as a subscript).
6. If any of the numbers are one there is no need to write them.

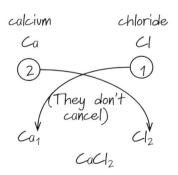

The chemical formula of a compound is not always immediately obvious, but if you follow the rules you will have no problems. To work it out you need to know the combining powers of the elements that make up the compound. How are these obtained from the periodic table? In fact there is a simple

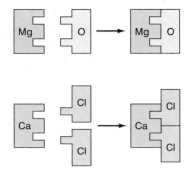

relationship between the group number of the periodic table and the combining power. Groups are the vertical columns in the periodic table.

Group number	1	2	3	4	5	6	7	8(or 0)
Combining power	1	2	3	4	3	2	1	0

The rules are as follows:

- Groups 1–4   combining power = group number;
- Groups 5–8   combining power = 8 – (group number).
- If an element is not in one of the main groups, its combining power will be included in the name of the compound containing it. For example: copper(II) oxide. Copper is a transition metal and is in the middle block of the periodic table. In this example, copper has a combining power of 2.
- Sometimes an element does not have the combining power you would predict from its position in the periodic table. If this is the case its combining power will also be included in the name of the compound containing it. For example: phosphorus(V) oxide. phosphorus is in group 5 and so you would expect it to have a combining power of 3. In this compound its combining power is 5.
- The only exception is hydrogen. Hydrogen is often not included in a group nor is its combining power given in the name of compounds containing hydrogen. It has a combining power of 1.

The combining power is linked to the number of electrons in the atom of the element. This will be looked at in detail in the next chapter.

COMPOUNDS CONTAINING MORE THAN TWO ELEMENTS

Some elements commonly exist bonded together in what is called a **radical**. For example, in copper(II) sulphate, the sulphate part of the compound is a radical. There are a number of common radicals, each having its own combining power.

combining power = 1		combining power = 2		combining power = 3	
hydroxide	OH	carbonate	CO_3	phosphate	PO_4
hydrogen carbonate	HCO_3	sulphate	SO_4		
nitrate	NO_3				
ammonium	NH_4				

These combining powers cannot be worked out easily from the periodic table. They have to be learnt. Those which have been shaded in the table are the ones most commonly encountered at GCSE.

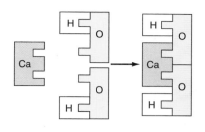

Fortunately the same rules apply to radicals as to elements.

copper(II) sulphate
Cu SO_4
2   2
    $CuSO_4$

potassium nitrate
K NO_3
1   1
    KNO_3

However, if the formula contains more than one radical unit the radical must be put in brackets.

calcium hydroxide
Ca   OH
2   1
    $Ca(OH)_2$

ammonium carbonate
NH_4 CO_3
1   2
    $(NH_4)_2CO_3$

The brackets are used in just the same way as in maths. The number outside the bracket multiplies everything in the brackets. There is a temptation to write calcium hydroxide as $CaOH_2$ rather than $Ca(OH)_2$. This is incorrect.

$CaOH_2$ contains one Ca, one O, two H   **✗**

$Ca(OH)_2$ contains one Ca, two O, two H   **✓**

Check yourself

QUESTIONS

Q1 Work out the chemical formulae of the following compounds:
a)   sodium chloride.
b)   magnesium fluoride.
c)   aluminium nitride.
d)   lithium oxide.
e)   potassium sulphide.
f)   hydrogen oxide.
g)   nitrogen hydride.
h)   carbon oxide (carbon dioxide).
i)   silicon bromide.
j)   calcium iodide.

Q2 Work out the chemical formulae of the following compounds:
a)   iron(III) oxide.
b)   phosphorus(V) chloride.
c)   chromium(III) bromide.
d)   sulphur(VI) oxide (sulphur trioxide).
e)   sulphur(IV) oxide (sulphur dioxide).
f)   copper(II) sulphide.

Q3 Work out the chemical formulae of the following compounds:
a)   sodium hydroxide.
b)   potassium carbonate.
c)   ammonium chloride.
d)   nitric acid (hydrogen nitrate).
e)   sulphuric acid (hydrogen sulphate).
f)   phosphoric acid (hydrogen phosphate).
g)   copper(II) nitrate.
h)   magnesium hydroxide.
i)   ammonium sulphate.
j)   aluminium sulphate.

ANSWERS

A1

a) $NaCl$
b) MgF_2
c) AlN
d) Li_2O
e) K_2S
f) H_2O
g) NH_3
h) CO_2
i) $SiBr_4$
j) CaI_2

A2

a) Fe_2O_3
b) PCl_5
c) $CrBr_3$
d) SO_3
e) SO_2
f) CuS

A3

a) $NaOH$
b) K_2CO_3
c) NH_4Cl
d) HNO_3
e) H_2SO_4
f) H_3PO_4
g) $Cu(NO_3)_2$
h) $Mg(OH)_2$
i) $(NH_4)_2SO_4$
j) $Al_2(SO_4)_3$

TUTORIAL

T1

a) Na (Gp 1) *combining power 1*, Cl (Gp 7) *combining power 1*
b) Mg (Gp 2) *combining power 2*, F (Gp 7) *combining power 1, 2, 1, cross over = Mg1 F2 = MgF₂*
c) Al (Gp 3) *combining power 3*, N (Gp 5) *combining power 3, 3, 3, cancel = Al1 N1 = AlN*
d) Li (Gp 1) *combining power 1*, O (Gp 6) *combining power 2*
e) K (Gp 1) *combining power 1*, S (Gp 6) *combining power 2*
f) H *combining power 1*, O (Gp 6) *combining power 2*
g) N (Gp 5) *combining power 3*, H *combining power 1*
h) C (Gp 4) *combining power 4*, O (Gp 6) *combining power 2, 4, 2, cancel = C1 O2 = CO₂*
i) Si (Gp 4) *combining power 4*, Br (Gp 7) *combining power 1*
j) Ca (Gp 2) *combining power 2*, I (Gp 7) *combining power 1*

T2

a) Fe *combining power 3*, O (Gp 6) *combining power 2*
b) P *combining power 5*, Cl (Gp 7) *combining power 1*
c) Cr *combining power 3*, Br (Gp 7) *combining power 1*
d) S *combining power 6*, O (Gp 6) *combining power 2, 6, 2, cancel = S1 O3 = SO₃*
e) S *combining power 4*, O (Gp 6) *combining power 2, 4, 2, cancel = S1 O2 = SO₂*
f) Cu *combining power 2*, S (Gp 6) *combining power 2, 2, 2, cancel = Cu1 S1 = CuS*

T3

a) Na *combining power 1*, OH *combining power 1*
b) K *combining power 1*, CO₃ *combining power 2*
c) NH₄ *combining power 1*, Cl (Gp 7) *combining power 1*
d) H *combining power 1*, NO₃ *combining power 1*
e) H *combining power 1*, SO₄ *combining power 2*
f) H *combining power 1*, PO₄ *combining power 3*
g) Cu *combining power 2*, NO₃ *combining power 1 don't forget the brackets*
h) Mg (Gp 2) *combining power 2*, OH *combining power 1 don't forget the brackets*
i) NH₄ *combining power 1*, SO₄ *combining power 2 don't forget the brackets*
j) Al (Gp 3) *combining power 3*, SO₄ *combining power 2 don't forget the brackets*

CHEMICAL EQUATIONS

Chemical equations are a means of summarising chemical reactions.
They tell you:

- what you start with (the **reactants**)
- how much of each chemical you need
- what you finish with (the **products**)
- how much of each chemical you produce.

WRITING CHEMICAL EQUATIONS

In a chemical equation the starting chemicals are called the **reactants** and the finishing chemicals are called the **products**. For example, when a lighted splint is put into a test tube of hydrogen the hydrogen burns with a 'pop'. In fact the hydrogen reacts with oxygen in the air (the reactants) to form water (the product). Starting with the **word equation**, the **chemical equation** for this reaction can be worked out.

1. Write down the word equation.
hydrogen + oxygen → water

2. Write down the symbols (for elements) and formulae (for compounds).
$H_2 + O_2 \rightarrow H_2O$

3. Balance the equation (to make sure there are the same number of each type of atom on each side of the equation).
$2H_2 + O_2 \rightarrow 2H_2O$

The reaction between hydrogen and oxygen produces a lot of energy as well as water – enough to launch a rocket.

This equation shows that for every two molecules of hydrogen which react, one molecule of oxygen is needed and two molecules of water are formed.

$$2H_2 \qquad + \qquad O_2 \qquad \rightarrow \qquad 2H_2O$$
2 molecules $\qquad$ one molecule $\qquad$ two molecules

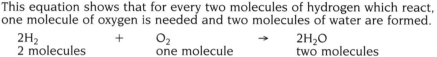

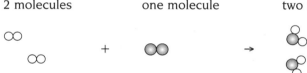

Note. Many of the elements are **diatomic**. They exist as molecules containing two atoms.

Element	Form in which it exists
hydrogen	H_2
oxygen	O_2
nitrogen	N_2
chlorine	Cl_2
bromine	Br_2
iodine	I_2

What is the equation when natural gas (methane) burns in air to form carbon dioxide and water?

1. Write down the word equation.
methane + oxygen → carbon dioxide + water

2. Write down the symbols (for elements) and formulae (for compounds).
$CH_4 + O_2 \rightarrow CO_2 + H_2O$

3. Balance the equation (to make sure there are the same number of each type of atom on each side of the equation).
$CH_4 + 2O_2 \rightarrow CO_2 + 2H_2O$

Methane is burning in the oxygen in the air to form carbon dioxide and water.

Balancing equations can be quite tricky. Basically it is done by trial and error. However, the golden rule is that *balancing numbers can only be put in front of the formulae*. Let us look at the balancing of the equation for the reaction between methane and oxygen.

	Reactants	Products
Start with the unbalanced equation.	$CH_4 + O_2$	$CO_2 + H_2O$
Count the number of atoms on each side of the equation.	1C ✓, 4H, 2O	1C ✓, 2H, 3O
There is a need to increase the number of H atoms on the products side of the equation. Put a '2' in front of the H_2O.	$CH_4 + O_2$	$CO_2 + 2H_2O$
Count the number of atoms on each side of the equation again.	1C ✓, 4H ✓, 2O	1C ✓, 4H ✓, 4O
There is a need to increase the number of O atoms on the reactant side of the equation. Put a '2' in front of the O_2.	$CH_4 + 2O_2$	$CO_2 + 2H_2O$
Count the atoms on each side of the equation again.	1C ✓, 4H ✓, 4O ✓	1C ✓, 4H ✓, 4O ✓

No atoms have been created or destroyed in the reaction. The equation is balanced!

$$CH_4 + 2O_2 \rightarrow CO_2 + 2H_2O$$

The number of each type of atom is the same on the left and right sides of the equation.

  →

In balancing equations involving radicals, you can use the same procedure only if the radical exists on both sides of the equation. For example, if lead(II) nitrate solution is mixed with potassium iodide solution, lead(II) iodide and potassium nitrate are produced.

1. Words	lead(II) nitrate	+	potassium iodide	→	lead(II) iodide	+ potassium nitrate
2. Symbols	$Pb(NO_3)_2$	+	KI	→	PbI_2	+ KNO_3
3. Balance the nitrates	$Pb(NO_3)_2$	+	KI	→	PbI_2	+ $2KNO_3$
3. Balance the iodides	$Pb(NO_3)_2$	+	2KI	→	PbI_2	+ $2KNO_3$

IONIC EQUATIONS

Ionic equations show reactions involving **ions**. An ion is a charged atom or molecule. The size of the charge on an ion is the same as the combining power – whether it is positive or negative depends on which part of the periodic table the element is placed in. If you have not come across ions before you will need to look at chapter 8. You will get further practice using ionic equations in chapter 12.

In many ionic reactions some of the ions play no part in the reaction. These ions are called **spectator ions**. In the reaction given above to produce lead(II) iodide, the potassium and nitrate ions play no part in the reaction – they are

This reaction occurs simply on mixing the solutions of lead(II) nitrate and potassium iodide. Lead iodide is an insoluble yellow compound.

spectators. The important ions are the lead(II) ions and the iodide ions. A simplified ionic equation can therefore be written:

$Pb^{2+}(aq) + 2I^-(aq) \rightarrow PbI_2(s)$

The equation must balance in terms of chemical symbols *and* charges.

	Reactants	**Products**
	$Pb^{2+}(aq) + 2I^-(aq)$	$PbI_2(s)$
symbols	1Pb ✓, 2I ✓	1Pb ✓, 2I ✓
charges	2+ and 2− = 0 ✓	0 ✓

In these equations state symbols are often used.

State	**State symbol**
solid	s
liquid	l
gas	g
solution	aq

The equation shows that *any* solution containing lead(II) ions will react with *any* solution containing iodide ions to form lead(II) iodide.

Another ionic equation can be written for the formation of copper(II) hydroxide:

$Cu^{2+}(aq) + 2OH^-(aq) \rightarrow Cu(OH)_2(s)$

	Reactants	**Products**
	$Cu^{2+}(aq) + 2OH^-(aq)$	$Cu(OH)_2(s)$
symbols	1Cu ✓, 2O ✓, 2H ✓	1Cu ✓, 2O ✓, 2H ✓
charges	2+ and 2− = 0 ✓	0 ✓

Any solution containing copper(II) ions and any solution containing hydroxide ions can be used to make copper(II) hydroxide, which appears as a solid.

HALF EQUATIONS

In electrolysis, the reactions at the electrodes can be shown as **half equations**. For example, when copper is deposited at the cathode the half equation can be written as:

$Cu^{2+}(aq) + 2e^- \rightarrow Cu(s)$

The symbol e^- stands for an electron. At the cathode positive ions gain electrons and become neutral. Once again the equation must balance in terms of symbols *and* charges.

A typical reaction at the anode during electrolysis would be:

$2Cl^-(aq) \rightarrow Cl_2(g) + 2e^-$

In this reaction two chloride ions combine to form one molecule of chlorine releasing two electrons. Further examples of half equations are given in chapter 10.

Check yourself

QUESTIONS

Q1 Magnesium reacts with hydrochloric acid to form magnesium chloride and hydrogen.
a) Write a word equation for the reaction.
b) Write a balanced symbol equation.

Q2 Sulphur dioxide (sulphur(IV) oxide) reacts with oxygen to produce sulphur trioxide (sulphur(VI) oxide).
a) Write a word equation for the reaction.
b) Write a balanced symbol equation.

Q3 Write symbol equations from the following word equations.
a) carbon + oxygen → carbon dioxide
b) iron + oxygen → iron(III) oxide
c) iron(III) oxide + carbon → iron + carbon dioxide
d) calcium carbonate + hydrochloric acid → calcium chloride + carbon dioxide + water

Q4 Balance the following equations.
a) $N_2 + H_2 \rightarrow NH_3$
b) $C_3H_8 + O_2 \rightarrow CO_2 + H_2O$
c) $H_2O + O_2 + SO_2 \rightarrow H_2SO_4$
d) $Cu(NO_3)_2 \rightarrow CuO + NO_2 + O_2$

Q5 Write ionic equations for the following reactions.
a) calcium ions and carbonate ions form calcium carbonate.
b) iron(II) ions and hydroxide ions form iron(II) hydroxide.
c) silver(I) ions and bromide ions form silver bromide.

Q6 Write half equations for the following reactions.
a) the formation of aluminium atoms from aluminium ions
b) the formation of sodium ions from sodium atoms
c) the formation of oxygen from oxide ions.
d) the formation of bromine from bromide ions.

REMEMBER! Cover the answers if you want to.

ANSWERS

A1
a) magnesium + hydrochloric acid → magnesium chloride + hydrogen
b) $Mg + 2HCl \rightarrow MgCl_2 + H_2$

A2
a) sulphur dioxide + oxygen → sulphur trioxide
b) $2SO_2 + O_2 \rightarrow 2SO_3$

A3
a) $C + O_2 \rightarrow CO_2$
b) $4Fe + 3O_2 \rightarrow 2Fe_2O_3$
c) $2Fe_2O_3 + 3C \rightarrow 4Fe + 3CO_2$
d) $CaCO_3 + 2HCl \rightarrow CaCl_2 + CO_2 + H_2O$

TUTORIAL

T1 Remember that hydrogen is diatomic, that is it forms H_2 molecules.

T2 Oxygen is also diatomic and exists as O_2 molecules.

T3
a) This doesn't need balancing!
b) Remember that balancing numbers must always go in front of symbols and formulae.
c) One maths trick to use here is to realise that the number of oxygen atoms on the right hand side must be even. Putting a '2' in front of Fe_2O_3 makes the oxygen on the left hand side even also.
d) Note that the carbonate radical does not appear on both sides of the equation.

ANSWERS

A4
a) $N_2 + 3H_2 \rightarrow 2NH_3$
b) $C_3H_8 + 5O_2 \rightarrow 3CO_2 + 4H_2O$
c) $2H_2O + O_2 + 2SO_2 \rightarrow 2H_2SO_4$
d) $2Cu(NO_3)_2 \rightarrow 2CuO + 4NO_2 + O_2$

A5
a) $Ca^{2+} + CO_3^{2-} \rightarrow CaCO_3$
b) $Fe^{2+} + 2OH^- \rightarrow Fe(OH)_2$
c) $Ag^+ + Br^- \rightarrow AgBr$

A6
a) $Al^{3+} + 3e^- \rightarrow Al$
b) $Na^+ + e^- \rightarrow Na$
c) $2O^{2-} \rightarrow O_2 + 4e^-$
d) $2Br^- \rightarrow Br_2 + 2e^-$

TUTORIAL

T4
a) *The number of hydrogen atoms on the left must always be even. Putting a 2 in front of the NH_3 makes the number of hydrogen atoms on the right even too.*
b) *Equations that appear to be complicated are often easy to balance. Leave the oxygen atoms until the last.*
c) *This equation shows how acid rain forms in the atmosphere.*
d) *Probably the hardest. Don't forget the '2' in front of the $Cu(NO_3)_2$ multiplies all the atoms in the formula, i.e. 2Cu, 4N, 12O.*

T5
The symbols and charges must balance on each side of the equation. State symbols could be used,
e.g. $Fe^{2+}(aq) + 2OH^-(aq) \rightarrow Fe(OH)_2(s)$
$Ag^+(aq) + Br^-(aq) \rightarrow AgBr(s)$

T6
At the cathode positive ions gain electrons. At the anode negative ions lose electrons. Remember that symbols and charges must balance.

EXAMINATION CHECKLIST

The facts and ideas that you should understand by studying this topic.

To be successful at Foundation GCSE Tier you should be able to:

- recognise the reactants and the products in a word equation
- construct word equations given the reactants and the products
- construct balanced symbol equations given the formulae of the reactants and the products.

In addition, to be successful at Higher GCSE Tier you should be able to:

- use the periodic table to work out the formulae of compounds
- construct balanced symbol equations
- write ionic equations by leaving out spectator ions
- write half equations.

Key Words		
Tick each word when you are sure of its meaning		
chemical equation	diatomic	radical
chemical formula	element	reactant
chemical symbol	half equation	spectator ion
combining power	ionic equation	word equation
compound	product	

CHAPTER 8

STRUCTURE AND BONDING

In order to understand the physical and chemical properties of substances we need to understand how substances are made up. This will help us to predict the properties of new substances. Fundamental to this knowledge is the idea of particles and more precisely ideas about atoms, molecules and ions.

SOLIDS, LIQUIDS AND GASES

All substances are either solid, liquid or gas. These are called the three **states of matter**. A substance in one state can be changed into another state by changing the temperature or the pressure. The names of the processes involved when changes of state occur are in the chart below:

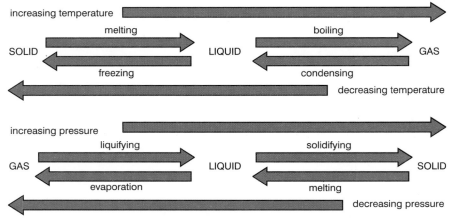

Property	Solid	Liquid	Gas
shape	definite shape	no shape – fills the container from the bottom	no shape – fills the whole container
volume	definite volume	definite volume	fills the whole container
density	high	medium	low
ease of flow	does not flow easily unless finely ground	flows easily	flows easily
ease of expansion	low	medium	high
ease of compression	very low	low	high
speed of diffusion	very slow	slow	fast

Solids, liquids and gases have quite different properties.

A PARTICLE MODEL

A simple particle model for solids, liquids and gases explains all the above properties. The ease with which gases can be compressed, compared to solids and liquids, can be explained in terms of the distances between the particles. Similarly the fixed shape of a solid can be explained in terms of the arrangements of the particles; density in terms of the distances between the particles; and ease of flow in terms of the forces between the particles.

Particle model	Solid	Liquid	Gas
Arrangement of particles	regular	random	random
Distance between particles	particles close together	particles quite close together	particles far apart
Forces between particles	strong	quite strong	very weak
Movement of particles	particles vibrate about a fixed point	particles move randomly and quite slowly	particles move randomly and quickly

Changes of state can also be explained using this simple particle model, or **kinetic theory** as it is called.

How changes of state relate to increasing temperature when a solid such as ice is heated

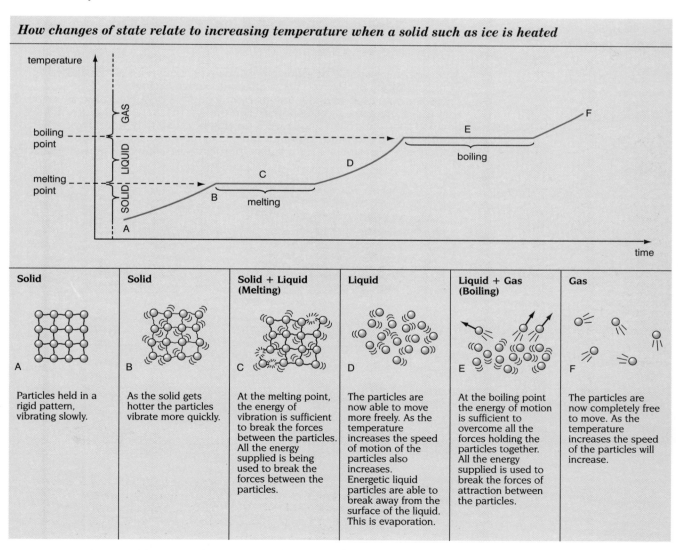

Solid	Solid	Solid + Liquid (Melting)	Liquid	Liquid + Gas (Boiling)	Gas
A	B	C	D	E	F
Particles held in a rigid pattern, vibrating slowly.	As the solid gets hotter the particles vibrate more quickly.	At the melting point, the energy of vibration is sufficient to break the forces between the particles. All the energy supplied is being used to break the forces between the particles.	The particles are now able to move more freely. As the temperature increases the speed of motion of the particles also increases. Energetic liquid particles are able to break away from the surface of the liquid. This is evaporation.	At the boiling point the energy of motion is sufficient to overcome all the forces holding the particles together. All the energy supplied is used to break the forces of attraction between the particles.	The particles are now completely free to move. As the temperature increases the speed of the particles will increase.

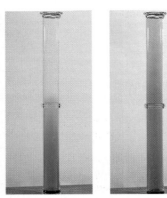

Diffusion is the process of the mixing and moving of particles. It is most common between gases. The two pictures above show bromine gas mixing with air. In about 5 minutes the bromine gas is completely mixed with the air. In the pictures to the right a crystal of potassium permanganate is shown dissolving in water. After a few hours the purple colour will have covered the whole dish.

DIFFUSION

Diffusion is a process in which particles mix together as a result of their constant movement. If you spill some perfume in the bathroom before long the smell will be all over the house! The particles of the perfume collide with particles of air and they mix or diffuse together. Gas particles move at about the speed of high velocity rifle bullets so unless they are held in a closed container they are always going to leave the container and mix with other gas particles. Though the particles move so quickly, it takes time for the mixing and spreading to occur because the gas particles in the air are also moving very rapidly. In the example of the perfume, the perfume particles have to 'battle' through rapidly moving air particles before they cover the whole house.

An important example of diffusion is known as **Brownian Motion** after the botanist Robert Brown who first noticed it. He was using a powerful microscope to look at pollen grains on the surface of some water. He noticed that the grains were moving about even though the water was quite still. He decided that the pollen grains were moving because they were being 'bumped' by rapidly moving water molecules. You may have seen Brownian motion in a 'smoke cell'. Here smoke particles are bombarded by air particles.

Dissolving is an example of diffusion.

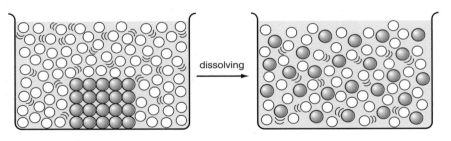

dissolving

Check yourself

QUESTIONS

Q1 The graph below shows how the temperature of a solid X changed as it was heated slowly.

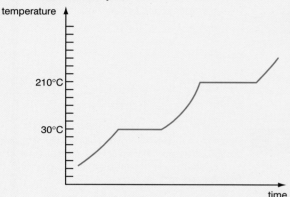

a) What was the melting point of X?
b) Explain the process of melting using a simple particle model.
c) What was the boiling point of X?
d) Explain the process of boiling using a simple particle model.

Q2 a) Use a simple particle model to explain evaporation.
b) Use a simple particle model to explain condensation.
c) What are the differences between the processes of boiling and evaporation?

Q3 Salt soon 'disappears' when sprinkled into warm water. Explain why *all* parts of the water then taste salty.

Q4 A dry glass tube is set up as shown in the diagram to demonstrate the diffusion of gases.

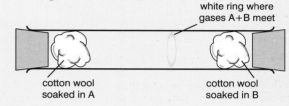

white ring where gases A+B meet

cotton wool soaked in A

cotton wool soaked in B

Hydrogen chloride gas is released from the cotton wool soaked in concentrated hydrochloric acid (B), ammonia gas is released from the cotton wool soaked in concentrated ammonia solution (A). Ammonium chloride is formed when these two gases meet.
a) What is diffusion?
b) Which gas particles move more quickly, ammonia or hydrogen chloride? Explain your answer.
c) It took about 10 minutes before the white ring of ammonium chloride formed. Explain why the ring didn't form immediately.

REMEMBER! Cover the answers if you want to.

ANSWERS

A1 a) 30 °C.
b) The particles have sufficient kinetic energy to break away from the solid structure and move into the liquid state. All the energy being supplied is being used to break the attractive forces between the particles and not to increase the speed of the particles.
c) 210 °C.
d) The particles have sufficient kinetic energy to break away from the surface of the liquid and form a gas. All the energy being supplied is being used to break the attractive forces between the particles in the liquid and not to increase the speed of the particles.

TUTORIAL

T1 *a) The first horizontal part of the heating curve.*
b) It is important to show that you understand that melting involves the change from solid to liquid. The other key point is that there is sufficient energy to overcome the attractive forces in the solid.
c) The second horizontal part of the heating curve.
d) It is important to show that you understand that boiling involves the change from liquid to gas. The temperature does not change as all the energy is going into breaking attractive forces rather than speeding up the particles.

ANSWERS

A2
a) Due to random collisions some particles gain sufficient energy to overcome the attractive forces in the liquid. These particles then leave the surface of the liquid as a vapour or gas.

b) Due to random collisions particles in the gas phase lose sufficient energy for attractions to form with other particles. The particles then remain in the liquid state.

c) A liquid can evaporate at any temperature. The higher the temperature the greater the rate of evaporation. A liquid boils at a fixed temperature (if the pressure is fixed).

A3
The water particles collide with the salt particles and break them away from the rigid structure of the solid. This is how the salt dissolves. Diffusion then occurs (salt particles mixing and moving with water particles) and the salt particles move slowly throughout the liquid.

A4
a) Diffusion is the mixing and moving of particles, in this case the mixing of hydrogen chloride and ammonia particles with air particles.

b) The ammonia moved quicker. It travelled further down the tube than the hydrogen chloride before the gases met and reacted.

c) The ammonia and hydrogen chloride particles were colliding with rapidly moving air particles.

TUTORIAL

T2
a) *Again show that you understand which change of state is involved, i.e. liquid to gas.*

b) *As with all these change of state questions you should mention which states are involved and how energy and forces of attraction are involved.*

c) *Both boiling and evaporation involve the loss of particles from the liquid state into the gas state.*

T3
*Diffusion is an important part of dissolving. Remember the term **diffusion**. Water currents are important in mixing solutions, but the mixing would happen by diffusion even with no currents.*

T4
a) *Always answer the question in the context of the experiment. In this case mention which gases are diffusing.*

b) *In fact from the distances travelled you can see that the ammonia must travel about twice as fast as the hydrogen chloride as it has travelled twice as far.*

c) *Don't forget that air plays a big part in the diffusion of gases.*

ATOMIC STRUCTURE

So far the components that make up substances have been referred to as particles. The smallest amount of an element that still behaves like that element is a single particle known as an **atom**. Each element has its own unique type of atom. Atoms are made up of smaller 'sub-atomic' particles.

THE SUB-ATOMIC PARTICLES

There are three main sub-atomic particles. These are the **proton**, **neutron** and **electron**. They are very small and have very little mass. However it is possible to compare their masses using a *relative* scale. It is also possible to compare their charges in a similar way.

The proton and neutron have the same mass and the proton and electron have opposite charges. Protons and neutrons are found in the centre of atoms

Sub-atomic Particle	Relative mass	Relative charge
proton	1	+1
neutron	1	0
electron	$\frac{1}{2000}$	−1

This central cluster of protons and neutrons is called the **nucleus**. The electrons form a series of 'shells' around the nucleus.

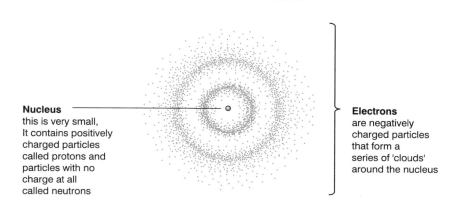

Nucleus
this is very small,
It contains positively
charged particles
called protons and
particles with no
charge at all
called neutrons

Electrons
are negatively
charged particles
that form a
series of 'clouds'
around the nucleus

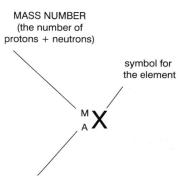

MASS NUMBER
(the number of
protons + neutrons)

symbol for
the element

M
A
X

ATOMIC NUMBER
(the number of protons
which equals the number
of electrons)

In order to describe the numbers of protons, neutrons and electrons in an atom, scientists use two numbers. These are called the **atomic number** and the **mass number**.

The atomic number is used to order the elements in the periodic table. The atomic structures of the first ten elements are shown in the table.

Hydrogen is the only atom that has no neutrons.

Element	Atomic number	Mass number	Number of protons	Number of neutrons	Number of electrons
Hydrogen	1	1	1	0	1
Helium	2	4	2	2	2
Lithium	3	7	3	4	3
Beryllium	4	9	4	5	4
Boron	5	10	5	5	5
Carbon	6	12	6	6	6
Nitrogen	7	14	7	7	7
Oxygen	8	16	8	8	8
Fluorine	9	19	9	10	9
Neon	10	20	10	10	10

Isotopes

Atoms of the same element with different numbers of neutrons are called **isotopes**. There are three isotopes of hydrogen:

Isotope	Symbol	Number of neutrons
normal hydrogen	^1_1H	0
deuterium	^2_1H	1
tritium	^3_1H	2

The three isotopes all contain the same number of protons and the same number of electrons. Isotopes have the same chemical properties but slightly different physical properties.

Electron arrangement

The electrons are arranged in shells around the nucleus. The shells do not all contain the same number of electrons. The shell nearest to the nucleus can only take two electrons whereas the next one further from the nucleus can take eight. Oxygen, with an atomic number of 8, has 8 electrons. Of these 2 will be in the first shell and 6 will be in the second shell. This arrangement is written 2, 6. A phosphorus atom with an atomic number of 15 has 15 electrons, arranged 2, 8, 5.

Electron Shell	Maximum number of electrons
1	2
2	8
3	8

The electron arrangements are very important as they determine the way that the atom reacts chemically.

Atom diagrams

The atomic structure of an atom can be shown simply in a diagram.

Atom diagrams for carbon and sulphur showing the number of protons, neutrons and the electron arrangements.

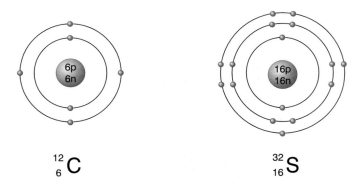

$^{12}_{6}C$ $^{32}_{16}S$

Check yourself

QUESTIONS

Q1 Explain the meanings of:
a)   atomic number;
b)   mass number.

Q2 Copy out and then complete the table below:

Atom	Number of protons	Number of neutrons	Number of electrons	Electron arrangement
$^{28}_{14}$Si				
$^{26}_{12}$Mg				
$^{32}_{16}$S				
$^{40}_{18}$Ar				

Q3 The table below shows information about the structure of six particles (A–F).

Particle	Protons (positive charge)	Neutrons (neutral)	Electrons (negative charge)
A	8	8	10
B	12	12	10
C	6	6	6
D	8	10	10
E	6	8	6
F	11	12	11

a)   In each of the questions (i) to (v) choose **one** of the six particles A–F above. Each letter may be used once, more than once or not at all.
Which particle:
   (i)   has a mass number of 12?
   (ii)   has the highest mass number?
   (iii) has no overall charge?
   (iv)   has an overall positive charge?
   (v)   is the same element as particle E?
b)   Draw an atom diagram for particle C.

ANSWERS

A1 a) The atomic number is the number of protons in an atom. (It is equal to the number of electrons.)

b) The mass number is the total number of protons and neutrons in an atom.

A2
Si	14	14	14	2, 8, 4
Mg	12	14	12	2, 8, 2
S	16	16	16	2, 8, 6
Ar	18	22	18	2, 8, 8

A3 a) (i) C, (ii) B
(iii) C or E or F
(iv) B
(v) C

b)

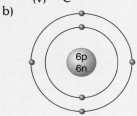

TUTORIAL

T1 a) Always define atomic number in terms of protons. However, as atoms are neutral there will always be equal numbers of protons and electrons.

b) Remember that electrons have very little mass and so mass number cannot refer to electrons.

T2 The top number is the mass number and the bottom (smaller) number is the atomic number. The difference between the two numbers is the number of neutrons.

T3 a) To have no overall charge the number of protons (positive charges) must equal the number of electrons (negative charges).
The atomic number (number of protons) is unique to a particular element. It is the atomic number that defines the element's position in the periodic table.

b) Remember that an atom diagram should show the number of protons and neutrons in the nucleus and the number and arrangement of electrons.

CHEMICAL BONDING

Atoms combine with other atoms in a chemical reaction to make a compound. For example sodium will react with chlorine to make sodium chloride or hydrogen will react with oxygen to make water. This reactivity is due to the electron arrangements in atoms. If atoms have incomplete electron shells they will usually react with other atoms. Only atoms with complete electron shells tend to be unreactive. Atoms in Group 8 (sometimes referred to as Group 0) of the periodic table, the **noble gases**, fall into this category.

When atoms combine they try to achieve full outer electron shells. They do this either by gaining electrons to fill the remaining gaps in the outer shell or by losing the remaining electrons in the outer shell to leave an inner complete shell.

There are two different ways in which atoms can bond together. The first way is called **ionic bonding** and the second is called **covalent bonding**.

IONIC BONDING

- Ionic bonding involves electron transfer.
- It occurs between metals and non-metals.
- Metals lose electrons from their outer shells and form positive **ions**.
- Non-metals gain electrons into their outer shells and form negative ions.
- The ions are held together by strong electrical (electrostatic) forces.
- Both metals and non-metals try to achieve complete outer electron shells.

Look at the reaction between sodium and chlorine as an example.

Sodium is a metal
It has an atomic number of 11 and so has 11 electrons, arranged 2, 8, 1.
Its atom diagram looks like this:

Chlorine is a non-metal.
It has an atomic number of 17 and so has 17 electrons, arranged 2, 8, 7.
Its atom diagram looks like this:

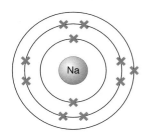
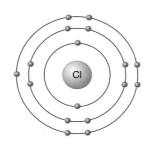

Sodium has one electron in its outer shell. It can achieve a full outer shell by losing this electron.
The sodium atom transfers its outermost electron to the chlorine atom.

Chlorine has seven electrons in its outer shell. It can achieve a full outer shell by gaining an extra electron.
The chlorine atom accepts an electron from the sodium.

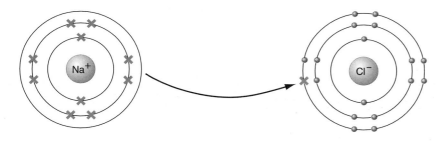

The sodium is no longer an atom. It does not have equal numbers of protons and electrons, it is no longer neutral. It is called an ion. It has one more proton than it has electrons.
It is therefore a positive ion with a charge of 1+.
The ion is written as Na^+.

The chlorine is no longer an atom. It does not have equal numbers of protons and electrons, it is no longer neutral. It is called an ion. It has one more electron than protons.
It is therefore a negative ion with a charge of 1−.
The ion is written as Cl^-.

Metals can transfer more than one electron to a non-metal. The example below shows how magnesium combines with oxygen to form magnesium oxide. The magnesium (electron arrangement 2, 8, 2) transfers two electrons to the oxygen (electron arrangement 2, 6). Magnesium therefore forms a Mg^{2+} ion and oxygen forms an O^{2-} ion. You can see this clearly in the following **dot-and-cross diagram**.

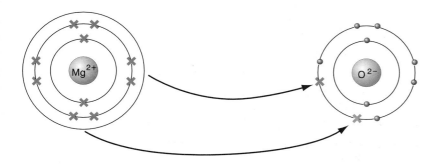

Aluminium has an electron arrangement 2, 8, 3. When it combines with fluorine with an electron arrangement 2, 7 three fluorine atoms are needed for each aluminium atom. The formula of aluminium fluoride is therefore AlF_3.

A dot-and-cross diagram of aluminium fluoride.

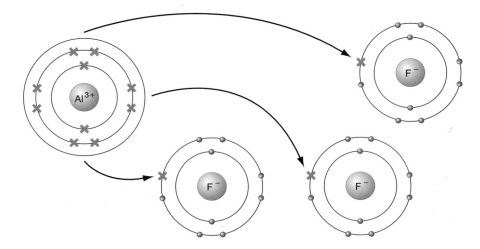

The ion formed by an element can be worked out from the position of the element in the periodic table. The elements in group 4 and group 8 (or 0) generally do not form ions.

group number	1	2	3	4	5	6	7	8 (or 0)
Ion charge	1+	2+	3+	X	3–	2–	1–	X

When a metal forms an ion it loses electrons. Losing electrons is an example of **oxidation**. For example, when a sodium atom forms the Na^+ ion it is oxidised. When a non-metal forms an ion it gains electrons. Gaining electrons is an example of **reduction**. For example, when a fluorine atom forms the F^- ion it is reduced.

COVALENT BONDING

- Covalent bonding involves electron sharing.
- It occurs between atoms of non-metals.
- It results in the formation of a **molecule**.
- The non-metal atoms try to achieve complete outer electron shells.

A single covalent bond is formed when two atoms each contribute one electron to a shared pair of electrons. For example, hydrogen gas exists as an H_2 molecule. Each hydrogen atom wants to fill its electron shell. They can do this by sharing electrons.

The dot-and-cross diagram and displayed formula of H_2.

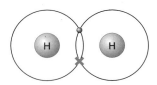

represented as

H — H

A single covalent bond can be represented by a single line. The formula of the molecule can be written as a **displayed formula**, H–H. The hydrogen and oxygen atoms in water are also held together by single covalent bonds.

Water contains single covalent bonds.

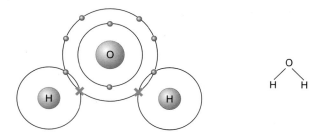

Some molecules contain double covalent bonds. An example is carbon dioxide. The carbon atom has an electron arrangement of 2, 4 and so needs an additional four electrons to complete its outer electron shell. It therefore needs to share its four electrons with four electrons from oxygen atoms (electron arrangement 2, 6). Two oxygen atoms are needed each sharing two electrons with the carbon atom.

Carbon dioxide contains double bonds.

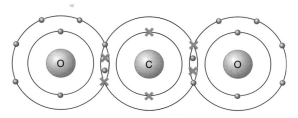

The number of covalent bonds a non-metal atom can form is linked to its position in the periodic table.

Remember that metals (Groups 1, 2, 3) do not form covalent bonds. The noble gases in Group 8 are unreactive and usually don't form covalent bonds.

Group in the periodic table	1	2	3	4	5	6	7	8 (or 0)
Covalent bonds formed	X	X	X	4	3	2	1	X

Compounds containing covalent bonds have very different properties to compounds that contain ionic bonds. These differences are considered in the next section.

Check yourself

QUESTIONS

Q1 For each of the following reactions say whether the compound formed is ionic or covalent:
(i) hydrogen and chlorine,
(ii) carbon and hydrogen,
(iii) sodium and oxygen,
(iv) chlorine and oxygen,
(v) calcium and bromine.

Q2 Write down the ions formed by the following elements:
(i) potassium,
(ii) aluminium,
(iii) sulphur,
(iv) fluorine.

Q3 Use the ion charges from question 2 to work out the formulae of the following compounds:
(i) potassium fluoride,
(ii) aluminium sulphide,
(iii) potassium sulphide.

Q4 Draw dot-and-cross diagrams to show the bonding in the following compounds:
a) methane, CH_4.
b) oxygen, O_2.
c) nitrogen, N_2.
d) potassium sulphide.
e) lithium oxide.

REMEMBER! Cover the answers if you want to.

ANSWERS

A1
(i) covalent,
(ii) covalent,
(iii) ionic,
(iv) covalent,
(v) ionic

A2
(i) K^+,
(ii) Al^{3+},
(iii) S^{2-}, (iv) F^-

A3
(i) KF,
(ii) Al_2S_3,
(iii) K_2S

A4 a)

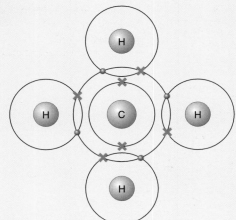

TUTORIAL

T1 *Remember that ionic bonding involves a metal and a non-metal; covalent bonding involves two or more non-metals. Hydrogen, chlorine, carbon, oxygen and bromine are non-metals. Only sodium and calcium in the list are metals.*

T2 *These can be worked out from the position of the element in the periodic table. Look back in the chapter for the rule if you have forgotten it. You will always be given a periodic table in the exam. Get used to using it, it will help to gain marks in the exam.*

T3 *In an ionic compound the number of positive and negative charges must balance out exactly, i.e. the formula must be neutral. For example $2K^+$ ions are needed to balance each S^{2-} ion. The formula is K_2S.*

T4 Notes:
1 *Remember ionic compounds contain a metal.*
2 *You can get the atomic number (needed to work out the number of electrons) from the periodic table.*
3 *With covalent compounds try drawing the displayed formula first. You will need to work out how many covalent bonds each atom can form. Again the periodic table helps here. For example, oxygen is in Group 6 so it can form two covalent bonds. Once you have got the displayed formula, each bond corresponds to a shared pair of electrons. Don't forget the first electron shell can hold 2 electrons, others hold 8 electrons. So don't miss any electrons out.*
4 *You will notice that nitrogen forms a triple bond, i.e. three pairs of electrons are shared.*
5 *In examples of ionic bonding don't forget to write the formula of the ions that are formed.*

ANSWERS

b)

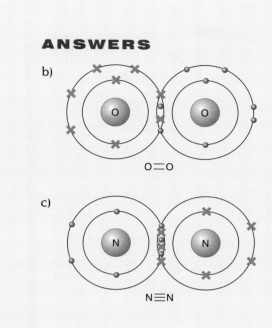

O=O

c)

N≡N

d)

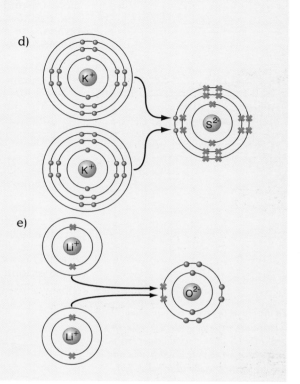

e)

STRUCTURES AND PROPERTIES

Ionic compounds

When sodium chloride is formed by ionic bonding the ions do not exist in pairs. Instead each sodium ion is surrounded by six chloride ions and each chloride ion is surrounded by six sodium ions. A **giant lattice structure**, as seen in the diagram (right), is formed.

The electrostatic attractions between the ions are very strong. The properties of sodium chloride can be explained using this model of its structure.

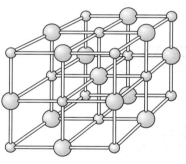

⬤ chloride ion   ⬭ sodium ion

Solid sodium chloride, the ions are held firmly in place – they are not free to move. Ionic compounds have giant ionic lattice structures like this.

Property	Explanation in terms of structure
hard crystals	strong forces between the ions
high melting point (801 °C)	strong forces between the ions
dissolves in water	the water is also able to form strong electrostatic attractions with the ions – the ions are 'plucked' off the lattice structure
does not conduct electricity when solid	strong forces between the ions prevent them from moving
does conduct electricity when molten or dissolved in water	the strong forces between the ions have been broken down and so the ions are able to move

Covalent compounds

Covalent bonds are also strong bonds. These bonds are formed *within* each molecule and are called **intramolecular bonds**. Other, much weaker, forces also exist *between* the individual molecules and are called **intermolecular bonds**. The properties of covalent compounds can be explained using a simple model involving these two types of bond or forces.

Properties	Explanation in terms of structure	
Hydrogen is a gas with a very low melting point (–259 °C).	The intermolecular forces between the molecules are weak.	
Hydrogen does not conduct electricity.	There are no ions or free electrons present. The covalent bond (intramolecular bond) is a strong bond and the electrons cannot be easily removed from it.	

Some covalently bonded compounds do not exist as **simple molecular structures** in the way that hydrogen does. Diamond, for example, exists as a giant structure with each carbon atom covalently bonded to four others. The bonding is extremely strong. Diamond has a melting point of about 3730 °C. Another form of carbon is graphite. Graphite has a different giant structure as seen in the diagram. Different forms of the same element are called **allotropes**.

In diamond (a) each carbon atom forms four strong covalent bonds. In graphite (b) each carbon atom forms only three strong covalent bonds and one weak intermolecular bond. This explains why graphite is flaky and can be used as a lubricant. The layers of hexagons can slide over each other. The electrical conductivity of graphite can be explained by the fact that each carbon atom has one electron which is not strongly bonded. These electrons can move through the structure.

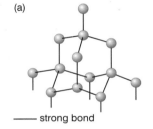

(a) —— strong bond

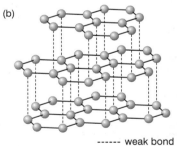

(b) ------ weak bond

Summary

The idea of giant and molecular structures is an important one. Structures can usually be identified as being giant or molecular from their melting points.

Structure	Atom	Molecule	Ion
Giant	Diamond, graphite, metals. High melting points.	Sand (silicon oxide molecules). High melting point.	All ionic compounds, e.g. sodium chloride. High melting points.
Simple molecular	Noble gases, e.g. helium. Low melting points.	Carbon dioxide, water. Low melting points.	None.

Check yourself

QUESTIONS

Q1 Explain why an ionic substance such as potassium chloride:
a) has a high melting point;
b) behaves as an electrolyte.

Q2 Use the structure of graphite to explain:
a) how carbon fibres can add strength to tennis rackets;
b) how graphite conducts electricity.

Q3 Explain why methane (CH_4), which has strong covalent bonds between the carbon atom and the hydrogen atoms, has a very low melting point.

Q4 An ionic compound exists as a *giant lattice structure*. What does this mean?

REMEMBER! Cover the answers if you want to.

ANSWERS

A1
a) Potassium chloride exists as a giant ionic lattice. There are very strong electrostatic forces holding the ions together. A considerable amount of energy is needed to overcome the forces of attraction so the melting point is high.
b) Potassium chloride is ionically bonded. When molten or dissolved in water the ions are free to move and carry an electric current.

A2
a) The carbon atoms are held by strong covalent bonds within the hexagonal layers – this can give a structure great strength.
b) Only three out of the four of carbon's outer shell electrons are used in forming covalent bonds. Each atom has an electron which is only loosely held to the atom. When a voltage is applied these electrons move forming an electric current.

A3 The bonds within the molecule are strong (the intramolecular bonds) but the bonds holding the methane molecules together are weak (intermolecular bonds). Therefore methane molecules require very little energy to be separated from each other.

A4 The ions do not exist as ion pairs. Instead each positive ion is surrounded by negative ions and each negative ion is surrounded by positive ions. This large and regular array of ions is known as a lattice.

TUTORIAL

T1
a) *The key ideas relating to melting point are: strong forces between particles; large amounts of energy needed to overcome these forces.*
b) *Remember that an electrolyte allows an electric current to flow through it when molten or dissolved in water – not when in the solid state.*

T2
a) *Have a look at the diagram of the structure of graphite. Remember covalent bonds are strong bonds.*
b) *Graphite has a structure similar to that of metals – there is a 'cloud' of electrons which are free to move. For more on metal structures see chapter 14.*

T3 *This is one of the ideas candidates find very confusing. Always try to be precise. There are bonds within a molecule as well as bonds between molecules. You wouldn't be expected to use 'intramolecular' and 'intermolecular' but if you do, be sure to get them the right way round!*

T4 *Have another look at the sodium chloride structure. The diagram shows just a small part of the lattice or giant array.*

EXAMINATION CHECKLIST

The facts and ideas that you should understand by studying this topic.

To be successful at Foundation GCSE Tier you should be able to:

- draw the arrangement of particles in a solid, liquid and a gas
- describe the changes in arrangement and motion of particles during freezing, melting, boiling and condensation
- explain that a solid dissolves as a result of solvent particles surrounding particles in the solid causing the solid structure to break down
- recall that atoms contain a positively charged nucleus surrounded by negatively charged electrons
- recall that the nucleus contain protons and neutrons and that electrons exist in shells
- work out atomic structure from the atomic and mass numbers
- recall that an ion is a charged atom and can be formed from an atom by it losing or gaining electrons
- distinguish between an atom, a molecule and an ion
- recall that non-metals combine together by sharing electrons (covalent bonding)
- recall that a metal and non-metal combine by transferring electrons (ionic bonding)
- recognise the structures of diamond and graphite and know their simple properties.

In addition, to be successful at Higher GCSE Tier you should be able to:

- use the kinetic theory to explain changes of state by appreciating the significance of the forces between particles
- explain the existence of isotopes through your understanding of the structure of the atom
- explain the relationship between an element's electronic structure and its position in the periodic table
- recall that the reactivity of an element relates to its electronic structure
- use dot-and-cross diagrams for simple ionic and covalent compounds
- draw simple displayed formulae
- explain how the properties of diamond and graphite relate to their structures
- explain that ionic lattices are held together by the attraction between oppositely charged ions
- explain some of the simple properties of ionic and molecular structures.

Key Words

Tick each word when you are sure of its meaning

allotrope

atomic number

Brownian motion

covalent bonding

diffusion

electron

giant structure

intermolecular

intramolecular

ionic bonding

isotope

kinetic theory

mass number

molecular structure

molecule

neutron

nucleus

proton

states of matter

EXAM PRACTICE

Sample Student's Answers & Examiner's Comments

1 **a)** Lithium has an atomic number of 3 and a mass number of 7. This is often represented by the symbol

$$^{7}_{3}\text{Li}$$

i) State the number of neutrons, protons and electrons which make up a neutral atom of this element.

Neutrons ....4.......... ✓

Protons ....3.......... ✓

Electrons ....3.......... ✓

ii) Sometimes atoms of the same element occur with a different number of neutrons in the nucleus. What do we call such atoms?

....isotopes.... ✓

iii) Using the symbolism at the start of the question, how would you represent an atom of lithium which contained only 3 neutrons?

$$^{6}_{3} \quad \text{Li}$$ ✓

iv) Atoms of lithium form ions by the loss of one electron. Write the formula of a lithium ion.

....Li^{+}.... ✓ (6)

b) **i)** Give the electronic structure of a single, uncombined atom of chlorine (atomic number + 17).

✓

ii) Draw, showing only the outer electrons of each atom, the structure of a chlorine molecule, Cl_2.

(3)

✗

c) Show what happens to the electronic structure of an atom of chlorine when it forms a chloride ion and say whether the atom has been oxidised, reduced or neither of these.

(2)

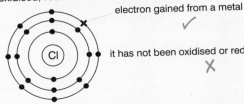

electron gained from a metal ✓

it has not been oxidised or reduced ✗

8/11

Cl^-

(Total 11 marks)

London Examinations

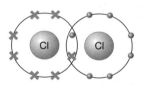

Questions to Answer

Answers to questions 2 – 3 can be found on page 364.

2 The diagram represents the arrangement of electrons in a magnesium atom.

a) Complete the table.

	protons	neutrons	electrons	electron arrangement
magnesium-24				2,8,2
oxygen-16		8	8	

(3)

b) Magnesium oxide is formed by ionic bonding. Explain fully, in terms of transfer of electrons and the formation of ions, the changes which occur when magnesium oxide is formed from magnesium and oxygen atoms.

(4)

c) Sodium chloride and magnesium oxide have similar crystal structures and both contain ionic bonding. The melting points of sodium chloride and magnesium oxide are 800 °C and 2800 °C respectively.

Suggest why the melting point of magnesium oxide is much higher than the melting point of sodium chloride. (Sodium chloride contains Na^+ and Cl^- ions.)

(3)

Midland Examining Group

3 a) The diagram shows part of a lithium atom.

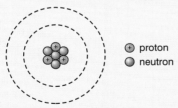

⊕ proton
◯ neutron

Complete the diagram by drawing in the correct number of electrons in each shell. Use an 'e' to represent an electron.

(1)

b) A fluorine atom can be represented as:

$$^{19}_{9}F$$

(i) How many electrons are there in a fluorine atom?

(ii) How many neutrons are there in a fluorine atom?

(2)

c) An isotope of an element X has the formula

$$^{41}_{19}X$$

Explain what is meant by the word *isotope*. (2)

d) Complete the table by filling in the blank spaces.

Name of substance	Formula	Type of bonding
Hydrogen	H_2	
Magnesium chloride		ionic
Water	H_2O	covalent

(2)

e) A fluorine molecule is made from two fluorine atoms which are held together by a covalent bond. Show this as a 'dot and cross' diagram for outer shell electrons only. (Atomic number of fluorine = 9.) (1)

f) Give **two** properties that ionic compounds, such as magnesium chloride, might be expected to have. (2)

Southern Examining Group

FUELS AND ENERGY

Many important chemicals used today are obtained from crude oil. Fuels and plastics are just two examples. Both play an important part in our daily lives but there are disadvantages associated with their use. Only by fully understanding the benefits and drawbacks can we be sure to use them sensibly and efficiently.

KEY POINTS

To make the most of fossil fuels, refineries use two processes:
- **fractional distillation** – sorts the crude oil into fractions
- **cracking** – converts long molecule fractions into more valuable short molecules

MAKING THE MOST OF FOSSIL FUELS

Crude oil was formed millions of years ago. At that time, large parts of the Earth were covered with sea containing a large variety of animal life. These animals died and decayed in the mud on the sea bed. Their remains were pressed together under layers of rock and very slowly changed into the thick, dark liquid known as crude oil.

Crude oil is usually found deep underground, trapped between layers of rock that it can't seep through. Rock that liquid can't move through is described as **impermeable**. Natural gas is often trapped in pockets above the crude oil.

Crude oil, natural gas and coal are fossil fuels. There is only a limited supply of them and it would take millions of years to replace the supplies we use. This is why they are called 'finite' or **non-renewable fuels**. They are an extremely valuable resource which must be used efficiently.

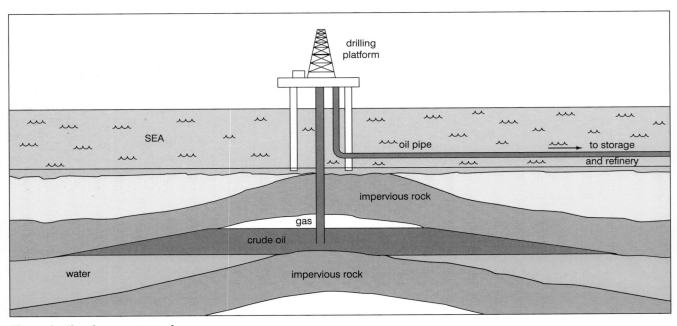

The crude oil and gas are trapped between layers of impermeable rock. An oil rig often has to drill through hundreds of metres of rock before finding the crude oil and gas.

SEPARATING THE FRACTIONS

Fossil fuels contain a large number of very useful chemicals. It is often important to separate these chemicals so that they are not wasted. For example, coal is often converted into coke by removing some of the chemicals in the coal. When the coke is burnt as a fuel these chemicals are not wasted. Crude oil is separated into a series of useful *fractions* by a process known as **fractional distillation**.

The crude oil is heated in a furnace and passed into the bottom of the fractionating column. Most of the fractions boil and turn into a vapour

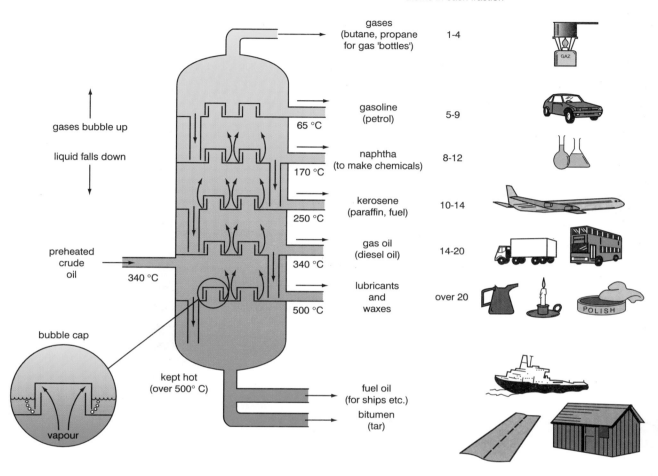

number of carbon
atoms in each fraction

Fraction	Carbon atoms
gases (butane, propane for gas 'bottles')	1-4
gasoline (petrol)	5-9
naphtha (to make chemicals)	8-12
kerosene (paraffin, fuel)	10-14
gas oil (diesel oil)	14-20
lubricants and waxes	over 20

gases bubble up

liquid falls down

preheated crude oil

340 °C

65 °C

170 °C

250 °C

340 °C

500 °C

bubble cap

kept hot (over 500° C)

vapour

fuel oil (for ships etc.)

bitumen (tar)

The fractionating column converts the crude oil into many useful fractions.

mixture. The vapour mixture rises up the column which is divided into different stages. Each stage is maintained at a different temperature and the temperature decreases as the vapour moves up the column. At each stage the vapour has to pass through bubble caps. If the temperature of these bubble caps is just below the boiling point of the fraction, the fraction will condense and form a liquid. These liquids can then be removed from the column. Fractions that do not condense move further up the column. All the liquid fractions eventually condense leaving the gases which pass out through the top of the column.

The fractions that come off near the top of the column are light-coloured runny liquids. Those removed near the bottom of the column are dark and treacle-like. Thick liquids that are not runny, such as these bottom-most fractions, are described as 'viscous'.

HOW DOES FRACTIONAL DISTILLATION WORK?

The process of fractional distillation is able to separate the components present in crude oil because they have different boiling points. Why they have different boiling points can be understood using a simple particle model.

Crude oil is a mixture of **hydrocarbon** molecules. A hydrocarbon is a molecule that contains only hydrogen and carbon atoms. The molecules are chemically bonded in similar ways but contain different numbers of carbon atoms. They form a series.

You will notice that each hydrocarbon in the series has one more carbon atom and two more hydrogen atoms than the one above it. Their formulae differ by CH₂.

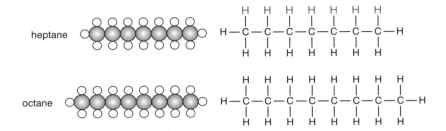

The bonds between the carbon and hydrogen atoms *within* the molecule are strong covalent bonds (see chapter 8). The bonds *between* the molecules are weak. It is these weak bonds between the molecules that have to be broken if the hydrocarbon is to boil.

The longer a hydrocarbon molecule is then the stronger the bonds that exist between the molecules. The stronger these bonds, the higher the boiling point as more energy needs to be supplied to overcome the larger forces.

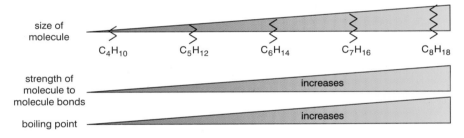

As well as having lower boiling points the smaller molecule hydrocarbons are more **volatile**, that is, they more readily form a vapour. Smaller molecules are more likely to gain sufficient energy to break away from the surface of the liquid and evaporate. For example, petrol (with molecules containing between 5 and 10 carbon atoms) smells much more than engine oil (with molecules containing between 14 and 20 carbon atoms) because it is more volatile. Another difference between the fractions is how easily they burn and how smoky their flames are.

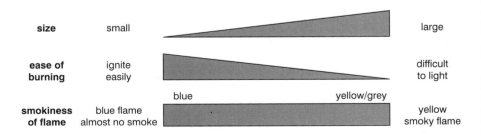

CRACKING THE OIL FRACTIONS

Unfortunately the amounts of the different fractions produced by fractional distillation do not match the amounts that are needed or usable. The following table shows the composition of a sample of crude oil from the

Middle East after fractional distillation. It also shows the demand (how much is needed) for each fraction.

Fraction (in order of increasing boiling point)	Percentage produced by fractional distillation	Percentage needed (demand)
liquefied petroleum gases (LPG)	3	4
petrol	13	22
naphtha	9	5
paraffin	12	7
diesel	14	24
heavy oils and bitumen	49	38

You will see that the lower boiling point fractions (e.g. petrol and gas) are in greater demand than the heavier ones (e.g. naphtha and paraffin). These larger molecules can be broken down into smaller ones by a process known as **cracking**. Cracking requires the following conditions:

- a high temperature
- a catalyst.

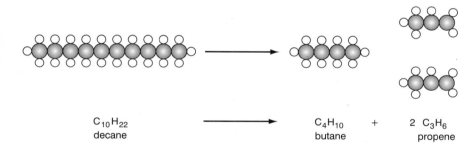

The decane molecule ($C_{10}H_{22}$) is converted into the smaller molecules butane (C_4H_{10}) and propene (C_3H_6).

$$C_{10}H_{22} \longrightarrow C_4H_{10} + 2\ C_3H_6$$

decane   butane   propene

The butane and propene formed in this example of cracking have different types of structures. These structures will be looked at in more detail in the next section.

Check yourself

QUESTIONS

Q1
a) How was crude oil formed?
b) Why is crude oil a non-renewable fuel?

Q2 The diagram shows part of a column used to separate the components present in crude oil.

d) Another component was collected between 230 °C and 300 °C. What would it be like?
e) Component A is used as a fuel in a car engine. Suggest why component C would not be suitable as a fuel in a car engine.

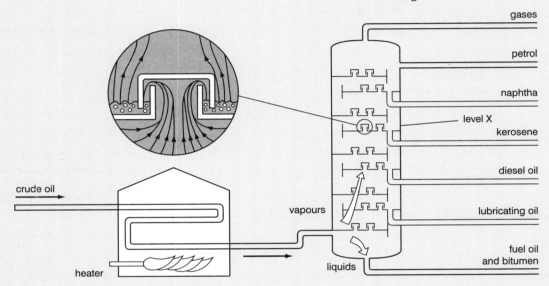

a) Name the process used to separate crude oil into fractions.
b) What happens to the boiling point of the mixture as it goes up the column?
c) The mixture of vapours arrives at level X. What now happens to the various parts of the mixture?

The table shows some of the properties of the crude oil components:

Q3 The cracking of decane molecules is shown by the equation $C_{10}H_{22} \rightarrow Y + C_2H_4$

a) Decane is a hydrocarbon. What is a hydrocarbon?
b) What conditions are needed for cracking?
c) Write down the molecular formula for hydrocarbon Y.

component	liquid temperature (°C)	how runny?	colour	how it burns
A	up to 70	very	colourless	easily, blue flame
B	70 to 150	fairly	pale yellow	fairly easily, a smoky flame
C	150 to 230	not very	dark yellow	difficult to light, a very smoky flame

ANSWERS

A1
a) Small sea creatures died, settled in the mud at the bottom of the oceans and decayed. They were compressed over a period of millions of years and slowly changed into crude oil.

b) It takes millions of years to form. Once supplies have been used up they cannot be replaced.

A2
a) Fractional distillation.

b) The boiling point of the fractions decreases.

c) Components which have boiling points just below the temperature of X will condense to a liquid. Components which have boiling points above the temperature of X will remain in the vapour state and continue up the column.

d) The component would be: not very runny (viscous); very dark yellow/orange in colour; very difficult to light; very smoky when burning.

e) It would not ignite easily. It would produce a lot of soot on burning.

A3
a) A molecule containing carbon and hydrogen atoms only.

b) High temperature, catalyst.

c) C_8H_{18}.

TUTORIAL

T1
Remember the process involves compression and takes place over millions of years. Do not confuse crude oil with coal which is formed from plant material.

A common mistake is to say 'it cannot be used again'. No fuel can be used again but some, such as trees (wood), can be re-grown quite quickly. Wood is therefore renewable.

T2
There is often a mark for 'fractional' and a mark for 'distillation'.
Remember the column is hotter at the bottom than at the top.
If a vapour is cooled below its boiling point then it will condense.

In a question like this, the clues to the answer will be in the table. Always give as full an answer as possible.
Four points can be scored using the last three columns of the table.
Again the clues are in the table. A fuel needs to ignite easily and be 'clean'.

T3
Don't miss out the word 'only'. A lot of compounds contain carbon and hydrogen but are not hydrocarbons, e.g. glucose $C_6H_{12}O_6$.
'Catalytic crackers' are used at oil refineries.
The equation must balance so the number of carbon atoms must be 10 − 2, the number of hydrogen atoms 22 − 4.

USING HYDROCARBONS

Most of the common fuels used today are hydrocarbons – substances that contain only carbon and hydrogen atoms.

COMBUSTION

When a hydrocarbon is burnt in a plentiful supply of air it reacts with the oxygen in the air to form carbon dioxide and water. This reaction is an example of **combustion**.

hydrocarbon + oxygen → carbon dioxide + water

For example, when methane (natural gas) is burnt:

$$CH_4(g) + 2O_2(g) \rightarrow CO_2(g) + 2H_2O(l)$$

KEY POINTS

Problems with burning hydrocarbon fuels:
- incomplete combustion
- acid rain
- Greenhouse Effect

Reactions:
- saturated unreactive alkanes
- unsaturated reactive alkenes
- alkenes make polymers

The complete combustion of methane in a plentiful supply of air.

The air contains only about of 20% oxygen by volume. When a hydrocarbon fuel is burnt there is not always enough oxygen to **oxidise** the carbon to carbon dioxide and the hydrogen to water. Instead **incomplete combustion** occurs and carbon or carbon monoxide are formed.

methane + oxygen → carbon monoxide + water

$$2CH_4(g) + 3O_2(g) → 2CO(g) + 4H_2O(l)$$

methane + oxygen → carbon + water

$$CH_4(g) + O_2(g) → C(s) + 2H_2O(l)$$

Incomplete combustion is both costly and dangerous. It is costly because the full energy content of the fuel is not being released and the formation of carbon or soot reduces the efficiency of the 'burner' being used. It can be dangerous as carbon monoxide is an extremely poisonous gas. Carbon monoxide molecules attach themselves to the haemoglobin of the blood preventing oxygen molecules from doing the same. Brain cells deprived of their supply of oxygen will quickly start to die.

The tell-tale sign that a fuel is burning incompletely is that the flame will be yellow. In contrast when complete combustion occurs the flame will be blue. You should be familiar with this from your use of a Bunsen burner.

Acid rain

Sulphur is an impurity in many fuels. When the fuel is burnt the sulphur is oxidised to sulphur dioxide. In the atmosphere, in the presence of oxygen and water, sulphuric acid is formed.

sulphur + oxygen → sulphur dioxide

$$S(s) + O_2(g) → SO_2(g)$$

sulphur dioxide + oxygen + water → sulphuric acid

$$2SO_2(g) + O_2(g) + 2H_2O(l) → 2H_2SO_4(aq)$$

Buildings, particularly those made of limestone and marble, are damaged by **acid rain**. Metal constructions are also attacked by the sulphuric acid. Acid rain can also damage trees and kill fish. Sulphur dioxide also forms a dry deposit on buildings, so that when it rains, strong acid is formed on the walls of the buildings and badly damages them. Power stations are now being fitted with 'Flue Gas Desulphurisation Plants' (FGD) to reduce the release of sulphur dioxide into the atmosphere.

Acid rain is also caused by solutions of the oxides of nitrogen. These are produced in the exhaust fumes of cars and aeroplanes. Catalytic converters are fitted to car exhaust systems so that the oxides of nitrogen can be converted into harmless nitrogen gas.

The Greenhouse Effect

The increase in carbon dioxide in the air since 1700.

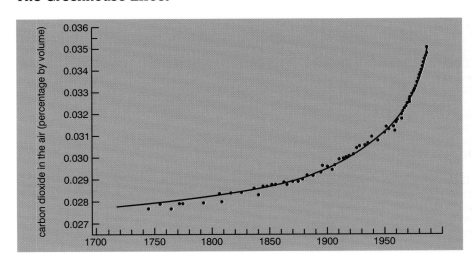

Carbon dioxide in the atmosphere enables energy from the Sun to be trapped as heat instead of reflecting straight back into space – this is the **greenhouse effect**. The worry is that if there is too much carbon dioxide in the Earth's atmosphere then the overall temperature of the planet will increase (global warming) causing the polar ice caps to melt and the sea level to rise.

There can be no doubt that the level of carbon dioxide in the atmosphere is rising. This is a consequence of the rapid increase in the use of fossil fuels at the same time as vast areas of forest have been cut down. The effect of these changes can be seen by looking at the **Carbon Cycle**. Many people feel that the best way to minimise the greenhouse effect is to reduce emissions of carbon dioxide. However, others feel that the problem may be caused by other greenhouse gases. Some scientists are not even sure that there is such a thing as global warming. Instead they point to recent rises in average temperatures as being no more than a climatic change, many of which have occurred on Earth over the last 1000 years. See chapter 5 for more on this.

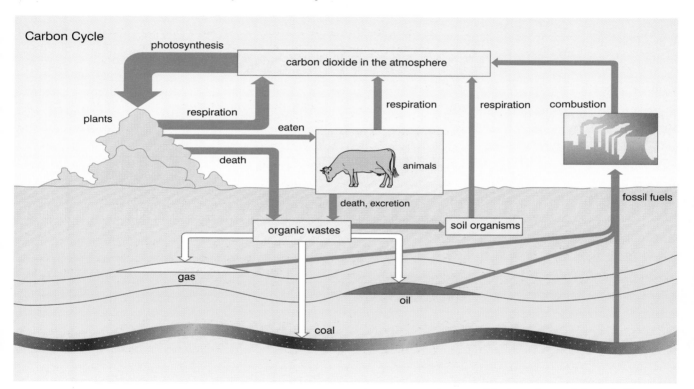

Carbon Cycle

ALKANES AND ALKENES

There are two common families of hydrocarbons, the **alkanes** and the **alkenes**. Members of a family have similar chemical properties, and physical properties that change gradually from one member to the next.

Many alkanes are obtained from crude oil by fractional distillation. The first members of the family are used extensively as fuels. Apart from burning, however, they are remarkably unreactive. Alkanes are made up of atoms joined by single covalent bonds, so they are known as **saturated** hydrocarbons.

The alkenes are also hydrocarbons and they are often formed in the cracking process. They contain a carbon–carbon double bond. Hydrocarbons with at least one double bond are known as **unsaturated** hydrocarbons. Alkenes burn well but, unlike the alkanes, they are reactive in other ways as well. Their reactivity is due to the carbon–carbon double bond in their structure.

The importance of plants in the Carbon Cycle is that they remove carbon dioxide from the atmosphere by photosynthesis. Carbon dioxide is also removed from the atmosphere as it dissolves in the oceans.

		Molecular formula	Displayed formula	Boiling point (°C)	State at room temperature and pressure
Alkanes	methane	CH_4		−162	gas
	ethane	C_2H_6		−89	gas
	propane	C_3H_8		−42	gas
	butane	C_4H_{10}		0	gas
	pentane	C_5H_{12}		36	liquid
Alkenes	ethene	C_2H_4		−104	gas
	propene	C_3H_6		−48	gas
	butene	C_4H_{18}		−6	gas
	pentene	C_5H_{10}		30	liquid

Alkenes can be distinguished from alkanes by adding bromine water to the hydrocarbon. There will be no reaction in the case of the alkane. In the case of the alkene the bromine will be decolorised. The type of reaction is known as an **addition** reaction.

ethene (colourless gas) + bromine (brown liquid) → 1,2-dibromoethane (colourless liquid)

POLYMERS

Alkenes are particularly important chemicals because they can be used to make **polymers**. Polymers are very large molecules made up of many identical smaller molecules called **monomers**. Alkenes are able to react with themselves. They join together into long chains like adding beads to a necklace. When the monomers add together like this the material produced is called an **addition polymer**. Poly(ethene) or polythene is made this way.

By changing the atoms or groups of atoms attached to the carbon–carbon double bond a whole range of different polymers can be made.

Although plastics have very many uses they are very difficult to dispose of. Most of them are not **biodegradable** – that is they cannot be decomposed by bacteria in the soil. Currently most waste plastic material is buried in landfill sites or it is burnt. Both methods have drawbacks. Burning plastics produces toxic fumes. Landfill sites are filling up, and it is difficult to find new sites. Some types of plastic can be melted down and used again. These plastics are called **thermoplastics**. Other types of plastic decompose when they are heated. These plastics are called **thermosetting plastics**. Recycling is difficult because the different types of plastic must be separated.

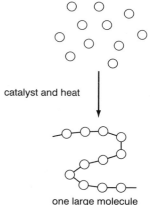

many small molecules

catalyst and heat

one large molecule

Ethene molecules link together to produce a long polymer chain of poly(ethene).

name of monomer	displayed formula of monomer	name of polymer	displayed formula of polymer	uses of polymer
ethene	H, H / C=C / H, H	poly(ethene)	(H H / -C-C- / H H)n	buckets, bowls, plastic bags
chloroethene (vinyl chloride)	H, H / C=C / Cl, H	poly(chloroethene) (polyvinylchloride)	(H H / -C-C- / Cl H)n	plastic sheets, artificial leather
phenylethene (styrene)	H, H / C=C / C_6H_6, H	poly(phenylethene) (polystyrene)	(H H / -C-C- / C_6H_5 H)n	yoghurt cartons, packaging
tetrafluoroethene	F, F / C=C / F, F	poly(tetrafluroethene) or PTFE	(F F / -C-C- / F F)n	non-stick coating in frying pans

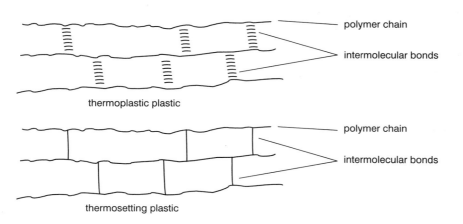

polymer chain

intermolecular bonds

thermoplastic plastic

polymer chain

intermolecular bonds

thermosetting plastic

In thermoplastics the intermolecular bonds are weak and break on heating. The plastic can be melted and re-moulded. In thermosetting plastics the intermolecular bonds are strong inter-linking covalent bonds. The whole structure breaks down when these bonds are broken by heating.

Check yourself

QUESTIONS

Q1
a) Petrol is a hydrocarbon. Write a word equation for the reaction when petrol burns in a plentiful supply of air.
b) When petrol is burned in a car engine carbon monoxide may be formed. What condition causes the formation of carbon monoxide?
c) Explain what effect carbon monoxide has on the body.

Q2
a) The early atmosphere of the Earth may have contained methane and carbon monoxide but very little oxygen. At that time the temperature was high and there were many electrical storms in the atmosphere. Why is it unlikely that oxygen would have been present in large amounts?
b) As early plant life began to develop, the amount of oxygen in the atmosphere tended to increase. (i) Name the process by which plants add oxygen to the atmosphere. (ii) What gas is removed from the atmosphere by the same process?
c) The amount of carbon dioxide in the atmosphere is partly controlled by the oceans. How does this happen?

Q3
a) Draw displayed formulae for (i) hexane, (ii) hexene.
b) Describe a test that could be used to distinguish between hexane and hexene.

Q4
a) Draw the displayed formula for propene.
b) Write an equation using displayed formulae to show how propene can be polymerised.
c) What is the name of the polymer formed in b)?
d) Explain why propane cannot form polymers as propene does.

Q5 What are the environmental problems associated with the disposal of plastics?

REMEMBER! Cover the answers if you want to.

ANSWERS

A1
a) petrol + oxygen → carbon dioxide + water
b) Shortage of oxygen.
c) The carbon monoxide combines with the haemoglobin in the blood, preventing oxygen from doing so. Supply of oxygen to the body is reduced. Specifically this can quickly cause the death of brain cells.

A2
a) The oxygen would have reacted with the methane and the carbon monoxide (oxidised them).
b) (i) Photosynthesis,
   (ii) carbon dioxide.
c) The carbon dioxide dissolves in the water of the oceans (forms carbonic acid).

TUTORIAL

T1
a) *All hydrocarbons produce carbon dioxide and water when burnt in a plentiful supply of air. It is much better to use oxygen rather than air in the equation.*
b) *Remember this is incomplete combustion.*
c) *A much more detailed answer is required than 'it causes suffocation'.*

T2
a) *Under conditions of high temperature and lightning strikes oxidation would have occurred.*
b) *If you missed this look back at the Carbon Cycle diagram.*
c) *This process was not shown on the Carbon Cycle! It needs to be remembered though.*

ANSWERS

A3 a) (i)

hexane

(ii)

hexane

b) Add bromine water. It will be decolorised by hexene but not by hexane.

A4 a)

b)

c) Poly(propene).
d) Propane is saturated and doesn't have a carbon–carbon double bond to undergo addition reactions.

A5 Most plastics are not biodegradable. If disposed of by burning they produce toxic fumes. Recycling is difficult as the different types of plastics need to be separated from each other.

TUTORIAL

T3 a) *Remember 'hex' means 6. The ending 'ane' means only single bonds whereas 'ene' means there is a carbon-carbon double bond.*
b) *This is the standard test for all alkenes. You should give the test reagent and what you would observe.*

T4 a) *'Prop' means 3 carbon atoms.*
b) *If you had problems here look back at how ethene polymerises. Then replace one H atom by a CH_3 group.*
c) *Just put 'poly' in front of the name of the monomer.*
d) *The carbon–carbon double bond is the reactive part of the molecule.*

T5 *Currently incineration and landfill are the two main ways of disposing of plastics.*

ENERGY TRANSFERS

In the majority of reactions energy is transferred to the surroundings. These reactions are **exothermic**. In a minority of cases energy is absorbed from the surroundings as a reaction takes place. These reactions are **endothermic**. All reactions involving the combustion of fuels are exothermic.

Type of reaction	Energy change	Temperature change of reaction and surroundings
exothermic	energy transferred *to* the surroundings	temperature increases
endothermic	energy absorbed *from* the surroundings	temperature decreases

KEY POINTS

Energy in reactions:
- **breaking bonds in the reactants takes in energy – endothermic**
- **making bonds in the products releases energy – exothermic**

Energy transfers in a wide range of chemical reactions can be measured using a polystyrene cup as a calorimeter. If a lid is put on the cup, very little energy is transferred to the air and quite accurate results can be obtained.

temperature goes up
EXOTHERMIC

temperature goes down
ENDOTHERMIC

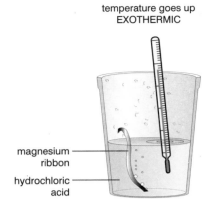

magnesium ribbon

hydrochloric acid

sodium hydrogen carbonate

hydrochloric acid

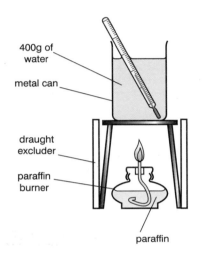

400 g of water

metal can

draught excluder

paraffin burner

paraffin

Measuring the energy produced on burning a liquid fuel.

When choosing a fuel one of the first considerations is how much energy the fuel can produce. The energy transferred when a fuel burns can be measured using a calorimetric technique. The energy released by the burning fuel is transferred to some water and the temperature rise of the water is measured.

This technique will not give a very accurate answer because considerable amounts of energy will be transferred to the surrounding air. Nevertheless, if the same technique is used to burn different fuels the energy they release can be compared. The rise in temperature of the water is a measure of the energy transferred to the water. It is often worth calculating the exact amount of energy transferred using the following equation:

$$\text{energy transferred to the water} = \text{mass of water} \times \text{specific heat capacity of water} \times \text{rise in temperature of water}$$

The 'specific heat capacity' of water is 4.2 J/g/°C (or 4200 J/kg/°C). As the density of water is 1 g/cm^3, the mass of water in g is the same as the volume of water in cm^3.

Example 2.0 g of paraffin were burned in a spirit burner under a metal can containing 400 cm^3 of water. The temperature of the water rose from 20 °C to 70 °C. Calculate the energy produced by the paraffin.

Equation	Energy	= mass of water × 4.2 × temperature rise
Substitute values	E	= 400 × 4.2 × 50
Calculate	E	= 84 000 J per 2 g of paraffin
		= 42 000 J per g of paraffin

If this energy value is to be compared with those produced by other fuels, one way is to use the energy produced by 1 g of each fuel.

WHERE DOES THE ENERGY COME FROM?

When a fuel is burnt the reaction can be considered to take place in two stages. In the first stage the covalent bonds between the atoms in the fuel molecules and the oxygen molecules are broken. In the second stage the atoms combine together and new covalent bonds are formed. For example, consider the combustion of propane.

propane + oxygen → carbon dioxide + water
$C_3H_8(g)$ + $5O_2(g)$ → $3CO_2(g)$ + $4H_2O(l)$

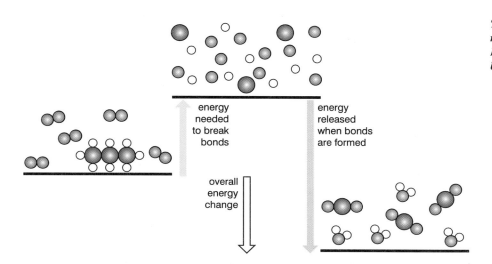

The two stages of the chemical reaction between propane and oxygen. Existing bonds are broken and new bonds are formed.

energy needed to break bonds

energy released when bonds are formed

overall energy change

Stage 1   Energy is needed (absorbed from the surroundings) to break the bonds. This process is endothermic.

Stage 2   Energy is released (transferred to the surroundings) as the bonds form. This process is exothermic.

The overall reaction is exothermic because more energy is released when bonds are formed than is needed initially to break the bonds. A simplified energy level diagram showing the exothermic nature of the reaction is shown on the right.

If the alkanes are compared as fuels you will see that the larger the molecule the more energy is released on combustion. This is because although more bonds have to be broken in the first stage of the reaction, more bonds are formed in the second stage. The increase in energy from one alkane to the next is almost constant. This constant increase is due to the extra CH_2 unit in the molecule. In this comparison the energy released on combustion has been worked out per mole of alkane. In this way the comparison can be made when the same number of molecules of each alkane is burnt. (More information on moles is given in chapter 13.)

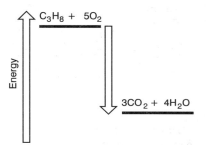

Energy

$C_3H_8 + 5O_2$

$3CO_2 + 4H_2O$

Alkane		Energy of combustion (kJ/ mole)
methane	CH_4	882
ethane	C_2H_6	1542
propane	C_3H_8	2202
butane	C_4H_{10}	2877
pentane	C_5H_{12}	3487
hexane	C_6H_{14}	4141

Check yourself

QUESTIONS

Q1 A 0.2 g strip of magnesium ribbon is added to 40 cm^3 of hydrochloric acid in a polystyrene beaker. The temperature rises by 32 °C. (The specific heat capacity of the hydrochloric acid can be assumed to be the same as that of water, i.e. 4.2 J/g/°C). Calculate (i) the energy released in the reaction, and (ii) the energy released per gram of magnesium.

Q2 Calcium oxide reacts with water as shown in the equation.

$$CaO(s) + H_2O(l) \rightarrow Ca(OH)_2(s)$$

An energy level diagram for this reaction is shown below.

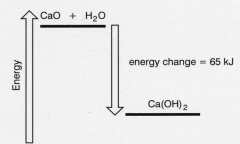

a) What does the energy level diagram tell us about the type of energy change which takes place in this reaction?

b) What does the energy level diagram indicate about the amounts of energy required to break bonds and form new bonds in this reaction.

Q3 Chlorine(Cl_2) and hydrogen(H_2) react together to make hydrogen chloride (HCl). The equation can be written as:

$$H—H + Cl—Cl \rightarrow H—Cl + H—Cl$$

When this reaction occurs energy is transferred to the surroundings. Explain this in terms of the energy transfer processes taking place when bonds are broken and when bonds are made.

Q4 Many reactions proceed by the breaking of covalent bonds followed by the formation of new bonds. Use this information to explain the following.

a) The reaction between hydrogen and oxygen is very exothermic.

b) Hydrogen and oxygen do not react when left together at room temperature.

REMEMBER! Cover the answers if you want to.

ANSWERS

A1
a) Energy = 40 × 4.2 × 32 = 5376 J
b) Energy = 5376/0.2 = 26 880 J/g

A2
a) The reaction is exothermic.
b) More energy is released when bonds are formed than is used to break the bonds in the first place.

A3 More energy is released when the H–Cl bonds are made than is needed to break the H–H and Cl–Cl bonds. The overall energy change is therefore exothermic.

A4
a) Much more energy is released on forming the bonds in water than is needed to break the H–H and O=O bonds.
b) Some energy is needed initially to start to break the H–H and O=O bonds.

TUTORIAL

T1 *Use the formula, energy = mass × 4.2 × temp. change.*

T2 *In an energy level diagram if the products are lower in energy than the reactants the reaction is exothermic. Remember that the overall energy change = energy needed to break bonds – energy released on forming bonds.*

T3 *Again this refers to the balance between bond breaking and bond forming. Always refer to the specific bonds involved.*

T4
a) Again this refers to the balance between bond breaking and bond forming. Always refer to the specific bonds involved.
b) Many reactions need energy (often called the activation energy) to start them off. This is useful otherwise fuels would immediately catch fire when in contact with the air. Heat or a spark can start the reaction.

EXAMINATION CHECKLIST

The facts and ideas that you should understand by studying this topic.

To *be successful at* **Foundation** *GCSE Tier you should be able to*:

- explain how crude oil is formed
- explain why crude oil is non-renewable
- recall that petrol, diesel, propane, butane and paraffin are made from fossil fuels
- describe the process of the fractional distillation of crude oil
- describe the conditions used for cracking oil fractions
- recall that crude oil contains hydrocarbons and that hydrocarbons are molecules made up of carbon and hydrogen only
- describe the differences in structure between alkanes and alkenes
- interpret displayed formulae
- explain that the combustion of a hydrocarbon produces carbon dioxide and water
- explain how a shortage of oxygen during combustion leads to the formation of carbon or carbon monoxide
- recall that gases other than carbon dioxide can cause a greenhouse effect
- recall some of the effects of acid rain
- describe one possible theory for the evolution of the atmosphere
- describe the process of polymerisation
- explain the terms exothermic and endothermic
- describe a simple calorimetric method for comparing energy transfers in different reactions.

In *addition, to be successful at* **Higher** *GCSE Tier you should be able to*:

- explain how crude oil can be separated by fractional distillation
- explain that cracking converts saturated hydrocarbons into more useful products
- explain how unsaturated hydrocarbons can be converted into saturated hydrocarbons
- describe some of the positive and negative factors in the use of fossil fuels
- explain the 'trapping' of solar radiation in the greenhouse effect and the controversy surrounding it
- describe the effect acid rain has on stone and metal
- describe how the atmosphere has evolved
- explain how the carbon cycle helps to maintain the composition of the atmosphere
- relate the properties of plastics to simple models of their structures
- describe chemical reaction in terms of the energy transfers associated with the breaking and making of chemical bonds
- calculate energy changes for simple calorimetric experiments.

Key Words

Tick each word when you are sure of its meaning

acid rain
addition
alkanes
alkenes
biodegradable
combustion
cracking
endothermic
exothermic
fractional distillation
greenhouse effect
hydrocarbon
monomer
non-renewable
oxidise
polymer
saturated
thermoplastics
thermosetting
unsaturated
volatile

EXAM PRACTICE

Sample Student's Answers & Examiner's Comments

EXAMINER'S COMMENTS

a) The correct responses have been chosen.

b) (i) The correct response has been given. Notice that 2 marks have been awarded. One for fractional and one for distillation.
(ii) The candidate has missed this altogether. This often happens when answers have to be written on a chart or diagram given previously. The lower boiling point liquids (or molecules with the smaller size) are found higher up the column. Hence petrol could have been written on the top line, diesel on the lower line.
(iii) The candidate has correctly appreciated that the compounds are hydrocarbons.

---

1 a) Crude oil was formed in the Earth from marine (sea) deposits over millions of years. Possible conditions are listed below. Tick the two conditions which are needed to change the deposits into crude oil.

low temperatures	
water	
high temperatures	✓
air	
high pressures	✓
a magnetic field	

(2)

b) The diagram shows how crude oil can be separated into useful products.

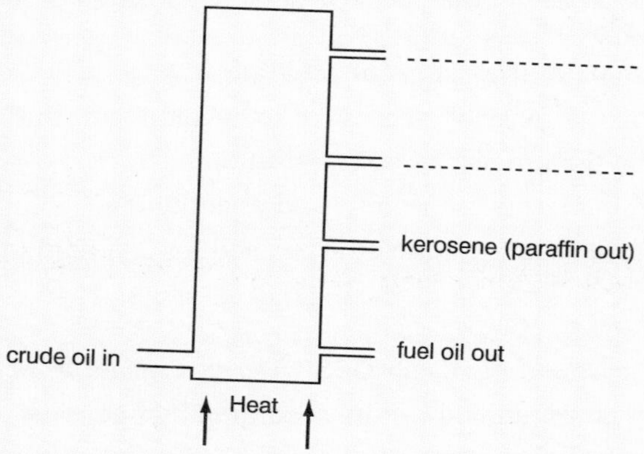

(i) Name the process of separation shown in the diagram.
  fractional distillation ✓✓

(ii) Write the names of TWO more products in the correct spaces the in diagram.

(iii) Two elements are present in compounds in all the products obtained.

Name these elements.

carbon ✓ and *hydrogen* ✓ (6)

c) When oil has been refined in this way there is often too much fuel oil. The larger molecules of fuel oil can be broken down into smaller, more useful molecules. This is shown in the diagram below.

$$\boxed{\text{FUEL OIL}} \xrightarrow{\text{PROCESS } \mathbf{X}} \boxed{\begin{array}{c}\text{COMPOUNDS WITH}\\ \text{SMALLER MOLECULES}\end{array}}$$

(i) What is the name of process X?

cracking ✓

(ii) Give TWO conditions used in process X.

heating to high temperature ✓

high pressure ✗ (3)

d) Diesel fuel has a much higher boiling point than petrol.

(i) Which of these fuels is more **volatile**?

(ii) Which of these fuels has larger molecules?

diesel ✓

(iii) Which fuel poses the greater fire hazard? Explain your answer.

petrol — it has the smallest molecules

and a lower boiling point ✓ ⑩/₁₅ (4)

(Total 15 marks)

London Examinations

EXAMINER'S COMMENTS

c) **(i)** *The correct answer has been given.*
(ii) *Only one of the conditions is correct, i.e. high temperature. The condition missed is that a catalyst is required. You would not be expected to give the name of the catalyst.*

d) **(i)** *The candidate left this part blank. Not a policy to be recommended especially as there was a 50:50 chance of gaining the mark. Always have a go. The word volatile means 'readily forms a vapour'. The fuel with the smaller molecules will be the more volatile, i.e. petrol.*
(ii) *The correct answer was given here.*
(iii) *Petrol is the correct answer. One mark has been gained for the reason given relating to boiling point. (A similar answer in terms of it having a higher volatility would have been even better.) For the second mark the candidate needed to say that vapour mixes more readily with oxygen than liquid does.*

● *A mark of 10 out of 15 corresponds to a grade C on this foundation tier question.*

Questions to Answer

Answers to questions 2 – 3 can be found on page 365.

2 Many power stations burn fossil fuels. The fossil fuels are normally oil, gas or coal. Two power stations in the United Kingdom are trying out a new fossil fuel. It is called Orimulsion.
- It is a mixture of tar and water.
- Orimulsion is about half the price of coal.
- When Orimulsion burns it transfers energy.
- The tar in Orimulsion is not pure because it contains traces of sulphur compounds.

a) Sulphur dioxide is one of the gases made by the power station. It can cause acid rain. Give one effect of acid rain. (1)

b) Carbon dioxide is also made by the power station. Carbon dioxide in the upper atmosphere causes a greenhouse effect. Give three reasons why it is difficult to prove that the increasing amount of carbon dioxide produced by human action is responsible for global warming. (3)

c) Orimulsion contains hydrocarbons. Look at the four displayed formulae shown below. Decane is a hydrocarbon found in Orimulsion.

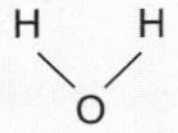

Decane burns and reacts with the oxygen in the air. Water and carbon dioxide are formed.

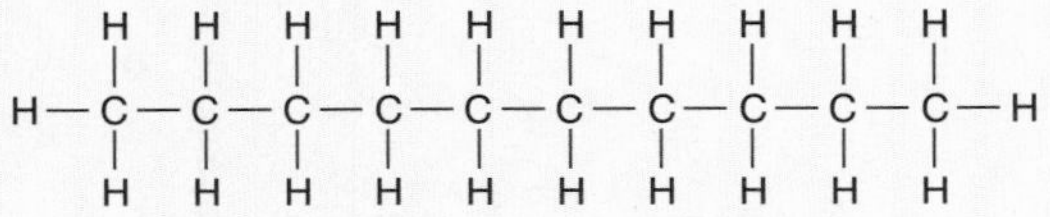

(i) What is the molecular formula for decane? (1)

(ii) To which series of hydrocarbons does decane belong? (1)

(iii) Energy is transferred to the surroundings when decane burns. What is the name of the type of reaction that transfers energy to the surroundings? (1)

(iv) Some bonds are broken when decane burns. Mark with a circle on the displayed formulae two different bonds that are broken. (1)

(v) Some bonds are also made when decane burns. What type of energy transfer process takes place when bonds are made? (1)

Midland Examining Group

3 Crude oil is a mixture of many compounds. Most of these compounds are hydrocarbons. The structure of one of these compounds is shown in the diagram.

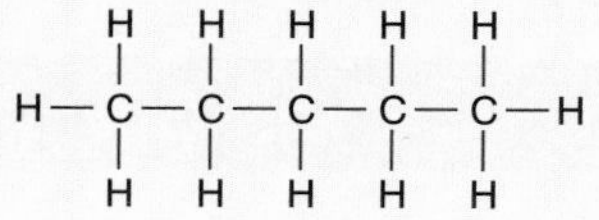

$$H-\underset{\underset{H}{|}}{\overset{\overset{H}{|}}{C}}-\underset{\underset{H}{|}}{\overset{\overset{H}{|}}{C}}-\underset{\underset{H}{|}}{\overset{\overset{H}{|}}{C}}-\underset{\underset{H}{|}}{\overset{\overset{H}{|}}{C}}-\underset{\underset{H}{|}}{\overset{\overset{H}{|}}{C}}-H$$

a)   What is a hydrocarbon?   (1)

b)   What is the chemical formula of the molecule shown in the diagram?   (1)

c)   Crude oil consists of a large number of different compounds. Fractional distillation and cracking can be used to produce useful compounds from crude oil. Describe in as much detail as you can how these two processes produce alkanes that are useful as fuels.   (5)

d)   Ethene is an unsaturated hydrocarbon. What is meant by the term 'unsaturated'?   (1)

e)   Complete the following equation to show how three ethene molecules join together to form part of a poly(ethene) molecule.   (2)

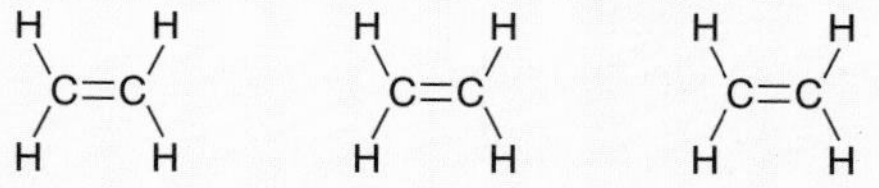

$$\underset{\underset{H}{/}}{\overset{\overset{H}{\backslash}}{C}}=\underset{\underset{H}{\backslash}}{\overset{\overset{H}{/}}{C}}\qquad \underset{\underset{H}{/}}{\overset{\overset{H}{\backslash}}{C}}=\underset{\underset{H}{\backslash}}{\overset{\overset{H}{/}}{C}}\qquad \underset{\underset{H}{/}}{\overset{\overset{H}{\backslash}}{C}}=\underset{\underset{H}{\backslash}}{\overset{\overset{H}{/}}{C}}$$

f)   (i)   Suggest one property of poly(ethene) which makes it suitable as a material for food containers.   (1)

(ii)   Thermosetting plastics cannot be remoulded after they have been heated and allowed to cool. Explain why.   (2)

Northern Examinations and Assessment Board

CHAPTER 10

ROCKS AND METALS

The Earth's crust is made up of rocks which contain chemicals called minerals. Many of these minerals contain metals. In order to find a particular mineral you need to know something about the geology of the Earth's crust. The challenge is then to obtain a pure sample of the metal from its mineral.

GEOLOGICAL CHANGE

Until the beginning of this century scientists thought that the Earth's crust had remained unchanged for millions of years. Then in 1912 the theory of **plate tectonics** was proposed.

PLATE TECTONICS

Evidence obtained by monitoring the shock waves produced by earthquakes suggests that the Earth is made up of layers. The layers are a thin rocky **crust**, the **mantle** and the **core**.

The layered structure of the Earth.

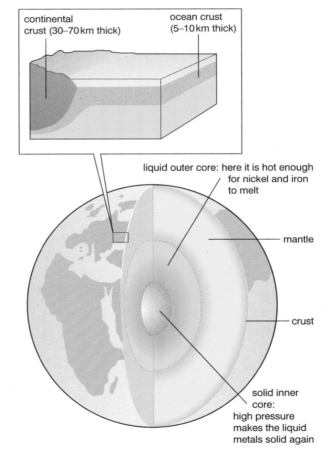

continental crust (30–70km thick)

ocean crust (5–10km thick)

liquid outer core: here it is hot enough for nickel and iron to melt

mantle

crust

solid inner core: high pressure makes the liquid metals solid again

The centre of the Earth is hot due to the decay of radioactive material in the core. Due to the large amount of energy released, the outer core and the mantle are in a liquid state. So the outer crust is really 'floating' on the liquid material of the mantle, which is called **magma**. Energy from the hot core is transferred out through the magma by the process of **convection**. This results in the formation of huge convection currents in the magma.

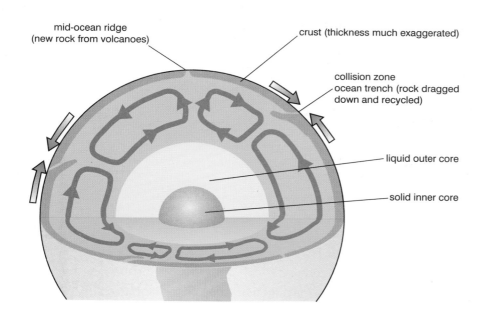

mid-ocean ridge
(new rock from volcanoes)

crust (thickness much exaggerated)

collision zone
ocean trench (rock dragged
down and recycled)

liquid outer core

solid inner core

The huge convection currents in the mantle cause the plates to move across the surface of the Earth.

The crust is thought to be made up of large plates with boundaries between the different plates. The convection currents in the mantle cause these plates to move by a few centimetres each year. If this theory is correct the plates must have been moving for millions of years. Evidence from the study of rock types and fossils does support the theory and suggests that millions of years ago the continents were joined together. The movement of these tectonic plates can be used to explain earthquakes and volcanoes.

Some plates carry continents (**continental plates**) – others carry oceans (**oceanic plates**). Where the convection currents cause the plates to grind together, earthquakes occur. Where the convection currents cause the plates to move apart, a weakness or gap forms in the crust. Magma forces its way up through this gap forming a volcano or, if the gap is in the ocean, a mid-ocean

The main tectonic plates of the Earth.

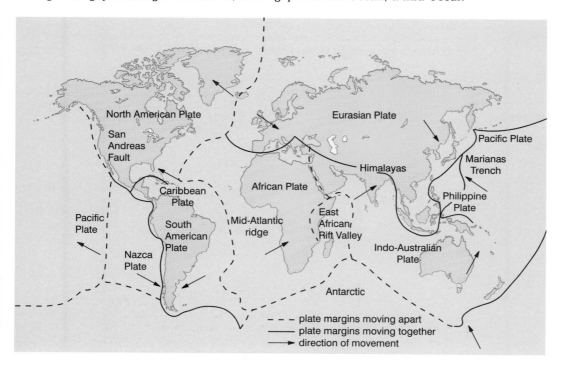

North American Plate

San Andreas Fault

Eurasian Plate

Pacific Plate

Marianas Trench

Himalayas

Caribbean Plate

African Plate

Philippine Plate

Pacific Plate

South American Plate

Mid-Atlantic ridge

East African Rift Valley

Nazca Plate

Indo-Australian Plate

Antarctic

- - - plate margins moving apart
——— plate margins moving together
——→ direction of movement

175

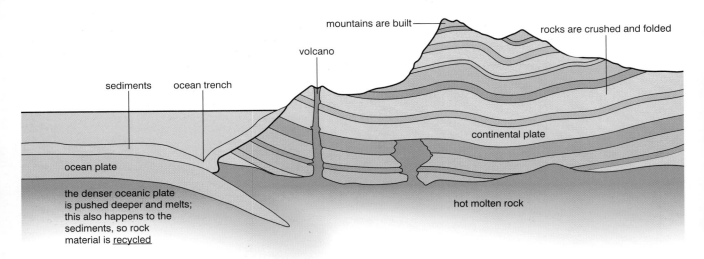

mountains are built

volcano

rocks are crushed and folded

sediments   ocean trench

continental plate

ocean plate

the denser oceanic plate
is pushed deeper and melts;
this also happens to the
sediments, so rock
material is <u>recycled</u>

hot molten rock

*What happens when plates collide.
This shows the process of subduction –
'drawing under'. This is happening
along the west coast of America.*

ridge. You will see from the diagram that some plates move towards each other and collide. If these are continental plates this can result in the formation of mountain ranges. The Himalayas are thought to have been formed in this way. If the collision is between an oceanic plate and a continental plate, the denser oceanic plate is pushed down into the mantle forming an ocean trench. This process is known as **subduction**.

By identifying areas of volcanic action and the occurrence of ocean trenches, as well as areas prone to earthquakes, scientists have been able to identify the separate plates and their boundaries.

THE ROCK CYCLE

Rocks are constantly being broken down and new rocks are constantly forming. Molten rocks in the magma cool and form **igneous rocks**. **Weathering** and **erosion** produce small sediments of rock which are transported (**transportation**) by water into streams and rivers. Here they are ground into even smaller particles which become compressed by those forming above them, eventually forming **sedimentary rock**. This process can take millions of years.

*Weathering, erosion and
transportation – all important
processes in the formation of
sedimentary rocks.*

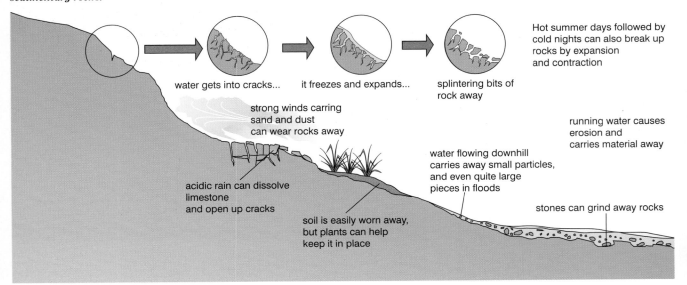

water gets into cracks...   it freezes and expands...   splintering bits of rock away

Hot summer days followed by
cold nights can also break up
rocks by expansion
and contraction

strong winds carring
sand and dust
can wear rocks away

running water causes
erosion and
carries material away

acidic rain can dissolve
limestone
and open up cracks

water flowing downhill
carries away small particles,
and even quite large
pieces in floods

stones can grind away rocks

soil is easily worn away,
but plants can help
keep it in place

Type of rock	Method of formation	Appearance	Examples
Igneous	Cooling of hot, liquid rock (lava from a volcano or underground magma).	Hard, containing crystals of different minerals. Crystal size depends on how quickly the molten rock crystallised. Large crystals are produced on slow cooling; small crystals are formed when the cooling is quick.	Granite, basalt.
Sedimentary	Layers of mud, sand or the shells and bones of living creatures are compressed under high pressure.	The rocks exist in layers with newer layers forming on top of older layers. Fossils are commonly found.	Limestone, sandstone.
Metamorphic	Formed from igneous and sedimentary rocks under conditions of high temperature and pressure.	Grains and crystals are often distorted and fossils are rarely present.	Marble (made from limestone), slate.

When sedimentary rocks are subjected to conditions of high temperature and pressure **metamorphic rocks** are formed. If the sedimentary or metamorphic rocks are forced into the magma in the process of subduction, igneous rocks are formed and the rock cycle is completed.

The evidence obtained from the structure of rocks (known as the rock record) has provided information on the age of the Earth and about the changes that have taken place over billions of years. When sedimentary rocks are formed the lower layers are older than the upper layers. The thickness of the layers and the fossils they contain have provided a lot of information about the evolution of plants and animals.

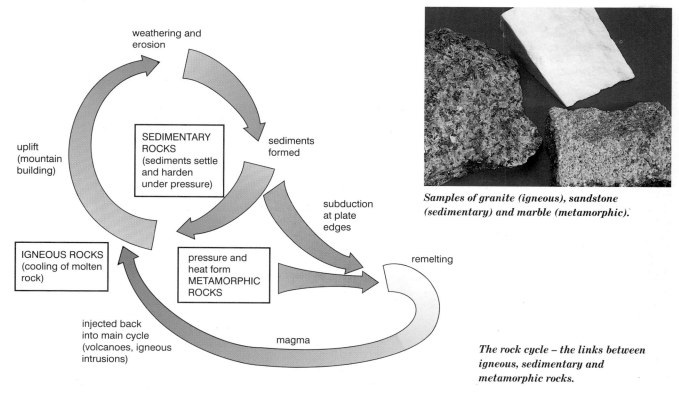

Samples of granite (igneous), sandstone (sedimentary) and marble (metamorphic).

weathering and erosion

uplift (mountain building)

SEDIMENTARY ROCKS (sediments settle and harden under pressure)

sediments formed

subduction at plate edges

IGNEOUS ROCKS (cooling of molten rock)

pressure and heat form METAMORPHIC ROCKS

remelting

injected back into main cycle (volcanoes, igneous intrusions)

magma

The rock cycle – the links between igneous, sedimentary and metamorphic rocks.

Check yourself

QUESTIONS

Q1 In the diagram A, B and C represent plates of the Earth's crust moving in the direction of the arrows.

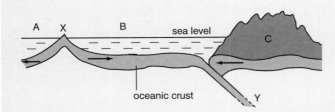

a) Write down one natural occurrence that could happen at the boundary between plates.
b) Explain how igneous rock is formed at the plate boundary X.
c) Metamorphic rock forms at Y. What two conditions are needed for the formation of metamorphic rock?

Q2 The crust of the Earth is divided into several tectonic plates.
a) What causes the tectonic plates to move?
b) Describe what happens to tectonic plates when they collide.

Q3 Sedimentary rock can be made from igneous rock. Three processes are involved. These are weathering, erosion and transportation. Explain how these three processes lead to the formation of sedimentary rock.

Q4 The diagram shows a cross-section of a type of volcano.

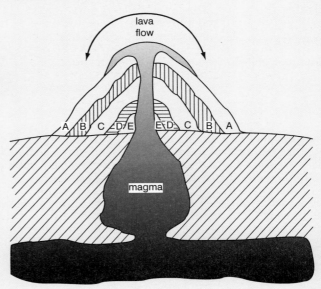

a) Which of the layers A, B, C, D or E is the oldest?
b) Would you expect the crystals on the outer surface of rock layer A to be larger or smaller than those in the middle of layer A? Explain your answer.

Q5 Use your ideas of the rock cycle to explain how a metamorphic rock can be changed into an igneous rock.

REMEMBER! Cover the answers if you want to.

ANSWERS

A1
a) Earthquakes (or volcanoes).
b) As the two plates move apart a weakness forms in the crust. Magma from the mantle is forced up, cools and forms igneous rock.
c) High temperature and high pressure.

TUTORIAL

T1
a) *Remember that earthquakes are formed when two plates rub together.*
b) *Igneous rocks are formed when the lava from a volcano cools. Volcanoes are not always formed, sometimes the magma doesn't break through the crust.*
c) *As the oceanic plate is forced into the mantle the rocks are squashed and heated.*

ANSWERS

A2 a) Convection currents in the liquid mantle.

b) If two continental plates collide, mountains form. If a continental plate collides with an oceanic plate then an ocean trench forms.

A3 Weathering – water gets into cracks in the rocks and then freezes and expands. As the ice melts, fragments of rock break away. Erosion – running water and strong winds carrying sand can wear rocks away . Acid rain dissolves limestone. Transportation – rain water carries particles into streams and rivers.

A4 a) E.

b) Smaller. On the outer surface the lava will cool more quickly than in the middle. Quicker cooling forms smaller crystals.

A5 If metamorphic rocks are pushed into the mantle (during subduction for example) they will melt and form magma. If the magma then returns to the surface where two plates are moving apart, igneous rock will be formed as the magma cools.

TUTORIAL

T2 a) *Energy from the core is transferred to the mantle causing convection currents.*

b) *You need to distinguish between the two types of plate. Look back at the diagram on page 176 if you are not sure about this.*

T3 *It is important that you are clear about these three processes. If not, look back at the diagram on page 176. The differences between weathering and erosion are slight and so you shouldn't worry if you had trouble separating them. This is a very common examination question.*

T4 a) *As lava comes out of the volcano it forms layers on top of previously solidified lava. Layer E was the first to form.*

b) *The link between crystal size and rate of cooling is an important one. Where the cooling is slow, the particles have time to form into large regular arrays (crystals).*

T5 *The collision between continental and oceanic plates causes the denser oceanic plate to be pushed into the mantle. The intense heat in the mantle is sufficient to melt the rock. Try to use the term subduction.*

THE REACTIVITY SERIES OF METALS

Metals are found in the form of **ores**. The ores contain **minerals** mixed with waste rock. In almost all cases the mineral is a compound of the metal, not the pure metal. One exception is gold which can be found in a pure state. Extracting a metal from its ore usually involves two steps:

- the mineral is physically separated from unwanted rock

- the mineral is chemically broken down to obtain the metal.

The chemical method chosen to break down a mineral depends on the **reactivity** of the metal. As a general rule, the more *reactive* a metal is the harder it is to break down its compounds. The less reactive metals can be obtained by heating their oxides with carbon. The more reactive metals are obtained from their minerals by a process known as **electrolysis**.

Metals have very different levels of reactivity. Some, such as potassium, can react as soon they come into contact with the air. Others, such as gold, do not react with the air over hundreds of years. The common metals can be placed in a **reactivity series** which can then be used to summarise the reactions of metals. It can also be used to predict how different metals will react with the air, with water and with acids.

Metal	Reactivity	Reaction with air	Reaction with water	Reaction with acids e.g. dilute hydrochloric acid.
potassium	most	burn in air or oxygen to form a metal oxide	react with cold water to form a metal hydroxide and hydrogen	react violently to form a metal chloride and hydrogen
sodium				
calcium				
magnesium			react with steam to form a metal oxide and hydrogen	react to form a metal chloride and hydrogen
aluminium				
zinc				
iron				
tin		react with air and oxygen but do not burn	no reaction with water or steam	no reaction with dilute acid
lead				
copper				
silver		do not react with air or oxygen		
gold	least			

Note – Aluminium rarely appears as reactive as it should be from its position in the reactivity series. This is because it rapidly reacts with oxygen to form aluminium oxide. The aluminium oxide forms a layer on the surface of the metal, preventing further reaction.

REACTIONS OF METALS WITH AIR AND OXYGEN

The reactions shown in the table can be summarised using word or symbol equations.

potassium + oxygen → potassium oxide
$4K(s)$ + $O_2(g)$ → $2K_2O(s)$

magnesium + oxygen → magnesium oxide
$2Mg(s)$ + $O_2(g)$ → $2MgO(s)$

copper + oxygen → copper(II) oxide
$2Cu(s)$ + $O_2(g)$ → $2CuO(s)$

Rusting involves the reaction of iron with oxygen in the presence of water. Rusting can be prevented if the iron is coated with, or is in contact with, a more reactive metal. Zinc is often used in this way. The metal panels and frames of cars are often made from galvanised steel – steel coated with zinc. The zinc, being more reactive than iron, reacts with oxygen and water before the iron does. This protects the iron.

REACTIONS OF METALS WITH WATER

These reactions are like a competition reaction. The metals are competing with the hydrogen for the oxygen in water. Metals that are more reactive than hydrogen will succeed and combine with the oxygen.

The most reactive metals at the top of the reactivity series react with water forming a metal hydroxide and hydrogen:

sodium   +   water   →   sodium hydroxide   +   hydrogen
$2Na(s)$ + $2H_2O(l)$ → $2NaOH(aq)$ + $H_2(g)$

Those metals in the middle of the reactivity series react with steam rather than cold water and this time a metal oxide is formed as well as hydrogen.

zinc   +   steam   →   zinc oxide   +   hydrogen
$Zn(s)$ + $H_2O(g)$ → $ZnO(s)$ + $H_2(g)$

REACTIONS OF METALS WITH ACIDS

This time the metals are competing with the hydrogen in the acid. In the case of hydrochloric acid, the metal and the hydrogen compete for the 'chloride' part of the acid. If the metals are more reactive than hydrogen they will succeed and form a metal chloride and hydrogen:

magnesium   +   hydrochloric acid   →   magnesium chloride   +   hydrogen
$Mg(s)$ + $2HCl(aq)$ → $MgCl_2(aq)$ + $H_2(g)$

The order of reactivity of metals with dilute sulphuric acid is the same as with dilute hydrochloric acid. A typical reaction would be:

zinc   +   sulphuric acid   →   zinc sulphate   +   hydrogen
$Zn(s)$ + $H_2SO_4(aq)$ → $ZnSO_4(aq)$ + $H_2(g)$

DISPLACEMENT REACTIONS

Metals can also compete with each other. For example, if magnesium metal is added to copper(II) sulphate solution, the magnesium displaces (takes the place of) the copper as it is a more reactive metal. This type of reaction is known as a **displacement reaction**.

magnesium   +   copper(II) sulphate   →   magnesium sulphate   +   copper
$Mg(s)$ + $CuSO_4(aq)$ → $MgSO_4(aq)$ + $Cu(s)$

In this reaction the blue colour of the copper(II) sulphate solution fades as the reaction proceeds (magnesium sulphate solution is colourless) and the magnesium ribbon becomes coated in copper. If copper is added to magnesium sulphate solution no reaction is observed. The copper is less reactive and so cannot 'take' the sulphate radical from the more reactive magnesium.

copper + magnesium sulphate → no reaction

Before reaction:
The copper and sulphate are quite 'happy' together – nothing is there to compete.

When magnesium is added:
Magnesium is more reactive than copper. Magnesium wins the competition and copper is displaced or 'thrown out' on its own.

THE VOLTAGE OF SIMPLE CELLS

If two different metals are placed in a conducting solution, a voltage is produced. A conducting solution contains free moving ions and is called an **electrolyte**. The size of the voltage produced does not depend on the electrolyte, it only depends on the difference in the reactivity of the metals. So for example, the voltage obtained when silver and copper are used is about 0.5 volts whereas when copper and magnesium are used the voltage is nearer to 2.7 volts. This is further evidence for the order of the reactivity series of metals and shows that the difference in reactivity between magnesium and copper is much greater than between copper and silver.

A simple electrical cell can be made by dipping magnesium and copper into a solution of an electrolyte such as sodium chloride.

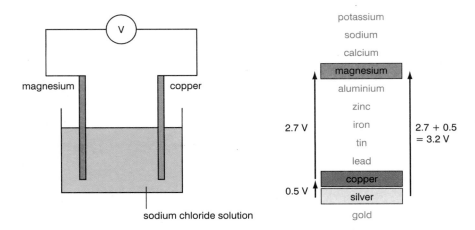

The voltages between different metals add up in a simple way. The voltage created by silver and magnesium is equal to the voltage created by silver and copper added to the voltage created by copper and magnesium.

Check yourself

QUESTIONS

Q1 The table below shows the reactions of some metals with oxygen and with water.

Metal	Reaction with oxygen	Reaction with water
sodium	burns easily	very rapid reaction with cold water
copper	does not burn but surface quickly turns black	no reaction
zinc	burns, but only at high temperatures	no reaction with cold water but reacts quickly with steam
A	burns easily when heated	
B	does not burn and remains unchanged on heating	

a) Place the five metals in order of reactivity, with the most reactive first.

b) Predict the reactions of metals A and B with water.

Q2 Four metals were tested to see which would react in different metal chloride solutions. The table shows the results. (A tick means a reaction occurred; a cross means there was no reaction).

	magnesium chloride	lead(II) chloride	zinc chloride	nickel(II) chloride
magnesium	✗	✓	✓	✓
lead	✗	✗	✗	✗
zinc	✗	✓	✗	✓
nickel	✗	✓	✗	✗

a) Put the metals in order of reactivity with the most reactive first.

b) What would you expect to see in one of the reactions in the table?

c) Magnesium reacts with nickel(II) chloride. (i) Write a word equation for this reaction. (ii) Write a symbol equation for the reaction.

Q3 Iron or steel food cans are sometimes coated with tin. When they are scratched the iron rusts very quickly. On other occasions the iron or steel cans are coated with zinc. When these cans are scratched the iron does not rust. Use ideas of the reactivity series to explain these differences.

Q4 A student wants to make an electrical cell. She puts two rods into a solution and connects a voltmeter as shown in the diagram.

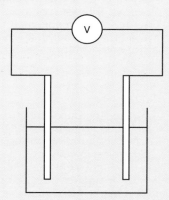

QUESTIONS

Q4
a) The rods must be electrical conductors. Name a non-metallic element that could be used for one of the rods.

b) The solution must be an electrolyte. (i) What is an electrolyte? (ii) Name a substance that could be used as the electrolyte in this cell.

c) The table on the right shows some voltages obtained in different cells.

   (i) Explain why the zinc and copper combination produces the highest voltage.

   (ii) Estimate the voltage you would expect for the zinc–iron combination. Explain how you made your estimate.

Rod 1	Rod 2	Voltage/volts
zinc	copper	0.60
lead	copper	0.02
iron	copper	0.30
zinc	iron	

REMEMBER! Cover the answers if you want to.

ANSWERS

A1
a) Sodium (most reactive), A, zinc, copper, B (least reactive).

b) A – reacts with cold water (but less rapid than sodium). B – no reaction with cold water or steam.

A2
a) Magnesium (most reactive), zinc, nickel, lead (least reactive).

b) A coating forming on the metal, a darkening in colour.

c) (i) Magnesium + nickel(II) chloride → magnesium chloride + nickel

   (ii) $Mg(s) + NiCl_2(aq) \rightarrow MgCl_2(aq) + Ni(s)$

TUTORIAL

T1
a) *The order can be obtained by looking at the reactions with oxygen. Don't forget to write the order with the most reactive first!*

b) *'A' must be a reactive metal and so will react with cold water. Its reaction with oxygen shows that it is less reactive than sodium. 'B' is less reactive than copper and copper doesn't react with water or steam.*

T2
a) *A more reactive metal will displace a less reactive metal from a solution of one of its compounds. Magnesium displaces all three other metals, zinc displaces two, nickel displaces one, lead displaces none.*

b) *The exact observation will depend on the actual metal and solution used.*

c) *Magnesium displaces the nickel and gains the chloride radical. The rules for writing formulae are given in chapter 7. The equation does not need balancing.*

ANSWERS

A3 The order of reactivity is zinc, iron, tin. When iron is in contact with zinc the zinc will react in preference to the iron. So the iron will be protected for as long as the zinc lasts. Because tin is less reactive than iron the iron will rust in preference to the tin.

A4
a) Carbon.
b) (i) An electrolyte is a substance that when molten or dissolved in water conducts an electric current.
(ii) Sodium chloride solution (any salt solution or acid)
c) (i) These two metals have the greatest difference in reactivity.
(ii) 0.30 volts. Zn/Cu is 0.6 V; Fe/Cu is 0.3 V therefore Zn/Fe should be 0.6 – 0.3 = 0.3 V.

TUTORIAL

T3 *This is an example of 'sacrificial protection'. A more reactive metal will always oxidise before a less reactive metal. The surprising thing is that sacrificial protection occurs even when the coating of zinc is broken.*

T4
a) Carbon is the only non-metal that is a relatively good conductor of electricity.
b) You will find out more about electrolytes in the next section when electrolysis is considered. Any solution containing an ionic compound (see chapter 8) will behave as an electrolyte.
c) With the activity series you have been given the greatest voltage would be obtained using sodium and gold rods. This would be very hazardous, however, as the sodium would react violently with the water in the electrolyte solution!

EXTRACTION OF METALS

Now that we have looked at the reactivity of metals we can look at how to extract those metals from their ores.

Metal	Extraction method
potassium	
sodium	
calcium	The most reactive metals are obtained using electrolysis.
magnesium	
aluminium	
(carbon)	
zinc	
iron	
tin	These metals are below carbon in the reactivity series and so can be obtained by heating their oxides with carbon.
lead	
copper	
silver	The least reactive metals are found as pure elements.
gold	

USING CARBON TO EXTRACT METALS

This method will only work for metals below carbon in the reactivity series. It involves the **reduction** of a metal oxide to the metal. Copper is extracted by heating the mineral malachite (copper(II) carbonate) with carbon. The reaction takes place in two stages:

Stage 1 – The malachite decomposes

copper(II) carbonate → copper(II) oxide + carbon dioxide
$CuCO_3(s)$ → $CuO(s)$ + $CO_2(g)$

Stage 2 – The copper(II) oxide is reduced by the carbon

copper(II) oxide + carbon → copper + carbon dioxide
$2CuO(s)$ + $C(s)$ → $2Cu(s)$ + $CO_2(g)$

The copper produced by this process is purified by electrolysis (see the next section).

Iron is also produced on a very large scale by this method of reduction using carbon. The reaction is undertaken in a huge furnace called a 'blast furnace'. Three important raw materials are put in the top of the furnace:

- iron ore (iron(III) oxide) – the source of iron
- coke – the source of carbon needed for the reduction;
- limestone – needed to remove the impurities as a 'slag'.

A blast furnace is used to reduce iron(III) oxide to iron.

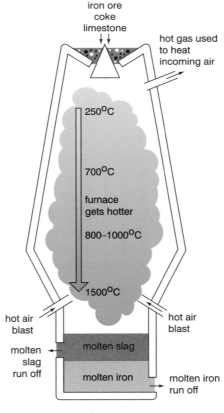

iron ore
coke
limestone

hot gas used to heat incoming air

250°C

700°C

furnace gets hotter

800–1000°C

1500°C

hot air blast

hot air blast

molten slag run off

molten slag

molten iron

molten iron run off

1 Iron ore, limestone and coke are fed into the top of the blast furnace

2 Hot air is blasted up the furnace from the bottom

3 Oxygen from the air reacts with coke to form carbon dioxide:
$C(s) + O_2(g) \longrightarrow CO_2(g)$

4 Carbon dioxide reacts with more coke to form carbon monoxide:
$CO_2(g) + C(s) \longrightarrow 2CO(g)$

5 Carbon monoxide is a reducing agent. Iron (III) oxide is reduced to iron:
⌐reduction = loss of oxygen ⌐
$Fe_2O_3(s) + 3CO(g) \longrightarrow 2Fe(l) + 3CO_2(g)$

6 Dense molten iron runs to the bottom of the furnace and is run off. There are many impurities in iron ore. The limestone helps to remove these as shown in 7 and 8.

7 Limestone is broken down by heat to calcium oxide:
$CaCO_3(s) \longrightarrow CaO(s) + CO_2(g)$

8 Calcium oxide reacts with impurities like sand (silicon dioxide) to form a liquid called 'slag':
$CaO(s) + SiO_2(s) \longrightarrow CaSiO_3(l)$
           impurity   slag
The liquid slag falls to the bottom of the furnace and is tapped off.

The reduction is slightly more complicated than in the reduction of copper(II) oxide because it happens in three stages, but the overall reaction is:

iron oxide   +   carbon   →   iron   +   carbon dioxide
$Fe_2O_3(s)$ + $3C$ → $2Fe$ + $3CO_2$

The three stages are:

Stage 1 – The coke (carbon) reacts with oxygen 'blasted' into the furnace.
carbon   +   oxygen   →   carbon dioxide
$C(s)$ + $O_2(g)$ → $CO_2(g)$

Stage 2 – The carbon dioxide is reduced by unreacted coke to form carbon monoxide.
carbon dioxide   +   carbon   →   carbon monoxide
$CO_2(g)$ + $C(s)$ → $2CO(g)$

Stage 3 – The iron(III) oxide is reduced by the carbon monoxide to iron.
iron(III) oxide   +   carbon monoxide   →   iron   +   carbon dioxide
$Fe_2O_3(s)$ + $3CO(g)$ → $2Fe(s)$ + $3CO_2(g)$

Lead and zinc are also extracted in large quantities by heating their oxides with carbon.

USING ELECTROLYSIS TO EXTRACT REACTIVE METALS

Electrolysis is the breakdown of a chemical compound by an electric current. Metals that are above carbon in the reactivity series cannot be obtained by heating their oxides with carbon, they are reduced by electrolysis.

What conditions are needed?

The substance being electrolysed is known as the **electrolyte**. It must contain **ions** and these ions must be free to move. In other words the substance must either be molten or dissolved in water.

A d.c. voltage must be used. The **electrode** connected to the positive terminal of the power supply is known as the **anode**. The electrode connected to the negative terminal of the power supply is known as the **cathode**. The electrical circuit can be drawn as follows:

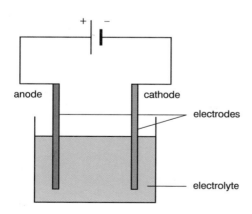

A typical electrical circuit used in electrolysis.

How does the electrolyte change?

The negative ions are attracted to the anode and release electrons. (Loss of electrons is oxidation.) For example:

chloride ions   →   chlorine molecules + electrons
$2Cl^-(aq)$ → $Cl_2(g)$ + $2e^-$

The positive ions are attracted to the cathode and gain electrons. (Gaining electrons is reduction.) For example:

copper ions + electrons → copper atoms
$Cu^{2+}(aq)$ + $2e^-$ → $Cu(s)$

The electrons move through the external circuit from the anode to the cathode.

Extracting aluminium

Aluminium is extracted from the ore bauxite (aluminium oxide) by electrolysis. The aluminium oxide is insoluble so it is melted to allow the ions to move when an electric current is passed through it. The anodes are made from carbon and the cathode is the carbon-lined steel case.

The extraction of aluminium is expensive. A mineral called cryolite is added to the aluminum oxide to lower the melting point and save energy costs.

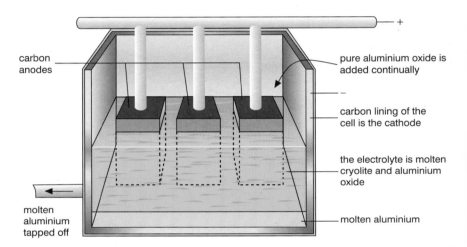

carbon anodes

pure aluminium oxide is added continually

carbon lining of the cell is the cathode

the electrolyte is molten cryolite and aluminium oxide

molten aluminium

molten aluminium tapped off

At *the cathode* – Aluminium is formed
aluminium ions + electrons → aluminium
$Al^{3+}(l)$ + $3e^-$ → $Al(l)$

At *the anode* – Oxygen is formed
oxide ions → oxygen molecules + electrons
$2O^{2-}(l)$ → $O_2(g)$ + $4e^-$

The oxygen reacts with the carbon anodes to form carbon dioxide. The rods constantly need to be replaced because of this.

USING ELECTROLYSIS TO PURIFY METALS

Copper is extracted from its ore by reduction with carbon. The copper produced is not pure enough for some of its uses, such as making electrical wiring. It can be purified using electrolysis. The impure copper is made the anode in a cell with copper(II) sulphate as an electrolyte. The cathode is made from a thin piece of pure copper.

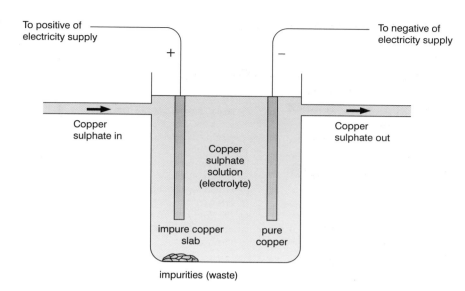

Copper is purified by electrolysis.

At *the anode* – The copper atoms dissolve forming copper ions.
copper atoms $\rightarrow$ copper ions $+$ electrons
$Cu(s)$ $\rightarrow$ $Cu^{2+}(aq)$ $+$ $2e^-$

At *the cathode* – The copper ions are deposited to form copper atoms.
copper ions $+$ electrons $\rightarrow$ copper atoms
$Cu^{2+}(aq)$ $+$ $2e^-$ $\rightarrow$ $Cu(s)$

Check yourself

QUESTIONS

Q1 Explain why lead can be obtained by heating lead oxide with carbon whereas sodium cannot be obtained by heating sodium oxide with carbon.

Q2 Iron is made from iron ore (iron oxide) in a blast furnace by heating with carbon.
a) Write a word equation for the overall reaction.
b) Is the iron oxide oxidised or reduced in this reaction? Explain your answer.
c) Why is limestone also added to the blast furnace?

Q3 Explain the following terms:
a) electrolysis
b) electrolyte
c) electrode
d) anode
e) cathode.

Q4 Aluminium is extracted from aluminium oxide (Al_2O_3) by electrolysis. Aluminium oxide contains Al^{3+} and O^{2-} ions. The aluminium oxide is heated until it is in a molten state.
a) Why is the electrolysis carried out on molten rather than solid aluminium oxide?
b) Which electrode does the aluminium form at?
c) Explain how aluminium atoms are formed from aluminium ions. Write an ionic equation for this change.
d) The carbon electrodes need to be replaced regularly. Explain why.

Q5 Electrolysis is used to purify copper.
a) Which electrode does the copper form on?
b) Write an ionic equation showing the formation of copper at this electrode.

ANSWERS

A1 Carbon is more reactive than lead, so it can remove the oxygen, reducing the lead oxide. Sodium is more reactive than carbon and will not lose its oxygen.

A2
a) Iron oxide + carbon → iron + carbon dioxide.
b) Reduced. Reduction is the loss of oxygen, the iron oxide has lost oxygen forming iron.
c) The limestone reacts with impurities forming a slag that floats to the top and is removed from the furnace.

A3
a) This is the breakdown (decomposition) of a compound using electricity.
b) This is a substance which when molten or dissolved in water allows an electric current to pass through it.
c) This is the substance which makes the electrical contact between the battery and the electrolyte.
d) This is the electrode connected to the positive terminal of the battery.
e) This is the electrode connected to the negative terminal of the battery.

A4
a) The ions must be free to move. In a solid the ions are held in a giant lattice structure.
b) Cathode.
c) Aluminium ions gain electrons and form aluminium atoms.
$Al^{3+} + 3e^- \rightarrow Al$
d) The oxygen that is produced at the anode oxidises the hot carbon electrodes forming carbon dioxide.

A5
a) The cathode.
b) $Cu^{2+} + 2e^- \rightarrow Cu$

TUTORIAL

T1 The important point here is the position of carbon in the reactivity series compared to the position of lead and sodium. Remember it is only metals with lower reactivity that can be obtained by heating an oxide with carbon.

T2
a) In the blast furnace much of the reduction is done by carbon monoxide.
b) Reduction and oxidation always occur together. In this reaction the carbon is oxidised.
c) Of the three raw materials in the blast furnace (coke, iron ore, limestone) this is the easiest one to forget.

T3
a) If you are still puzzled by electrolysis look back at page 187.
b) Electrolytes must contain ions and the ions must be free to move before a current will flow.
c) In a circuit there will always be two electrodes.
d) Ions that travel to the anode are called anions (negatively charged ions).
e) Ions that travel to the cathode are called cations (positively charged ions).

T4
a) Remember that the ions in a solid can be made free to move by either melting the solid or dissolving it in water. Aluminium oxide does not dissolve in water.
b) Aluminium ions are positively charged and so will be attracted to the negative electrode.
c) The ionic equation must balance in terms of symbols and charges.
d) This is one of the drawbacks of the method and also adds to the expense of producing aluminium.

T5
a) All metals ions (positively charged) are deposited as atoms on the cathode.
b) If you need more practice with ionic equations see chapter 7.

EXAMINATION CHECKLIST

The facts and ideas that you should understand by studying this topic.

To be successful at Foundation GCSE Tier you should be able to:

- recall the three main types of rock (igneous, metamorphic and sedimentary) and explain how they form
- recall that the rate of cooling of magma determines crystal size in igneous rocks
- recall that the Earth consists of a crust, mantle and core
- recall that the Earth's crust is composed of large interlocking plates (tectonic plates) that move slowly
- recall that the movement of tectonic plates causes volcanic activity and earthquakes at tectonic boundaries
- use the reactivity series to predict the reactions of metals with air, water and acids
- recall that an ore is a mixture of a mineral and surrounding rock
- recall that copper can be extracted from its mineral by heating with carbon
- recall that reduction is the loss of oxygen from an oxide
- explain how the reactivity series helps to predict which metals can be obtained by reduction of their oxides with carbon or carbon monoxide
- recall that reactive metals such as aluminium are usually extracted by electrolysis and that copper can be purified by electrolysis.

In addition, to be successful at Higher GCSE Tier you should be able to:

- explain the layered appearance and presence of fossils in sedimentary rocks and the distortion of crystals and fossils in metamorphic rocks
- explain the importance of the convection currents in the mantle
- use the theory of plate tectonics to explain the formation of mountains, mid-ocean ridges and the process of subduction
- identify the processes of oxidation and reduction from word and symbol equations
- recall the key features of the electrolysis of aluminium oxide in the production of aluminium
- explain how electrolysis is used to purify copper
- explain electrolysis reactions using half equations.

Key Words

Tick each word when you are sure of its meaning.

cathode
continental plate
displacement
electrolysis
electrolyte
erosion
magma
mantle
mineral
oceanic plate
ore
plate tectonics
reactivity series
reduction
sedimentary
subduction
transportation
weathering

EXAM PRACTICE

Sample Student's Answers & Examiner's Comments

a) The correct responses have been chosen.

b) **(i)** There is only one mark for this question and this is given for the reason. So simply guessing will not be enough for the mark. The reason given by the candidate is only just good enough. It would have been a better answer if 'it contains a mineral and waste rocks' had been added.

(ii) The word equation has been given correctly. You will notice that the names of the reactants (copper carbonate and carbon) and the products (copper and carbon dioxide) are given in the question. This is quite common in a Foundation Tier question. In a Higher Tier question you would not be given as much help.

---

1 This question is about rocks, ores and minerals.

Look at the list. It shows the three types of rock.

**igneous
metamorphic
sedimentary**

a) Finish the sentences by choosing one of the types of rock.

(i) One type of rock is formed when molten rock from a volcano changes into a solid.

This type of rock is called .....*igneous*.......... ✓ rock. (1)

(ii) Another type of rock is made when small bits of rock settle at the bottom of lakes and are squashed together.

This type of rock is called .....*sedimentary*.......... ✓ rock. (1)

b) Copper is a metal which is made from a mineral. The mineral comes from an ore found in the ground.

The ore is first purified to get the mineral out of the waste rocks.

(i) Is copper ore an element, a compound or a mixture?

Answer .*mixture*......

Write down a scientific reason for your answer.
.*contains more than one substance*...... ✓
.... (1)

(ii) The mineral in the copper ore is called copper carbonate.

Copper carbonate can be changed into copper metal.

The copper carbonate is heated with carbon. Copper metal and carbon dioxide are made.

Write down a word equation for this change.
.*copper carbonate + carbon → copper + carbon dioxide*. ✓ (1)

c) Look at the Table. It shows some information about some metals.

Name of mineral	Metal	Reaction of the metal with water
malachite	copper	no reaction with cold water or steam
magnesite	magnesium	slow reaction with hot water to form a colourless gas
rock salt	sodium	rapid reaction with cold water to form a colourless gas
zinc blende	zinc	only reacts with steam

(i) What is the name of the mineral of sodium?

Rock salt ✓ (1)

(ii) Use the table to write the metals in order of reactivity.

Most reactive sodium

Magnesium

Least reactive Zinc

Copper ✓ (1)

d) Aluminium can be made from a mineral called alumina. Aluminium oxide is the chemical name for alumina.

Describe how aluminium is made from alumina. Include in your answer the method used to break down the alumina and the conditions needed.

aluminium oxide is broken down to aluminium using electricity. The method is called electrolysis. ✓ An electric current is passed through the aluminium oxide forming aluminium and oxygen ✓

8/10

(4)

(Total 10 marks)

Midland Examining Group

EXAMINER'S COMMENTS

c) ***(i)*** *The correct mineral has been chosen from the table.*
(ii) *The order of reactivity can be worked out from the observations given in the last column of the table.*

d) *This question requires an extended answer. Questions like this must be included in examination papers. The candidate has scored just two of the four marks. One mark is obtained for the method used, 'electrolysis', and the second mark for giving the products as 'aluminium and oxygen'. Although prompted the candidate has not mentioned the conditions needed. Liquid or molten aluminium oxide would have scored a further mark. The final mark was awarded for stating that the aluminium would form at the cathode, the oxygen at the anode.*

- *A mark of 8/10 corresponds to a grade C on this Foundation Tier question.*

Questions to Answer

Answers to questions 2 – 3 can be found on page 366.

2 a)   Outline the formation of: (i) igneous rock, (ii) metamorphic rock.   (3)

b)   Explain how igneous rocks can become sedimentary rocks.
Give an indication of the time scale involved.   (4)

c)   Explain how sedimentary rock can be recycled as igneous rock.   (4)

London Examinations

3 a)   Aluminium is extracted from its ore by electrolysis.
The diagram shows the cell that is used.

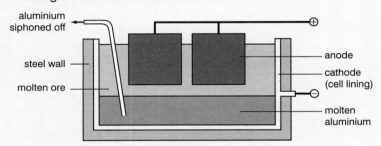

(i)   Name the ore from which the aluminium is extracted.   (1)

(ii)   What substance is the anode made from?   (1)

(iii)   Explain why the anodes must be replaced regularly.   (2)

b)   Copper was one of the earliest metals to be discovered.
Soon afterwards Man discovered how to get tin from tin ores.

Metal	First discovered
aluminium	155 years ago
copper	7000 years ago
gold	10000 years ago
iron	3000 years ago
lead	7000 years ago
magnesium	125 years ago
sodium	85 years ago
tin	6000 years ago
zinc	2000 years ago

Mixing and melting copper and tin made the 'metal' bronze. Bronze was not brittle and it could be sharpened. It was such an important discovery that it gave its name to a new age, the 'Bronze Age'.

(i) Before the Bronze Age, people only had stone and pottery to make things out of. How did the discovery of bronze improve the lives of the Bronze Age people? (2)

(ii) Part of the reactivity series is shown below.
Sodium (most reactive)
magnesium
aluminium
zinc
iron
tin
lead
copper
gold (least reactive)

What is the link between the position of a metal in the reactivity series and how long ago it was discovered? (1)

(iii) Explain why magnesium was not discovered until fairly recently. (2)

c) The diagram represents a blast furnace used for extracting iron from iron ore.

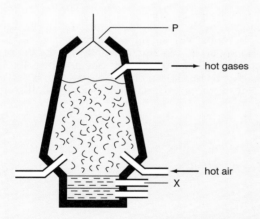

(i) Name **two** substances which are mixed with iron ore and fed into the blast furnace at point **P**. (2)

(ii) What substance is removed from the blast furnace at X? (1)

Southern Examining Group

CHAPTER 11

CHEMICAL REACTIONS

Chemical reactions, whether those involved in the ripening of tomatoes or in the production of important industrial chemicals such as ammonia and sulphuric acid, can be controlled. The speed of a reaction and the amount of product made can be carefully controlled.

KEY POINTS

Reaction rate:
- **high for quick reactions**
- **low for slow reactions**
- **depends on activation energy**
- **monitored by change in mass, volume or temperature.**

CHEMICAL CHANGE

When sugar is added to water the sugar molecules are separated from each other by water molecules. The sugar dissolves. This process does not change the sugar molecules themselves. If the sugar solution is heated, solid sugar will reform as the water evaporates. Dissolving is therefore a temporary physical change and does not involve a chemical change. A chemical change, or **chemical reaction**, is quite different. The features of a chemical reaction are summarised below:

- One or more new substances are produced. In many cases an observable change is apparent, for example a colour change or a gas is produced.

- An apparent change in mass occurs. This change is often quite small and difficult to detect unless accurate balances are used. (Mass *is* conserved in a chemical reaction – the apparent change in mass usually occurs because one of the reactants or products is a gas.)

- An energy change is almost always involved. In most cases energy is released and the surroundings become warmer. In some cases energy is absorbed from the surroundings and so the surroundings become colder. *Note:* Some temporary changes, such as evaporation, also produce energy changes.

Collision theory

For a chemical reaction to occur, the reacting particles (atoms, molecules or ions) must collide. There must also be enough energy involved in the collision to break the chemical bonds in the reacting molecules. If the energy transfer is not large enough the particles will just bounce off one another. A collision which does have enough energy to result in a chemical reaction is referred to as an **effective collision**. This approach to reactions is called **collision theory**.

Particles must collide with sufficient energy to make an effective collision.

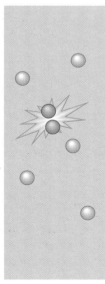
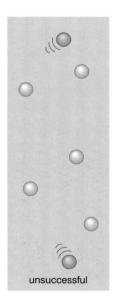

unsuccessful

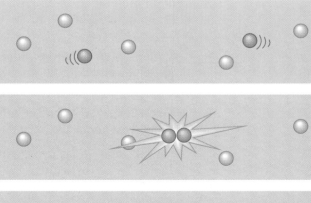

successful

Some chemical reactions occur extremely quickly. The explosive reaction between petrol and oxygen in a car engine is over in a fraction of a second. Other chemical reactions are much slower. It can be days before there is an obvious sign of a reaction when iron rusts. The reason why reactions occur at different speeds is because they have different **activation energies**. The activation energy is the minimum amount of energy required in a collision for a reaction to occur. As a general rule, the bigger the activation energy the slower the reaction will be at a particular temperature.

Speed, rate and time

A quick reaction takes place in a short time. It is said to have a high rate of reaction. It is easy to become confused about rate and time. As the time taken for a reaction to be completed increases, the rate of the reaction decreases. In other words:

$$\text{rate} \propto \frac{1}{\text{time}}$$

Speed	Rate	Time
Quick or Fast	High	Short
Slow	Low	Long

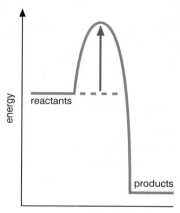

course of reaction

If the activation energy of a reaction is low, more of the collisions will be effective and the reaction will proceed quickly. If the activation energy is high a smaller proportion of collisions will be effective and the reaction will be slow.

MONITORING THE RATE OF A REACTION

When marble (calcium carbonate) reacts with hydrochloric acid the following reaction starts straight away:

calcium carbonate + hydrochloric acid → calcium chloride + carbon dioxide + water
$CaCO_3(s)$ + 2HCl(aq)   → $CaCl_2(aq)$ + $CO_2(g)$ + $H_2O(l)$

The reaction can be monitored as it proceeds either by measuring the volume of gas being formed or by measuring the change in mass of the reaction flask.

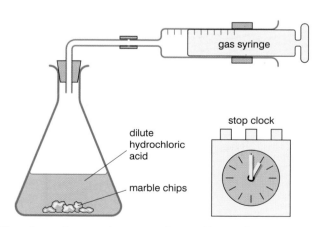

The volume of gas can be measured every 10 seconds.

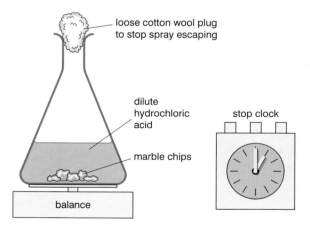

The change in mass can be measured every 10 seconds. The carbon dioxide produced in the reaction escapes into the air. The cotton wool plug is there to stop acid spray from escaping.

If the results obtained from both experiments are plotted on graph paper the two graphs have almost identical shapes. The rate of the reaction at any point can be calculated from the **gradient** of the curve. The shapes of the graphs can be divided into three regions:

1. The curve is the steepest (has the greatest gradient) and the reaction has its highest rate at this point. There are the maximum number of reacting particles present and the number of effective collisions per second is at its greatest.

2. The curve is not as steep (has a lower gradient) at this point and the rate of the reaction is lower. There are fewer reacting particles present and so the number of effective collisions per second will be less.

3. The curve is horizontal (has a zero gradient) and the reaction is complete. At least one of the reactants has been completely used up and so no further collisions can occur between the two reactants.

The rate of the reaction decreases as the reaction proceeds.

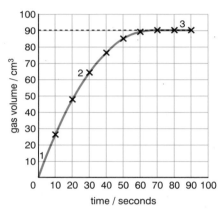

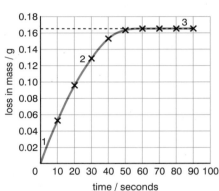

Check yourself

QUESTIONS

Q1 What are the three key features of a chemical reaction? Which of these features could also be found in a temporary or physical change?

Q2 For a chemical reaction to occur the reacting particles must collide. Why don't all collisions between the particles of the reactants lead to a chemical reaction?

QUESTIONS

Q3 The diagrams below show the activation energies of two different reactions A and B. Which reaction is likely to have the greater rate of reaction at a particular temperature?

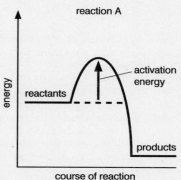

Q4 Look at the table of results obtained when dilute hydrochloric acid is added to marble chips.

Time/seconds	0	10	20	30	40	50	60	70	80	90
Volume of gas/cm³	0	20	36	49	58	65	69	70	70	70

a) What is the name of the gas produced in this reaction?
b) Use the results to calculate the volume of gas produced:
 (i) in the first 10 seconds
 (ii) between 10 and 20 seconds
 (iii) between 20 and 30 seconds
 (iv) between 80 and 90 seconds.
c) Explain how your answers to part (b) show that the rate of the reaction decreases as the reaction proceeds.
d) Use collision theory to explain why the rate of reaction decreases as the reaction proceeds.
e) In this experiment the rate of the reaction was followed by measuring the volume of gas produced every 10 seconds. What alternative measurement could have been used?

················ **REMEMBER! Cover the answers if you want to.** ················

ANSWERS

A1 The three features of a chemical reaction are: a new substance is produced; an apparent change in mass occurs; an energy change occurs. Energy changes and apparent mass changes are also features of many physical changes.

A2 Not all collisions provide enough energy for the reaction to take place.

TUTORIAL

T1 *The easiest chemical reactions to spot are those where the formation of a new substance is obvious, for example a white ash forming when magnesium ribbon burns in oxygen. The feature commonly missed by students is the change in mass. Energy changes are often noticeable. Think of the burning magnesium ribbon. But be careful though, some physical changes also have energy changes. For example, energy is absorbed from the surroundings when a liquid evaporates.*

T2 *Energy is needed to break chemical bonds so that the atoms, molecules or ions can rearrange and form new bonds. Remember that collisions that do have sufficient energy are referred to as **effective collisions**.*

ANSWERS

A3 Reaction A.

A4
a) Carbon dioxide.
b) (i) 20 cm^3, (ii) 16 cm^3, (iii) 13 cm^3, (iv) 0 cm^3.
c) The volume of gas produced in each 10 second interval decreases as the reaction proceeds. This means the rate of production of gas decreases.
d) As the reaction proceeds there are fewer particles of hydrochloric acid and calcium carbonate to collide with each other. Therefore there will be fewer collisions and hence fewer effective collisions.
e) Measuring the change in mass as the reaction proceeds.

TUTORIAL

T3 *Reaction A has a lower activation energy than reaction B. This means that there will be more effective collisions and so the rate of reaction will be higher. If you are still not clear about this idea of an 'energy barrier', look again at the explanation of activation energy in the* **collision theory** *section.*

T4
a) *Remember that marble has the chemical name, calcium carbonate. The carbon dioxide is released from the carbonate ion.*
b) *20 − 0 = 20; 36 − 20 = 16; 49 − 36 = 13; 70 − 70 = 0. Don't forget the units of volume.*
c) *The rate of reaction is measured in terms of the volume of gas produced in a certain amount of time. If you wanted to calculate rates in these time intervals you would need to divide the volume of gas produced by the time taken. So between 0 and 10 seconds, 20 cm^3 of gas was collected giving a rate of reaction of 20/10 = 2 cm^3/sec. Between 20 and 30 seconds the rate was 1.3 cm^3/sec.*
d) *Don't forget to mention the idea of 'effective collisions' (see page 196). This is a key idea in this chapter. If you want to impress the examiner use the correct names for the particles, e.g. in this reaction collisions occur between hydrogen ions (H$^+$ions) and carbonate ions (CO$_3{}^{2-}$) ions.*
e) *The carbon dioxide gas will escape causing a decrease in mass of the reaction container. The equipment you would need is shown on page 197.*

KEY POINTS

More collisions from:
- **increased concentration**
- **increased temperature**
- **increased surface area**
- **increased pressure**
- **a catalyst.**

More energy from:
- **increased temperature**
- **light.**

CONTROLLING THE RATE OF A REACTION

There are six key factors that can change the rate of a reaction:

- concentration (of a solution)
- temperature
- surface area (of a solid)
- pressure (of a gas)
- light
- a catalyst.

A simple collision theory can be used to explain how these factors affect the rate of a reaction. Remember the two important parts of the theory are:

1. The reacting particles must collide with each other.

2. There must be sufficient energy in the collision to overcome the activation energy.

Concentration

Increasing the concentration of a reactant will increase the rate of a reaction. If a piece of magnesium ribbon is added to a solution of hydrochloric acid the following reaction occurs:

magnesium	+	hydrochloric acid	→	magnesium chloride	+	hydrogen
Mg (s)	+	2HCl (aq)	→	MgCl$_2$ (aq)	+	H$_2$ (g)

As the magnesium and acid come into contact, the acid effervesces – hydrogen is given off. The graph below shows the volume of gas collected every 10 seconds when two different concentrations of hydrochloric acid are used. In experiment 1 the curve is steeper (has a greater gradient) than in experiment 2. In experiment 1 the reaction is complete after 20 seconds whereas in experiment 2 it takes 60 seconds. The rate of the reaction is higher with 2.0 M hydrochloric acid than with 0.5 M hydrochloric acid. In the 2.0 M hydrochloric acid solution the hydrogen ions are more likely to collide with the surface of the magnesium ribbon than in the 0.5 M hydrochloric acid.

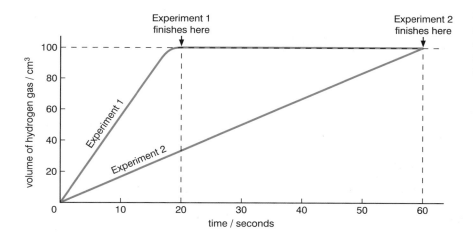

The same length of magnesium was used in each experiment. In experiment 1 the concentration of the acid was 2.0 M, in experiment 2 it was 0.5 M.

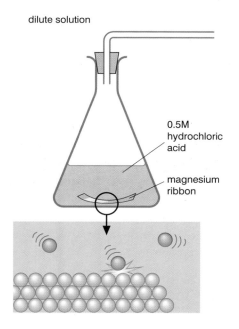

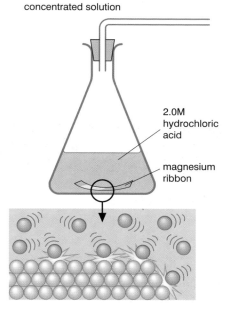

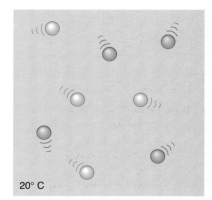

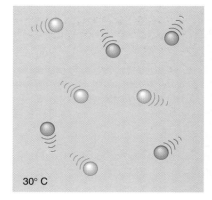

Temperature

Increasing the temperature will increase the rate of reaction. Warming a chemical transfers kinetic energy to the chemical's particles. More kinetic energy means that the particles move faster. As they are moving faster there will be more collisions each second. The increased energy of the collisions also means that the proportion of collisions that are effective will increase.

Increasing the temperature of a reaction such as that between calcium carbonate and hydrochloric acid will not increase the final amount of carbon dioxide produced. The same amount of gas will be produced in a shorter time.

The rates of the two reactions are different but the final loss in mass is the same.

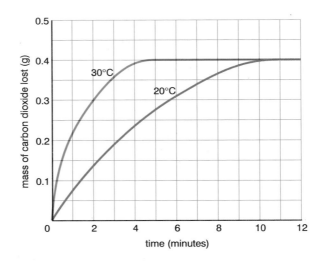

Surface area

Increasing the surface area of a solid reactant will increase the rate of a reaction. The reaction can only take place if the reacting particles collide. This means that the reaction takes place at the surface of the solid. The particles within the solid cannot react until those on the surface have reacted and moved away. Powdered calcium carbonate has a much larger surface area than the same mass of marble chips. A lump of coal will burn slowly in the air whereas coal dust can react explosively.

On a large lump of marble hydrochloric acid can only react with the outside surfaces. Breaking the lump into smaller pieces creates extra surfaces for the reaction.

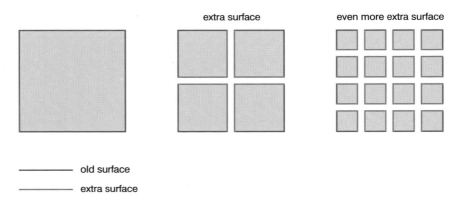

————————— old surface

————————— extra surface

Pressure

Increasing the pressure on the reaction between gases will increase the rate of the reaction. Increasing the pressure has the effect of reducing the volume of the gas and so moving the particles closer together. If the particles are closer together there will be more collisions and therefore more effective collisions.

The same number of particles are closer together in a smaller volume. There will be more effective collisions each second.

low pressure

high pressure

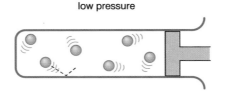
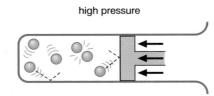

Light

Increasing the intensity of light will increase the rate of some reactions. This fact is important in photography. The photographic film is coated with chemicals that react when in contact with the light. Some laboratory chemicals, for example silver nitrate and hydrogen peroxide, are stored in brown glass bottles to reduce the effect of the light.

Catalysts

A **catalyst** is a substance that alters the rate of a chemical reaction without being used up. The mass of the catalyst remains unchanged throughout the reaction. Hydrogen peroxide decomposes slowly at room temperature into water and oxygen. This reaction is catalysed by manganese(IV) oxide.

$$\text{hydrogen peroxide} \xrightarrow{\text{manganese (IV) oxide}} \text{water} + \text{oxygen}$$

$$2H_2O_2 \xrightarrow{MnO_2} 2H_2O + O_2$$

Most catalysts work by providing an alternative 'route' for the reaction that has a lower activation energy 'barrier'. This increases the number of effective collisions each second. Some catalysts slow down reactions. These are called negative catalysts or **inhibitors**. Inhibitors are added to petrol to prevent 'pre-ignition' of the petrol vapour in the engine.

More examples of catalysts and a description of their key role in the manufacture of some important chemicals will be covered later on pages 209–210.

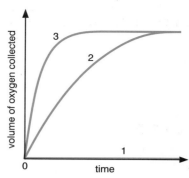

The manganese(IV) oxide has a dramatic effect on the rate of decomposition of hydrogen peroxide. The catalyst doesn't produce any extra oxygen but gives the same amount at a higher rate.

1 = no catalyst
2 = one spatula measure
3 = two spatula measures

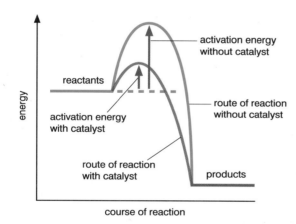

course of reaction

The catalyst provides a lower energy route from reactants to products.

Check yourself

QUESTIONS

Q1 Why does increasing the temperature increase the rate of a reaction?

Q2 The graph shows the results obtained in three different experiments. In each experiment marble chips were added to 50 cm^3 of 1 M hydrochloric acid (an excess) at room temperature. The same mass of marble was used each time but different sized chips were used in each experiment.

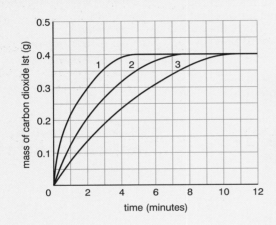

a) (i) In which experiment was the reaction the fastest?
   (ii) Give a reason for your answer.
b) (i) Which experiment used the largest marble chips?
   (ii) Give a reason for your answer?
c) (i) How long did it take for the reaction in experiment 2 to finish?
   (ii) Why did the reaction finish?
d) Why was the same mass of carbon dioxide lost in each experiment?
e) Experiment 3 was repeated at 50 °C rather than at room temperature. How would the results be different from those shown for experiment 3?

Q3 a) What is a catalyst?
b) How does a catalyst affect the rate of a reaction?

Q4 Which of the following types of reaction is most likely to be affected by changes in pressure:
a) a reaction involving a solid reactant and a liquid reactant
b) a reaction involving two gases? Explain your answer.

REMEMBER! Cover the answers if you want to.

ANSWERS

A1 As the temperature increases, the kinetic energy of the reacting particles also increases. The particles will therefore be moving faster and will collide more frequently. In addition, there will be more energy transferred in the collisions and so more are likely to be effective and lead to a reaction.

A2 a) (i) Experiment 1.
   (ii) The reaction finishes more quickly because the curve levels out soonest.
b) (i) Experiment 3.
   (ii) The largest chips have the smallest surface area and so the lowest rate of reaction.

TUTORIAL

T1 *Don't forget to mention the energy of the collision. Many students only mention the increased number of collisions.*

T2 a) *A better answer would mention the gradient of the curve, e.g. the gradient of the curve at the beginning of the reaction is the steepest.*
b) *The curve for experiment 3 has the lowest gradient at the beginning of the reaction. If you got this wrong check the section on **surface area** again.*

ANSWERS

c) (i) 7.5 minutes (approximately).
    (ii) The reaction finished because the marble was used up.
d) The same mass of marble was used in each experiment.
e) The curve would be steeper (have a greater gradient) at the beginning of the reaction but would reach the same plateau height.

A3
a) A catalyst is a substance that alters the rate of a chemical reaction but does not change itself.
b) A catalyst provides another route for the reaction with a different activation energy.

A4
b. Gases can be easily compressed. The particles can be pushed into a smaller volume where more collisions are likely. Increasing pressure has only a small effect on liquids and very little effect on solids. It is not easy to push the particles closer together.

TUTORIAL

c) Try and read off the graph as accurately as you can. In an examination you will be allowed a small margin of error. In this example 7.3 to 7.7 minutes would be acceptable. The question says that the hydrochloric acid was 'an excess'. This means that there was more than would be needed. Therefore, the reaction must have finished because the marble was used up. Always look for the phrase 'an excess' or 'in excess'.
d) As the acid is in excess the amount of marble will determine how much carbon dioxide is produced.
e) Increasing the temperature will increase the rate of the reaction but will not change the amount of carbon dioxide produced. If you are still in doubt about this read the section on **temperature** again.

T3
a) Remember that whilst most catalysts speed up reactions some slow them down.
b) Think of activation energy as an energy barrier that prevents reactants changing into products.

T4
If you had forgotten the particle models for solids, liquids and gases look again at chapter 8.

MAKING USE OF ENZYMES

An **enzyme** is a biological catalyst. Enzymes are protein molecules that control many of the chemical reactions that occur in living cells. The decomposition of hydrogen peroxide described on page 203 is also speeded up by an enzyme present in blood! This enzyme is called catalase and prevents the build up in the body of dangerous compounds called peroxides. Enzymes are used in a wide range of manufacturing processes – some of the more important ones are shown in the table.

Process	Enzyme Involved	Description
Brewing	Enzymes present in yeast	A mixture of sugar solution and yeast will produce ethanol (alcohol) and carbon dioxide. This process is called fermentation and is the basis of the beer and wine making industries.
Baking	Enzymes present in yeast	This also depends on the fermentation process. The carbon dioxide produced helps the bread dough rise. The ethanol evaporates during the baking process.
Washing using biological washing powders	Proteases	These enzymes break down proteins found in stains (e.g. blood, egg) on clothing.
Making cheese	Lipases	These enzymes speed up the ripening of cheese.

205

Increasing the temperature of a reaction will usually increase its rate. This is not the case in reactions where an enzyme is involved. The protein structure of an enzyme is affected by temperature. Above a certain temperature the protein becomes denatured and it will cease to function as a catalyst. Enzymes are also sensitive to pH. Inside cells, most enzymes work best in neutral conditions around pH 7. However, the enzymes in the stomach work best at about pH 2.

The effect of temperature and pH on the rate of a reaction involving enzymes.

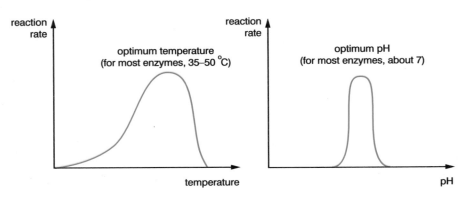

How enzymes work can best be understood by using a simple model. The enzyme provides a surface for the reaction to take place on. The surface of the enzyme molecule contains a cavity known as an **active site**. Reactant molecules become 'trapped' in the active site and so collide more frequently resulting in more effective collisions and a greater rate of reaction.

This model also explains why enzymes cease to function above a certain temperature. The protein molecule is a long folded chain. As the temperature is increased the folded chain jostles and reforms and so the shape of the active site changes. Above a certain temperature the active site has changed so much that it is no longer able to 'trap' reactant molecules. Such an enzyme is said to be **denatured**.

Check yourself

QUESTIONS

Q1 What is an enzyme?

Q2 In fermentation, glucose ($C_6H_{12}O_6$) is broken down to ethanol (C_2H_5OH) and carbon dioxide (CO_2).
a) What is used as a source of enzymes for this reaction?
b) Which two major industries make use of this reaction?
c) The optimum temperature for this reaction is about 40 °C. Why isn't a higher temperature used?
d) Write a word equation for the reaction.
e) Write a balanced symbol equation for the reaction.

Q3 a) What is an active site?
b) Use the active site model to explain why enzymes usually only act as catalysts for one reaction.

ANSWERS

A1 A biological catalyst.

A2
a) Yeast.
b) The brewing and baking industries.
c) At higher temperatures the enzymes present in the yeast are denatured and can no longer act as catalysts.
d) glucose → ethanol + carbon dioxide
e) $C_6H_{12}O_6 \rightarrow 2C_2H_5OH + 2CO_2$

A3
a) An active site is the part of an enzyme molecule in which reactant molecules are trapped.
b) The arrangement and shape of the active site in an enzyme will only 'fit' one specific type of molecule.

TUTORIAL

T1 *A really good answer would include the fact that they are protein molecules.*

T2
a) *Remember that the ethanol is the more important*
& b) *product in the brewing industry and that the carbon dioxide is more important in the baking industry.*
c) *A common mistake is to say that the 'enzymes are killed' at higher temperatures. Enzymes are not alive in the first place!*
e) *Remember when balancing an equation that the balancing numbers can only be placed in front of the formulae (see chapter 7).*

T3
a) *You can think of it as a hole or cavity.*
b) *This is sometimes referred to as a 'lock and key' model – only one key fits the lock.*

REVERSIBLE REACTIONS

When carbon burns in oxygen, carbon dioxide is formed:

carbon + oxygen → carbon dioxide
$C\ (s)$ + $O_2\ (g)$ → $CO_2\ (g)$

Carbon dioxide cannot be changed back into carbon and oxygen. The reaction cannot be reversed. When blue copper(II) sulphate crystals are heated a white powder is formed (anhydrous copper(II) sulphate) and water is lost as steam. If water is added to this white powder, blue copper(II) sulphate is re-formed. The reaction can be reversed:

copper(II) sulphate crystals $\rightleftharpoons$ anhydrous copper(II) sulphate + water
$CuSO_4.5H_2O\ (s)$ $\rightleftharpoons$ $CuSO_4\ (s)$ + $5H_2O\ (l)$

KEY POINTS

Many reactions go backwards and forwards.

Equilibrium when backwards rate and forwards rate are equal.

When copper(II) sulphate crystals are heated they turn from blue to white.

The reaction can then be reversed by adding water.

This is an example of a **reversible reaction**. The reaction can go from left to right or from right to left – notice the double-headed '⇌' arrow used when writing these equations. Another reversible reaction is the reaction between ethene and water to make ethanol. This is one of the reactions used industrially to make ethanol.

ethene + water ⇌ ethanol
C_2H_4 (g) + H_2O (g) ⇌ C_2H_5OH (g)

If ethene and water are heated in the presence of a catalyst in a sealed container ethanol is produced. As the ethene and water are used up, the rate of the forward reaction decreases. As the amount of ethanol increases the rate of the back reaction (the decomposition of ethanol) increases. Eventually the rate of formation of ethanol will be exactly equal to the rate of decomposition of ethanol. The amounts of ethene, water and ethanol will be constant. The reaction is said to be in **equilibrium**.

CHANGING THE POSITION OF EQUILIBRIUM

Reversible reactions can be a nuisance to an industrial chemist. You want to make a particular product but as soon as it forms it starts to change back into the reactants! Fortunately scientists have found ways of increasing the amount of product that can be obtained in a reversible reaction. They have found ways of moving the position of balance to favour the products rather than the reactants. The position of equilibrium can be changed in the following ways:

● changing concentrations

● changing pressure

● changing temperature.

Change in conditions of the reaction	Type of reversible reaction affected	Result of the change on position of equilibrium, reactants ⇌ products
Increase concentration of reactants	all	moves to the right
Decrease concentration of reactants	all	moves to the left
Increase concentration of products	all	moves to the left
Decrease concentration of products	all	moves to the right
Increase pressure	reactions involving gases	moves to the side with *fewer* molecules (no change if same number of molecules on each side)
Decrease pressure	reactions involving gases	moves to the side with the *greater* number of molecules (no change if same number of molecules on each side)
Increase temperature	forward reaction exothermic forward reaction endothermic	moves to the left moves to the right
Decrease temperature	forward reaction exothermic forward reaction endothermic	moves to the right moves to the left

Considering the production of ethanol again:

$$C_2H_4(g) + H_2O(g) \rightleftharpoons C_2H_5OH(g)$$

Note – the (g) indicates that the molecule exists in the gas state. The forward reaction is exothermic. To obtain the maximum conversion of ethene and steam into ethanol the conditions needed would be:

- high concentrations of ethene and steam
- low temperature
- high pressure.

In practice, the predictions about concentration and pressure are correct. Excess steam is used and 70 times normal atmospheric pressure is used. However, a high temperature of 300 °C is used. This is higher than you might have predicted. A higher temperature is used to provide a reasonable *rate* of reaction even though a *smaller conversion* to ethanol will be achieved. This is one example where the desire to change the position of equilibrium is balanced against the desire for the reaction to take place as quickly as possible. A temperature of 300 °C is therefore a compromise. A catalyst is also used in the manufacture of ethanol. The catalyst does not affect the position of the equilibrium but does increase the rate of the reaction.

THE HABER PROCESS

Ammonia is an important compound. It is used to make nitrogen-containing fertilisers. Ammonia is manufactured in the Haber Process from nitrogen and hydrogen. The conditions include an iron catalyst, a temperature of 450 °C and 200 times atmospheric pressure.

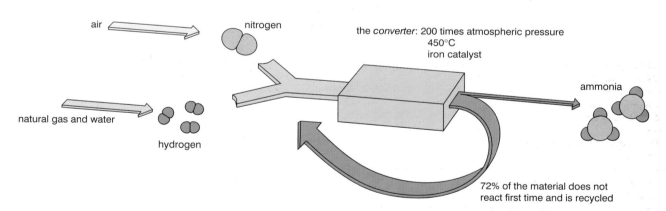

The Haber Process for making ammonia. The reactants have to be recycled to increase the amount of ammonia produced.

nitrogen + hydrogen $\rightleftharpoons$ ammonia
$$N_2(g) + 3H_2(g) \rightleftharpoons 2NH_3(g)$$

The forward reaction is exothermic. To drive the equilibrium to the right and obtain the maximum amount of ammonia the conditions would be:

- high concentrations of nitrogen and hydrogen
- low temperature
- high pressure.

The actual conditions used are shown in the diagram. The temperature of 450 °C is a compromise between rate and equilibrium requirements. If it were lower it would favour a greater conversion to ammonia but the reaction would be much slower. A catalyst is used to increase the rate of the reaction.

THE CONTACT PROCESS

Sulphuric acid is a very important starting material in the chemical industry. It is used in the manufacture of many other chemicals, from fertilisers to plastics. It is manufactured in a process known as the Contact Process – sulphur dioxide is oxidised to sulphur trioxide, which is then added to water to make sulphuric acid.

The conversion of sulphur dioxide to sulphur trioxide takes place at 450 °C in the presence of a vanadium(V) oxide catalyst.

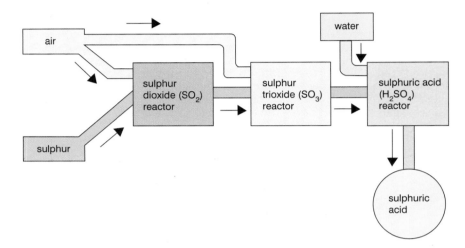

sulphur dioxide + oxygen $\rightleftharpoons$ sulphur trioxide
$2SO_2(g)$ + $O_2(g)$ $\rightleftharpoons$ $2SO_3(g)$

The forward reaction is exothermic. To drive the equilibrium to the right and obtain the maximum amount of sulphur trioxide the conditions needed would be:

- high concentrations of sulphur dioxide and oxygen
- low temperature
- high pressure.

Once again the temperature of 450 °C is a compromise between rate and equilibrium requirements. A catalyst is used to increase the rate of the reaction. High pressure is not used as this would be costly and, as 98% conversion can be achieved at normal atmospheric pressure, it is unnecessary.

Check yourself

QUESTIONS

Q1 When a chemical reaction is in **equilibrium** what does this mean?

Q2 This question is about the reaction shown below:

carbon monoxide + hydrogen ⇌ methanol
$CO(g) + 2H_2(g) \rightleftharpoons CH_3OH(g)$
The forward reaction is exothermic.

a) What does the symbol (g) stand for?
b) Complete the table below to show the conditions needed to drive this reversible reaction to the right. Answer using the words 'high' or 'low'. The first one has been done for you.

c) How do you think a catalyst would affect this reaction?

Q3 In the Haber process for making ammonia the forward reaction is favoured by a low temperature. Why is a temperature as high as 450 °C used?

Factor that affects the reaction	Condition needed to produce the most methanol (CH_3OH)
Concentration of carbon monoxide	high
Concentration of hydrogen	
Temperature	
Pressure	

.... **REMEMBER! Cover the answers if you want to.** ....

ANSWERS

A1 A reaction is in equilibrium when the number of reactant and product molecules do not change.

A2
a) The symbol stands for 'gas'.
b) high, low, high.
c) A catalyst would speed up the reaction but would not change the amount of methanol (product) formed.

A3 The higher temperature is used to increase the rate of the reaction.

TUTORIAL

T1 *You might remember that at equilibrium the rate of the forward reaction equals the rate of the backward reaction.*

T2
a) *It is called a state symbol. Other state symbols include (s) solid, (l) liquid, (aq) aqueous.*
b) *The concentration of the reactant should be high. If the forward reaction is exothermic, the temperature should be low. High pressure favours the side with the fewer molecules (there are 3 molecules on the left side and only 1 on the right side). These ideas are quite difficult. If you are in doubt look back at the table on page 208.*
c) *It is a very common mistake to say that a catalyst changes the position of equilibrium. A catalyst changes the rate of forward and backward reactions to the same extent – so that the mixture reaches the equilibrium position sooner (or later if it is an inhibitor). It does not change the position of equilibrium.*

T3 *The choice of temperature in industrial processes often balances how much can be formed (depends on the equilibrium position) and how fast it can be formed (depends on the rate of the reaction).*

Key Words

Tick each word when you are sure of its meaning.

activation energy

active site

catalyst

collision

concentration

enzyme

equilibrium

gradient

particle

product

rate

reactant

reversible

surface area

temperature

EXAMINATION CHECKLIST

The facts and ideas that you should understand by studying this topic.

To be successful at Foundation GCSE Tier you should be able to:

- recall that chemical reactions make new materials and usually involve an energy transfer
- recall that reactions stop when one of the reactants is used up
- recall that a collision between particles must take place for a chemical reaction to happen
- recall that an increase in concentration makes particles more crowded
- recall that small amounts of catalyst can be used to increase the rate of reactions
- explain why the rate of reaction changes as the reaction proceeds
- interpret graphs of reaction rates
- recall that an enzyme is a catalyst that controls biological processes
- explain what a reversible reaction is
- interpret graphs showing the effect of temperature and pressure on a reversible reaction
- recall how ammonia is manufactured in the Haber Process.

In addition, to be successful at Higher GCSE Tier you should be able to:

- explain the relationship between rate and time
- recall how the size of the activation energy determines the rate of the reaction
- explain how increasing the frequency of effective collisions between reacting particles increases the rate of reaction
- explain how changing concentration, temperature, surface area and pressure influences the rate of chemical reactions
- recall how catalysts alter the rate of a reaction
- recall that light can alter the rates of certain reactions
- recall how enzymes are affected by pH and temperature
- explain how the yield of products in a reversible reaction depends on the conditions.

EXAM PRACTICE

Sample Student's Answers & Examiner's Comments

1 A student was investigating the effects of acid rain on marble statues. The student found out that the acid rain contains two acids. One of these acids is dilute nitric acid.

The student set up the following apparatus.

Look at the diagram.

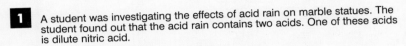

gas syringe

dilute nitric acid — marble

The student measured the volume of carbon dioxide formed every minute.

The results are shown on the graph.

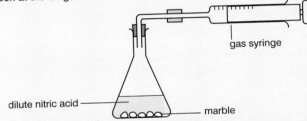

At the end of the reaction there will be some marble present in the flask.

EXAMINER'S COMMENTS

a) The chemical formulae have been written down correctly to match the word equation but the candidate has not balanced the equation. The equation should have been:
$CaCO_3(s) + 2HNO_3(aq) \rightarrow Ca(NO_3)_2(aq) + H_2O(l) + CO_2(g)$

b) The intercept from the graph has been read correctly. In this question the unit was given. If the unit had not been given the candidate would be expected to give the unit as well.

c) The curve has been drawn correctly. One mark is given for drawing the curve steeper (with a greater gradient) than the original curve. This is because the rate of reaction with the powder will be greater due to its increased surface area. The second mark is given for drawing the curve with the same final volume of carbon dioxide. The marble is in excess and so the volume of carbon dioxide depends on the amount of acid used. This remains constant.

d) The candidate has described and explained the changes as the question asks. However, only two of the four marks have been scored: one for saying that the reaction would be quicker with the warm acid; one for explaining that this is due to the hydrogen ions in the acid having more energy. The key points missed are: there will be more collisions each second (greater collision frequency) (1 mark) resulting in more effective collisions each second (1 mark).

● Note: The candidate has shown a good understanding of the reaction and has correctly referred to **hydrogen ions** in the acid rather than simply **acid particles**.

● A mark of 5/8 corresponds to a good grade C answer on this higher tier question.

---

a) The equation for the reaction taking place is written below.

calcium carbonate + nitric acid → calcium nitrate + water + carbon dioxide

Look at the table. It shows the chemical formula of each substance in the reaction.

Substance	Formula
calcium carbonate	$CaCO_3$
nitric acid	HNO_3
calcium nitrate	$Ca(NO_3)_2$
water	H_2O
carbon dioxide	CO_2

Write down a balanced equation for the reaction.

$CaCO_3 + HNO_3 \rightarrow Ca(NO_3)_2 + H_2O + CO_2$ ✗ (1)

b) What was the total volume of carbon dioxide formed after 2 minutes?

Answer 30 ✓ cm³ (1)

c) The experiment was repeated.

The only change was that marble powder was used instead of pieces.

Draw on the graph the curve you would expect. (2)

d) The experiment was repeated a second time.

This time the only change was that warm nitric acid was used instead of cold nitric acid.

Describe and explain, using ideas about reacting particles, the changes you would expect in the reaction.

The reaction of the marble with warm nitric acid would be quicker than the reaction with cold nitric acid. This is because ✓ the hydrogen ions in the acid will be moving faster and have more energy. ✓

(4)

5/8

(Total 8 marks)

Midland Examining Group

Questions to Answer

Answers to questions 2 – 3 can be found on pages 367.

2 Holes in car bodies can be filled using a product called the 'Fastglass System'. This product comes in three parts.

1   resin – this is a liquid

2   filler – this is a powdered solid

3   paste – this is a gel

The three parts are mixed together and then put into the hole in the car body.

The mixture becomes warm and then sets solid.

The time it takes for the mixture to turn solid is called the 'setting time'.

Look at the Table. it shows the results of an investigation into the setting times of different mixes. All experiments were carried out at 15 °C.

Experiment number	volume of resin/cm^3	mass of filler/g	volume of paste/cm^3	setting time/min
1	10	20	2	6
2	10	20	0	600
3	10	20	4	3
4	20	20	2	8

a)   What evidence is there that a chemical reaction occurs when the mixture is made?   (1)

b)   Use the information in the Table to explain the purpose of the paste.   (2)

c)   Experiment 1 is repeated. This time a temperature of 25 °C rather than 15 °C was used. Explain **how** and **why** the setting time will change. Use ideas about the movement of particles.   (4)

Midland Examining Group

3 Ammonia is used in the manufacture of fertilisers. Ammonia is made in the Haber process.

The equation shows the reaction that takes place in the Haber process.

hydrogen + nitrogen $\rightleftharpoons$ ammonia

$3H_2$ (g) + N_2 (g) $\rightleftharpoons$ $2NH_3$ (g)

a)   What does the sign $\rightleftharpoons$ represent? (1)

b)   The rate of the reaction makes a lot of difference to the cost of making ammonia.

   The greater the rate of reaction the cheaper it is to make ammonia.

   Suggest **two** ways of increasing the rate of reaction. (2)

c)   The large scale manufacture of ammonia by the Haber process has a number of economic, social and environmental effects.

   Write about some of the local and global economic, social and environmental effects. (4)

d)   The percentage of ammonia formed in the Haber process depends on the pressure and the temperature.

   Look at the graphs.

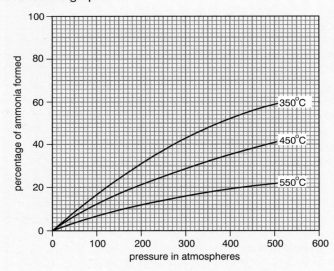

(i)   What is the percentage of ammonia formed at 350 °C and 200 atmospheres pressure? (1)

(ii) How does the percentage of ammonia change as the temperature is increased? (1)

(iii) How does the percentage of ammonia change as the pressure is increased? (1)

Midland Examining Group

CHAPTER 12

THE PERIODIC TABLE

Elements are the building blocks from which all materials are made. Over one hundred elements have been identified so far and each element has its own properties and reactions. This makes the study of elements very difficult as there appears to be so much to learn. To overcome this problem scientists have arranged the elements in a particular pattern called the **periodic table**. The periodic table used today has been developed from one first produced in 1869. It puts elements with similar reactions and properties close together. In this way it becomes much easier to understand and learn the properties and reactions of the elements.

GENERAL CHARACTERISTICS OF THE PERIODIC TABLE

The periodic table arranges the elements in order of increasing **atomic number**. The atomic number is the number of protons in the atom. As atoms are neutral, the atomic number also gives the number of electrons. The elements are then arranged in rows called **periods** and columns called **groups**.

Periods:

● Rows of elements arranged in increasing atomic number from left to right.

● Numbered from 1 to 7.

● The first period only contains two elements, hydrogen and helium.

● The elements in the middle block of the periodic table in periods 4, 5 and 6 are called the **transition metals**.

Groups:

● Columns of elements with the atomic number increasing down the column.

● Numbered from 1 to 8 (Group 8 is often referred to as Group 0).

● Elements in a group have similar properties – they are a 'chemical family'.

● Some groups have family names – the **alkali metals** (group 1), the **halogens** (group 7) and the **noble gases** (group 8/0).

Part of the modern periodic table showing periods and groups.

metal | non metal | transition metal | metalloids

The fact that elements in the same group have similar reactions can be explained in terms of the electron structures of their atoms (see chapter 8). If elements have the same number of electrons in their outer shells they will have similar chemical properties. The relationship between the group number and the number of electrons in the outer electron shell of the atom is shown in the table.

Group number	1	2	3	4	5	6	7	0 (8)
Electrons in the outer electron shell	1	2	3	4	5	6	7	2 or 8 (full)

METALS AND NON-METALS

Most elements can be classified as either **metals** or **non-metals**. In the periodic table the metals are arranged on the left-hand side and in the middle, the non-metals are on the right-hand side. Some of the elements close to the boundary between metals and non-metals have some properties of metals and some of non-metals. These elements, for example silicon and germanium, are called **metalloids**.

Metals and non-metals have quite different physical and chemical properties.

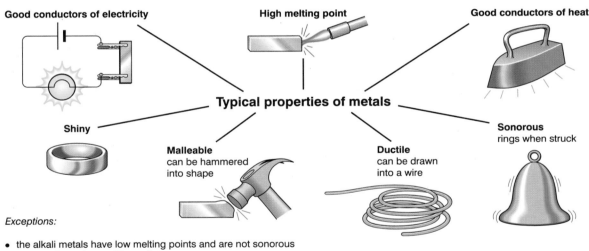

Good conductors of electricity

High melting point

Good conductors of heat

Shiny

Typical properties of metals

Sonorous
rings when struck

Malleable
can be hammered
into shape

Ductile
can be drawn
into a wire

Exceptions:

- the alkali metals have low melting points and are not sonorous
- mercury has a low melting point

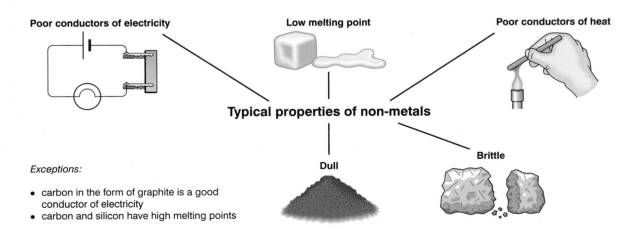

Poor conductors of electricity

Low melting point

Poor conductors of heat

Typical properties of non-metals

Brittle

Dull

Exceptions:

- carbon in the form of graphite is a good conductor of electricity
- carbon and silicon have high melting points

The oxides of elements can often be made by heating the element in air or oxygen. For example a metal such as magnesium burns in oxygen to form magnesium oxide.

magnesium + oxygen → magnesium oxide
2Mg(s) + O$_2$(g) → 2MgO(s)

The magnesium oxide forms as a white ash. If distilled water is added to the ash and the mixture is tested with universal indicator the pH will be greater than 7 – the oxide has formed an alkaline solution.

If sulphur is burned in oxygen, sulphur dioxide gas is formed. If this is dissolved in water and then tested with universal indicator solution the pH will be less than 7 – the oxide has formed an acidic solution.

sulphur + oxygen → sulphur dioxide
S(s) + O$_2$(g) → SO$_2$(g)

The oxides of most elements can be classified as **basic oxides** or **acidic oxides**. Some elements form neutral oxides, for example water is a neutral oxide. Basic oxides or bases that dissolve in water are called **alkalis**. Bases and alkalis react with acids to form **salts** in reactions known as **neutralisation** reactions. A typical neutralisation reaction occurs when sodium hydroxide (an alkali) reacts with hydrochloric acid. The salt formed is sodium chloride, common salt:

alkali + acid → salt + water
sodium hydroxide + hydrochloric acid → sodium chloride + water
NaOH(aq) + HCl(aq) → NaCl(aq) + H$_2$O(l)

Oxide	Type of oxide	pH of solution	Other reactions of the oxide
Metal oxide	basic	more than 7 (alkaline)	reacts with an acid to form a salt + water
Non-metal oxide	acidic	less than 7 (acidic)	reacts with a base to form a salt + water

Check yourself

QUESTIONS

Q1 Look at the diagram representing the periodic table. The letters stand for elements.

QUESTIONS

a) Which element is in group 4?
b) Which element is in the second period?
c) Which element is a noble gas?
d) Which element is a transition metal?
e) Which element is a metalloid?
f) Which elements are non-metals?
g) Which element is most likely to be a gas?

Q2 Why do elements in the same group react in similar ways?

Q3 Metals are usually malleable and ductile.
a) What do the terms malleable and ductile mean?
b) Give three other characteristic properties of metals.
c) In what way does graphite (carbon) behave as if it were a metal?

Q4 Explain the following terms:
a) base
b) alkali
c) salt.

Q5 Look at the table of experimental results at the end of the question.
a) Which of the oxides is/are acidic? Explain how you decided.
b) Which of the oxides is/are basic? Explain how you decided.
c) Copper(II) oxide reacts with sulphuric acid (H_2SO_4).
   (i) What is the name given to this type of reaction?
   (ii) Write a word equation for the reaction.
   (iii) Write a symbol equation for the reaction.
   (iv) Which oxide A to D is most likely to be copper(II) oxide?

Oxide of element	pH of solution	Does it react with an acid?	Does it react with an alkali?
A	7	✓	✗
B	3	✗	✓
C	7	✗	✗
D	10	✓	✗

REMEMBER! Cover the answers if you want to.

ANSWERS

A1
a) b
b) a
c) d
d) c
e) b
f) d and f
g) d

A2 They contain the same number of electrons in the outer electron shell.

TUTORIAL

T1
a) Group 4 is the fourth column from the left.
b) This is the second row.
c) The noble gas family is group 0 or 8.
d) Transition elements are in the middle block.
e) Metalloids are found near the 'staircase' on the right side of the periodic table which separates metals from non-metals.
f) These are on the right of the 'staircase'. If you included b you would not be penalised.
g) Gases are non-metals and so have to be on the right side. f is a possibility but in groups 5, 6 and 7 elements near the bottom of the group are solids.

T2 Remember that how an atom reacts depends on its outermost electrons.

ANSWERS

A3

a) Malleable means can be beaten into shape. Ductile means can be drawn into wires.
b) Good conductors of heat, good conductors of electricity, high melting point, shiny, sonorous. (Any three.)
c) Graphite is a good conductor of electricity.

A4

a) A base is a substance that will react with an acid to form a salt. If the base dissolves in water the solution formed will be alkaline.
b) An alkali is a soluble base.
c) A salt is formed when an acid reacts with a base.

A5

a) B. Its pH is less than 7.
b) A and D. Both react with an acid.
c) (i) Neutralisation.

   (ii) Copper oxide + sulphuric acid → copper sulphate + water.
   (iii) $CuO(s) + H_2SO_4(aq) \rightarrow CuSO_4(aq) + H_2O(l)$.
   (iv) A.

TUTORIAL

T3

a) *These are characteristic properties of metals.*
b) *Remember that these properties do not apply to all metals, e.g. mercury and the alkali metals.*
c) *Graphite is the only non-metal that is a good conductor of electricity.*

T4

a) *A base is in fact the oxide or hydroxide of a metal.*
b) *The common alkalis are sodium hydroxide and calcium hydroxide (limewater).*
c) *This type of reaction is known as a neutralisation.*

T5

a) *Remember that the acidic oxide will also react with an alkali.*
b) *'A' has a pH of 7 because it doesn't dissolve in water.*
c) *Copper oxide is a base. The reaction of a base with an acid is neutralisation.*
d) *Remember that base + acid → salt + water.*
e) *No extra balancing is required here.*
f) *Copper(II) oxide is insoluble in water and so the mixture will have a pH of 7.*

THE METALS

Most elements are metals. Some metals are highly reactive whilst others are almost completely unreactive. These types of metals are found in different parts of the periodic table.

GROUP 1 – THE ALKALI METALS

This is a group of very reactive metals. They all have only one electron in their outer electron shell. This electron is readily given away when the metal reacts with non-metals. The more electrons a metal atom has to lose in a reaction, the more energy is needed to start the reaction. This is why the Group 2 elements are less reactive – they have to lose two electrons when they react (see chapter 8). Reactivity also increases down the group due to the fact that as the atom gets bigger, the outer electron is further away from the nucleus and so is removed more easily.

| Li lithium 3 |
| Na sodium 11 |
| K potassium 19 |
| Rb rubidium 37 |
| Cs caesium 55 |

INCREASING REACTIVITY

Group 1 elements become more reactive as you go further down the group.

Name	Symbol	Melting point/°C	Density/ g/cm³
Lithium	Li	180	0.53
Sodium	Na	98	0.97
Potassium	K	64	0.86
Rubidium	Rb	39	1.53
Caesium	Cs	29	1.90

Sodium being cut under oil.

These elements have the following physical properties:

- soft to cut
- shiny when cut, but quickly tarnish in the air
- very low melting points compared to most metals
- very low densities compared to most metals (lithium, sodium and potassium will float on water)
- react very easily with air, water and other elements such as chlorine. The alkali metals are so reactive that they are stored in oil to prevent reaction with air and water.

Reaction	Observations	Equations
Air or oxygen	The metals burn easily with coloured flames: ● lithium – red ● sodium – orange ● potassium – lilac. A white solid oxide is formed.	lithium + oxygen → lithium oxide $4Li(s) + O_2(g) → 2Li_2O(s)$ sodium + oxygen → sodium oxide $4Na(s) + O_2(g) → 2Na_2O$ potassium + oxygen → potassium oxide $4K(s) + O_2(g) → 2K_2O$
Water	The metals react vigorously. They float on the surface; moving around rapidly; the heat of the reaction melts the metal so it forms a sphere; effervescence occurs and a gas is given off; the metal rapidly dissolves. With the more reactive metals (e.g. potassium) the hydrogen gas produced burns. The resulting solution is alkaline.	lithium + water → lithium hydroxide + hydrogen $2Li(s) + 2H_2O(l) → 2LiOH(aq) + H_2(g)$ sodium + water → sodium hydroxide + hydrogen $2Na(s) + 2H_2O(l) → 2NaOH(aq) + H_2(g)$ potassium + water → potassium hydroxide + hydrogen $2K(s) + 2H_2O(l) → 2KOH(aq) + H_2(g)$
Chlorine	The metals react easily burning in the chlorine to form a white solid.	lithium + chlorine → lithium chloride $2Li(s) + Cl_2(g) → 2LiCl(s)$ sodium + chlorine → sodium chloride $2Na(s) + Cl_2(g) → 2NaCl(s)$ potassium + chlorine → potassium chloride $2K(s) + Cl_2(g) → 2KCl(s)$

The compounds of the alkali metals are widely used:

- lithium carbonate – in light sensitive lenses for glasses
- lithium hydroxide – removes carbon dioxide in air conditioning systems
- sodium chloride – table salt
- sodium carbonate – a water softener
- sodium hydroxide – used in paper manufacture
- mono-sodium glutamate – a flavour enhancer
- sodium sulphite – a preservative
- potassium nitrate – a fertiliser also used to make explosives.

THE TRANSITION METALS

The transition metals are found in the centre of the periodic table. Compared to the alkali metals they are much less reactive and so are more 'everyday' metals. They have much higher melting points and densities. Iron, cobalt and nickel are the *only* magnetic elements.

Element	Symbol	Melting Point/°C	Density/ g/cm³
Chromium	Cr	1890	7.2
Iron	Fe	1535	7.9
Cobalt	Co	1492	8.9
Nickel	Ni	1453	8.9
Copper	Cu	1083	8.9

All the transition metals have more than one electron in their outer electron shell which is why they have a lower chemical reactivity than the alkali metals in Group 1. They react much more slowly with water and with oxygen. Some, like iron, will react with dilute acids – others, like copper, show no reaction. They are widely used as construction metals, particularly iron, and they are frequently used as **catalysts** in the chemical industry.

Property	Group 1 Metal	Transition Metal
Melting point	low	high
Density	low	high
Colours of compounds	white	coloured
Reactions with water/air	vigorous	slow or no reaction
Reactions with acid	violent (dangerous)	slow or no reaction

The compounds of the transition metals are usually coloured. The colours of gem stones such as emerald, sapphire and amethyst are due to traces of transition metal compounds. Copper compounds are usually blue or green; iron compounds tend to be either green or brown. The fact that transition metal compounds are coloured can be used to identify them during analysis. If sodium hydroxide solution is added to the solution of a transition metal compound, a precipitate of the metal hydroxide will be formed. The colour will help to identify the metal. For example:

copper sulphate + sodium hydroxide → copper(II) hydroxide + sodium sulphate
$CuSO_4(aq)$ + $2NaOH(aq)$ → $Cu(OH)_2(s)$ + $Na_2SO_4(aq)$

This can be written as an ionic equation (see chapter 7):
$Cu^{2+}(aq)$ + $2OH^-(aq)$ → $Cu(OH)_2(s)$

Colour of metal hydroxide	Likely metal present
blue	copper
green	nickel
green turning to brown	iron
greenish/blue	chromium

Flame tests can also be used in analysis. Transition metal compounds and other metal compounds produce characteristic colours.

Metal ion	Flame colour
lithium	red
sodium	orange
potassium	lilac
copper	turquoise
calcium	brick red
strontium	crimson
barium	green

The colour of the flame can be used to identify the metal ions present.

Check yourself

QUESTIONS

Q1 This question is about the Group 1 elements.
a) Which is the most reactive of the elements?
b) Why are the elements stored in oil?
c) Which element will be the easiest to cut?
d) Why do the elements tarnish quickly when they are cut?
e) Why is the group known as the alkali metals?
f) Why does sodium float when added to water?
g) Write word equations and symbolic equations for the following reactions:
   (i) rubidium and oxygen
   (ii) caesium and water
   (iii) potassium and chlorine.

Q2 This question is about the transition metals.
a) Give two differences in the physical properties of the transition metals compared to the alkali metals.
b) Transition metals are used as catalysts. What is a catalyst?
c) Suggest why the alkali metals are more reactive than the transition metals.

Q3 Look at the table of observations below:
a) Identify the metal present in each of the three compounds.
b) Explain why C could not contain a metal from Group 1 of the periodic table.
c) Write an ionic equation for the reaction of a solution of B with sodium hydroxide solution.

Compound tested	colour of compound	colour produced in a flame test	effect of adding sodium hydroxide solution to a solution of the compound
A	white	orange	no change
B	green	turquoise	blue precipitate formed
C	white	brick red	white precipitate formed

REMEMBER! Cover the answers if you want to.

ANSWERS

A1
a) Caesium.
b) To prevent the metal reacting with air and water.
c) Caesium.
d) On cutting, the metal is exposed to the air and rapidly oxidises.
e) When added to water they react to produce an alkali.
f) Sodium is less dense than water.

TUTORIAL

T1
a) Remember that the reactivity of metals increases down a group.
b) These are highly reactive metals. They oxidise rapidly without heat being needed.
c) The melting point gives a measure of hardness. Melting point decreases down the group.
d) Tarnishing is another word for oxidation.
e) The metal hydroxides formed in the reactions are alkalis.
f) Density increases down the group. The more reactive metals would not float but their reaction is so violent that the metal usually flies out of the water!

ANSWERS

g) (i) rubidium + oxygen → rubidium oxide.
$$4Rb(s) + O_2(g) → 2Rb_2O(s)$$
(ii) caesium + water → caesium hydroxide + hydrogen
$$2Cs(s) + 2H_2O(l) → 2CsOH(aq) + H_2(g)$$
(iii) potassium + chlorine → potassium chloride
$$2K(s) + Cl_2(g) → 2KCl(s)$$

A2 a) They are harder, have higher densities, higher melting points and are sonorous. (Any two.)
b) A catalyst is a substance that changes the speed of a chemical reaction without being changed itself.
c) The alkali metals only have one electron in the outer electron shell. Transition metals have more than one electron in the outer electron shell.

A3 a) A – sodium, B – copper, C – calcium.
b) The hydroxides of the Group 1 metals are soluble in water and would not form a precipitate.
c) $$Cu^{2+}(aq) + 2OH^-(aq) → Cu(OH)_2(s)$$

TUTORIAL

g) Rubidium behaves in the same way as sodium but more violently.

The reaction of the Group 1 metals with water produces the metal hydroxide (an alkali) and hydrogen.

Potassium chloride is a salt with an appearance very similar to sodium chloride, common salt.

T2 a) If you couldn't answer this look back at the summary given on page 218.
b) Catalysts, and their effect on rates of reaction, are covered in chapter 11.
c) Reactivity is principally related to the number of electrons in the outer electron shell. Metals lower down a group are also more reactive than those higher up, but this is of secondary importance.

T3 a) The sodium and calcium can be identified from the flame test information. Only one of the compounds contains a transition metal (forms a coloured compound).
b) Calcium is in Group 2. Calcium hydroxide is only sparingly soluble in water and so a precipitate forms.
c) The important part of the compound is the copper ion. The non-metal ion or radical does not play a part in the reaction – it is a spectator ion.

THE NON-METALS

There are only about 20 non-metal elements. As with metals there is a wide range of reactivity between different groups of non-metals. The most reactive non-metals are found in Group 7, the least reactive are found in the next group, Group 0.

GROUP 7 – THE HALOGENS

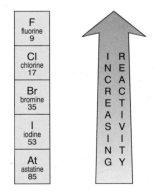

The elements become more reactive as you go further up the group.

This is a group of very reactive non-metals. The term 'halogen' means 'salt-maker' and the halogens react with most metals to make salts. The halogen elements have seven electrons in their outermost electron shell. Their high reactivity is due to the fact that they only need to gain one electron to obtain a full outer electron shell. The halogens react with metals gaining an electron and forming a singly charged negative ion (see ionic bonding in chapter 8). The reactivity of the elements decreases down the group due to the fact that as the atom gets bigger the extra electron will be further from the attractive force of the nucleus. This means it is harder for this electron to be gained.

Name	Symbol	Melting Point/°C	Boiling Point/°C	State at room temperature
Fluorine	F	−220	−188	gas
Chlorine	Cl	−101	−35	gas
Bromine	Br	−7	59	liquid
Iodine	I	114	184	solid

In some respects these elements appear very different, though they are chemically very similar.

Differences:

- appearances: fluorine is pale yellow gas; chlorine is a yellow/green gas; bromine is a brown liquid; iodine is a black solid.

Similarities:

- all have 7 electrons in their outermost electron shell
- all exist as **diatomic** molecules, i.e. molecules containing two atoms such as F_2, Cl_2, Br_2, I_2
- all react with water and react quickly with metals to form salts
- all undergo **displacement** reactions.

Reaction	Observations	Equations
Water 	The halogens dissolve in water and also react with it forming solutions that behave as bleaches. Chlorine solution is pale yellow. Bromine solution is brown. Iodine solution is brown.	chlorine + water → hydrochloric acid + hypochlorous acid (bleach) $Cl_2(g) + H_2O(l) → HCl(aq) + HClO(aq)$
Metals 	The halogens will form salts with all metals. For example, gold leaf will catch fire in chlorine without heating. With a metal such as iron, brown fumes of iron(III) chloride form.	iron + chlorine → iron (III) chloride $2Fe(s) + 3Cl_2(g) → 2FeCl_3(s)$ Fluor*ine* forms salts called fluor*ides* Chlor*ine* forms salts called chlor*ides* Brom*ine* forms salts called brom*ides* Iod*ine* forms salts called iod*ides*
Displacement 	A more reactive halogen will displace a less reactive halogen from a solution of a salt. Chlorine displaces bromine from sodium bromide solution. The colourless solution (sodium bromide) will turn brown as the chlorine is added due to the formation of bromine. Chlorine displaces iodine from sodium iodide solution. The colourless solution (sodium iodide) will turn brown as the chlorine is added due to the formation of iodine.	chlorine + sodium bromide → sodium chloride + bromine $Cl_2(g) + 2NaBr(aq) → 2NaCl(aq) + Br_2(aq)$ chlorine + sodium iodide → sodium chloride + iodine $Cl_2(g) + 2NaI(aq) → 2NaCl(aq) + I_2(aq)$

The halogens and their compounds have a wide range of uses:

- fluorides – in toothpaste help to prevent tooth decay
- fluorine compounds – make plastics like teflon (the non-stick surface on pans)
- chloro-fluorocarbons – propellants in aerosols (now being phased out due to effect on the ozone layer)
- chlorine – a bleach
- chlorine compounds – kill bacteria in drinking water and are used in antiseptics such as TCP
- hydrochloric acid – widely used in industry
- bromine compounds – make pesticides
- silver bromide – the light sensitive film coating in photography
- iodine solution – an antiseptic.

GROUP 0 – THE NOBLE GASES

This is a group of very unreactive non-metals. They used to be called the inert gases as it was thought that they didn't react with anything! Scientists have more recently managed to produce fluorine compounds of some of the noble gases. As far as the school laboratory is concerned, however, they are completely unreactive. This can be explained in terms of their electronic structures. The atoms all have complete outer electron shells. They don't need to lose electrons (as metals do), nor do they need to gain electrons (as most non-metals do).

Name	Symbol	Boiling point/°C
Helium	He	−269
Neon	Ne	−246
Argon	Ar	−186
Krypton	Kr	−152
Xenon	Xe	−108
Radon	Rn	−62

Similarities:

- full outer electron shells
- very unreactive
- gases
- exist as single atoms – they are **monatomic**, i.e. He, Ne, Ar, Kr, Xe, Rn.

Uses:

- helium – in balloons
- neon – in 'red lights'
- argon – in light bulbs
- krypton – in some lasers.

Check yourself

QUESTIONS

Q1 This question is about the Group 7 elements.
a) Which is the most reactive of the elements?
b) Which of the elements exists as a liquid at room temperature and pressure?
c) Which of the elements exist as a solid at room temperature and pressure?
d) Why are halogens such reactive non-metals?
e) Write word and symbol equations for the following reactions:
(i) sodium and chlorine
(ii) magnesium and bromine
(iii) hydrogen and fluorine.

a) What are the colours of the following solutions:
(i) aqueous chlorine (chlorine water),
(ii) aqueous bromine
(iii) aqueous iodine, (iv) aqueous sodium bromide?
b) What would be *observed* in the reaction between aqueous chlorine and sodium bromide solution?
c) Complete the table of results. Use a ✓ or ✗ as appropriate.
d) Write a word equation and symbol equation for the reaction between bromine and sodium iodide.

Q2 The table below records the results of some reactions.

	sodium chloride	sodium bromide	sodium iodide
chlorine	✗		✓
bromine		✗	✓
iodine			✗

(a ✓ indicates a reaction occurred, a ✗ indicates no reaction occurred.)

Q3 Chlorine exists as a diatomic molecule whereas neon is monatomic. Explain what is meant by the words diatomic and monatomic.

Q4 Explain why the noble gases are so unreactive.

Q5 Write down one use of each of the following noble gases: helium, neon, argon and krypton.

REMEMBER! Cover the answers if you want to.

ANSWERS

A1
a) Fluorine.
b) Bromine.
c) Iodine.
d) They only need to gain one electron in order to have a full outer electron shell. Other non-metals need to gain more than one electron.
e) (i) sodium + chlorine → sodium chloride
$2Na(s) + Cl_2(g) → 2NaCl(s)$
(ii) magnesium + bromine → magnesium bromide
$Mg(s) + Br_2(l) → MgBr_2(s)$

TUTORIAL

T1
a) Unlike groups of metals, reactivity decreases down the group.
b) Bromine is the only liquid non-metal.
c) Astatine would be expected to be a solid but it is radioactive with a very short half-life so it is difficult to confirm this prediction.
d) Remember that reactivity depends on the number of electrons in the outer electron shell.
e) The halogen elements react with all metals. They form salts. The salts are called fluorides, chlorides, bromides and iodides.

ANSWERS

(iii) hydrogen + fluorine → hydrogen
fluoride
$H_2(g) + F_2(g) → 2HF(g)$

A2 a) (i) Very pale yellow.
(ii) Brown.
(iii) Brown.
(iv) Colourless.
b) On mixing the pale yellow solution
with the colourless solution, a brown
solution is formed.
c) ✗ ✓ ✓
✗ ✗ ✓
✗ ✗ ✗
d) Bromine + sodium iodide → iodine +
sodium bromide
$Br_2(aq) + 2NaI(aq) → I_2(aq) +$
$2NaBr(aq)$

A3 A diatomic molecule contains two atoms
whereas a monatomic means a single atom.
Chlorine exists as Cl_2, neon exists as Ne.

A4 Noble gases have full outer electron shells
and so do not need to gain or lose
electrons in a reaction.

A5 Helium, used in weather balloons. Neon,
used in lighting, produces a red colour.
Argon, used as the gas in light bulbs.
Krypton, used in some lasers.

TUTORIAL

*Hydrogen fluoride will have very similar properties
to the more familiar hydrogen chloride. Its solution
in water, hydrofluoric acid is highly corrosive.*

T2 a) *Bromine is more soluble in water than iodine but
it is very difficult to distinguish between dilute
solutions of the two. Sometimes a solvent like
cyclohexane is added. The bromine is brown in
the cyclohexane, the iodine is violet.*
b) *Observations are what you see, smell or hear. You
don't see bromine, you see a brown solution.*
c) *The more reactive halogen will displace (take the
place of) the less reactive halogen. Chlorine will
displace bromine from sodium bromide because it
is more reactive. Iodine cannot displace chlorine or
bromine as it is less reactive than both of them.*
d) *Bromine is more reactive than iodine and so will
displace it from the solution. In this reaction it
would be difficult to see a change as both the
bromine and iodine solutions are brown. Adding
cyclohexane would confirm that a reaction had
taken place.*

T3 *Most of the non-metals exist as diatomic molecules, e.g.
H_2, O_2, N_2.*

T4 *Remember that metals want to lose electrons and form
positive ions, whereas non-metals often try to gain
electrons and form negative ions. Details of ionic
bonding are given in chapter 8.*

T5 *They may be unreactive but this property makes them
very suited to certain uses.*

EXAMINATION CHECKLIST

The facts and ideas that you should understand by studying this topic.

To be successful at Foundation GCSE Tier you should be able to:

- recall that elements are arranged in the periodic table in order of increasing atomic number
- recall that the columns are known as groups and the rows as periods
- identify the position in the periodic table of the alkali metals, the transition metals, the halogens and the noble gases
- recall the characteristic physical properties of metals and non-metals
- recall that metal oxides are basic and that most non-metal oxides are acidic
- recall that alkalis are soluble bases
- recall that the reactivity of an element depends on its electron structure
- recall that the alkali metals become more reactive as you go down the group, react with air and water and are stored under oil
- recall that an alkali metal reacts with water to form a metal hydroxide (an alkali) and hydrogen
- recall some of the uses of alkali metal compounds
- recall that transition metals have higher melting points and densities than the alkali metals
- recall that transition metals are used as catalysts and their compounds are coloured
- recall some of the flame colours of alkali metal and transition metal compounds
- recall that the halogens are a group of very reactive non-metals
- recall that the halogens react with most metals to form salts
- recall that the reactivity of the halogens decreases down the group
- recall some uses of the halogens and their compounds
- write word equations to represent simple reactions.

In addition, to be successful at Higher GCSE Tier you should be able to:

- explain neutralisation as the reaction between an acid and a base to form a salt and water
- explain how the periodic table is constructed in terms of electronic structure
- explain differences in reactivity of the alkali metals in terms of the ease with which electrons can be lost or transferred
- explain differences in the reactivity of the halogens in terms of the ease with which an electron can be gained
- explain the differences in reactivity between groups in terms of the differences in the electron arrangements
- explain displacement reactions of the halogens in terms of the differing reactivities of the elements
- write symbolic equations for the reactions encountered
- write ionic equations for the precipitation of transition metal hydroxides.

Key Words

Tick each word when you are sure of its meaning.

acidic oxide
alkali
alkali metals
atomic number
basic oxide
diatomic
displacement
ductile
group
halogens
malleable
metal
metalloid
monatomic
neutralisation
noble gases
non-metal
period
periodic table
salt
transition metals

231

EXAM PRACTICE

Sample Student's Answers & Examiner's Comments

EXAMINER'S COMMENTS

a) The mark would have been given for saying that all the elements have the same number of electrons in their outer shell. The candidate has gone further and correctly stated that they all have one electron in the outer shell.

b) **(i)** The correct answer has been given.
(ii) This answer lacks precision. The mark would be awarded for either saying that energy was transferred from the reaction to the surroundings or that the reaction was exothermic.
(iii) The candidate has not scored both marks. Apart from stating that the reaction would be more rapid than that with lithium, observations have not been given. The products have been correctly named but these are not what you would observe. Marks would be awarded for: the sodium floats on the water/ it melts into a sphere/ a colourless gas is given off/ an alkaline solution is formed/ the temperature of the solution increases.
(iv) The correct answer and explanation have been given.

c) The correct flame colour has been given. Yellow would also be allowed.

● A mark of 5/7 corresponds to a grade B on this higher tier question.

---

1 Lithium (Li), sodium (Na) and potassium (K) are in Group 1 of the periodic table.

a) These elements have similar chemical properties.

Explain why, using ideas about electronic structures.

All the elements have one electron in their outer electron shell ✓ (1)

b) Lithium reacts with water to form a solution of lithium hydroxide and a colourless gas. During this reaction the temperature of the water increases.

(i) What is the name of the colourless gas that is produced?

hydrogen ✓ (1)

(ii) Why does the temperature of the water increase?

the reaction between lithium and water produces heat ✗ (1)

(iii) Describe what you would observe if a small piece of sodium is added to water.

✗ *The sodium forms sodium hydroxide and hydrogen gas is given off. The reaction would be more rapid than the lithium reaction.* ✓ (2)

(iv) Caesium (Cs) is another Group 1 metal.

Is caesium more or less reactive than lithium?

Answer *more reactive*

Give a reason for your answer

Reactivity in group 1 increases down the group. ✓ (1)

c) The presence of a sodium compound can be tested using a flame test.

What is the flame test colour for a sodium compound?

orange ✓ (1)

(Total 7 marks)

5/7

Midland Examining Group

Questions to Answer

Answers to questions 2 – 3 can be found on pages 367–368.

2 a) (i) In the Periodic Table the elements are arranged in an increasing order.
What is it that increases? (1)

(ii) The table shows some information about six elements.

symbol	Sn	Sb	Te	I	Xe	Cs
Atomic number	50	51	52	53	54	55
Relative atomic mass	119	122	128	127	131	133

What is unusual about the data for Te and I?
Suggest an explanation. (3)

b) (i) How does the reactivity of the elements change going down Group 1? (1)

(ii) Name the products of the reaction between any member of Group 1 and water. (1)

(iii) What happens to the electrons in the outer shell of the atoms when elements of Group 1 react? (1)

c) (i) Write down the symbol and electronic structure of an atom in Group 7. (1)

(ii) How does the reactivity of the elements change going down Group 7? (1)

(iii) Suggest why the reactivity of the elements changes down Group 7. (1)

Midland Examining Group

3 Use your periodic table and your knowledge of more familiar elements to **suggest** answers to the following questions.

You are not expected to **know** about the chemistry of rubidium, iodine and selenium.

a) Rubidium, Rb, has the atomic number 37.
(i) In what group of the periodic table is rubidium? (1)

(ii) Name the two products you would expect when rubidium reacts with water. (2)

(iii) Suggest TWO observations you might SEE during the reaction. (2)

b) Iodine, I, has the atomic number 53. It reacts with hydrogen to form a gas, hydrogen iodide.

$$H_2 + I_2 \rightarrow 2HI$$

Hydrogen iodide is very soluble in water.

(i) Name a common laboratory liquid you would expect to behave like hydrogen iodide solution. (1)

(ii) What TWO products would you expect to be formed when magnesium reacts with hydrogen iodide solution? (2)

(iii) Suggest TWO observations you might SEE during the reaction. (2)

c) Selenium, Se, has the atomic number 34. Selenium trioxide reacts with water to form a new compound.

$H_2O(l) + SeO_3(s) \rightarrow H_2SeO_4(aq)$

(i) What effect would the solution have on universal indicator? (1)

(ii) Suggest TWO observations you might SEE when this solution was added to sodium carbonate. (2)

(iii) Name one of the compounds formed in the reaction with sodium carbonate. (1)

London Examinations

In the chemical industry knowing *what* is produced in a particular reaction is obviously very useful. Equally important, though, is to know *how much* can be produced. Chemists have developed methods of calculating these quantities. This chapter looks at some of these methods.

ATOMIC MASSES AND THE MOLE

Atoms are far too light to be weighed. Instead scientists have developed a **relative atomic mass** scale. Initially the hydrogen atom, the lightest atom, was chosen as the unit that all other atoms were weighed in terms of. On this scale, a carbon atom weighs the same as 12 hydrogen atoms, so carbon's relative atomic mass was given as 12.

Element	Symbol	Relative Atomic Mass		
Hydrogen	H	1		
Carbon	C	12		
Oxygen	O	16		
Magnesium	Mg	24		
Sulphur	S	32		
Calcium	Ca	40		
Copper	Cu	64		

Using this relative mass scale you can see, for example, that:

- 1 atom of magnesium has 24 × the mass of 1 atom of hydrogen
- 1 atom of magnesium has 2 × the mass of 1 atom of carbon
- 1 atom of copper has 2 × the mass of 1 atom of sulphur.

Recently, the reference point has been changed to carbon and the relative atomic mass is defined as:

the mass of an atom on a scale where the mass of a carbon atom is 12 units.

This change does not really affect the work that is done at GCSE. The relative atomic masses are not changed.

MOLES OF ATOMS

The **mole** is a number. It is a very large number, approximately 6×10^{23}. That is

600 000 000 000 000 000 000 000

What makes this number special is that:

- 6×10^{23} atoms of hydrogen have a mass of 1 g
- 6×10^{23} atoms of carbon have a mass of 12 g
- 6×10^{23} atoms of magnesium have a mass of 24 g

So the relative atomic mass (R.A.M.) of an element expressed in grams contains one mole of atoms. This means that the number of atoms of an element can be worked out by weighing.

these all
contain
1 mole of
atoms:

| 12 g | 24 g | 32 g | 56 g | 64 g |
| Carbon | Magnesium | Sulphur | Iron | Copper |

1 mole of carbon atoms has a mass of 12 g. 10 moles of carbon atoms would have a mass of 120 g. Calculations can be done using the simple equation:

$$\text{moles of atoms} = \frac{\text{mass}}{\text{R.A.M.}}$$

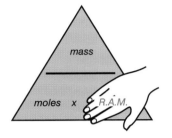

Put your finger over the quantity you are trying to work out. The triangle will then tell you whether to multiply or divide the other quantities.

Worked example

How many moles of atoms are there in 72 g of magnesium? (R.A.M. of magnesium = 24.)

1. Write down the formula: $\text{moles} = \dfrac{\text{mass}}{\text{R.A.M.}}$

2. Rearrange if necessary: (None needed)

3. Substitute the numbers: $\text{moles} = \dfrac{72}{24}$

4. Write the answer and units: moles = 3 moles

Worked example

What is the mass of 2 moles of sulphur atoms? (R.A.M. of sulphur = 32.)

1. Write down the formula: $\text{moles} = \dfrac{\text{mass}}{\text{R.A.M.}}$

2. Rearrange if necessary: mass = moles × R.A.M.

3. Substitute the numbers: mass = 2 × 32

4. Write the answer and units: mass = 64 g

Worked example

What is the mass of 0.1 moles of carbon atoms? (R.A.M. of carbon = 12)

1. Write down the formula: $\text{moles} = \dfrac{\text{mass}}{\text{R.A.M.}}$

2. Rearrange if necessary: mass = moles × R.A.M.

3. Substitute the numbers: mass = 0.1 × 12

4. Write the answer and units: mass = 1.2 g

WORKING OUT CHEMICAL FORMULAE

A chemical formula shows the number of atoms of each element that combine together.

H_2O A water molecule contains 2 hydrogen atoms and 1 oxygen atom.

Alternatively:

H_2O 1 mole of water molecules is made from 2 moles of hydrogen atoms and 1 mole of oxygen atoms.

The formula of a compound can be calculated if the number of moles of the combining elements are known.

Worked example

What is the simplest formula of a hydrocarbon that contains 60 g of carbon combined with 20 g of hydrogen? (R.A.M.s: H = 1, C = 12.)

	C	H
1. Write down the mass of each element:	60	20
2. Work out the number of moles of each element:	$\frac{60}{12} = 5$	$\frac{20}{1} = 20$
3. Find the simplest ratio (divide by the smaller number):	$\frac{5}{5} = 1$	$\frac{20}{5} = 4$
4. Write the formula showing the ratio of atoms:	CH_4	

Worked example

What is the simplest formula of magnesium oxide if 1.2 g of magnesium burned in oxygen to form 2.0 g of magnesium oxide? (O = 16, Mg = 24.)

	Mg	O
1. Write down the mass of each element:	1.2	$2.0 - 1.2 = 0.8$
2. Work out the number of moles of each element:	$\frac{1.2}{24} = 0.05$	$\frac{0.8}{16} = 0.05$
3. Find the simplest ratio:	$\frac{0.05}{0.05} = 1$	$\frac{0.05}{0.05} = 1$
4. Write the formula showing the ratio of atoms: MgO		

Worked example

What is the simplest formula of calcium carbonate if it contains 40% calcium, 12% carbon and 48% oxygen? (C = 12, O = 16, Ca = 40.)

	Ca	C	O
1. Write down the mass of each element:	40	12	48
2. Work out the number of moles of each element:	$\frac{40}{40} = 1$	$\frac{12}{12} = 1$	$\frac{48}{16} = 3$
3. Find the simplest ratio:	(Already in the simplest ratio)		
4. Write the formula showing the ratio of atoms:	$CaCO_3$		

MOLES OF MOLECULES

You can also refer to a mole of molecules. A mole of water molecules will be 6×10^{23} water molecules. The **relative molecular mass** of a molecule can be worked out by simply adding up the relative atomic masses of the atoms in the molecule. For example:

Water, H_2O (H = 1, O = 16)
The relative molecular mass (R.M.M.) = 1 + 1 + 16 = 18.

Carbon dioxide, CO_2 (C = 12, O = 16)
The R.M.M. = 12 + 16 + 16 = 44.
(Note: The '2' only applies to the oxygen atom.)

A similar approach can be used for any formula, including ionic formulae. As ionic compounds do not exist as molecules the **relative formula mass** (R.F.M.) can be worked out. For example:

Sodium chloride, NaCl (Na = 23, Cl = 35.5)
The relative formula mass (R.F.M.) = 23 + 35.5 = 58.5

Potassium Nitrate, KNO_3 (K = 39, N = 14, O = 16)
The R.F.M. = 39 + 14 + 16 + 16 + 16 = 101
(Note: The '3' only applies to the oxygen atoms.)

Calcium hydroxide, $Ca(OH)_2$ (Ca = 40, O = 16, H = 1)
The R.F.M. = 40 + (16 + 1)2 = 40 + 34 = 74
(Note: The '2' applies to everything in the bracket.)

Magnesium nitrate, $Mg(NO_3)_2$ (Mg = 24, N = 14, O = 16)
The R.F.M = 24 + (14 + 16 + 16 + 16)2 = 24 + (62)2 = 24 + 124 = 148

An equation can be written that can be used with atoms, molecules and ionic compounds.

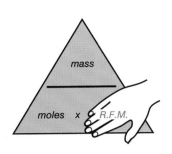

Put your finger over the quantity you are trying to work out.

$$\text{number of moles} = \frac{\text{mass}}{\text{R.F.M.}}$$

Check yourself

QUESTIONS

Q1 Calculate the number of moles in the following:
a) 56 g of silicon (Si = 28)
b) 3.1 g of phosphorus (P = 31)
c) 72 g of carbon (C = 12)
d) 11 g of carbon dioxide, CO_2 (C = 12, O = 16)
e) 68 g of hydrogen sulphide, H_2S (H = 1, S = 32)
f) 50 g of calcium carbonate, $CaCO_3$ (Ca = 40, C = 12, O = 16).

Q2 Calculate the mass of the following:
a) 2 moles of magnesium atoms (Mg = 24)
b) 0.2 moles of calcium atoms (Ca = 40)
c) 1.5 moles of carbon atoms (C = 12)
d) 2 moles of hydrogen molecules, H_2 (H = 1)
e) 0.1 moles of sulphuric acid, H_2SO_4 (H = 1, O = 16, S = 32)
f) 3 moles of sodium hydroxide, NaOH (H = 1, O = 16, Na = 23).

Q3 Chlorine exists in the form of two isotopes. 75% of atoms have an atomic mass of 35 and 25% have a mass of 37. Explain why the relative atomic mass of chlorine is 35.5.

Q4 Titanium chloride contains 25% titanium and 75% chlorine by mass. Work out the simplest formula of titanium chloride. (Ti = 48, Cl = 35.5)

Q5 8.0 g of copper(II) oxide are reduced and 6.4 g of copper remains. Calculate the formula of copper(II) oxide. (Cu = 64, O = 16)

ANSWERS

TUTORIAL

A1
a) 2.
b) 0.1.
c) 6.
d) 0.25.
e) 2.
f) 0.5.

T1
a) $\text{Moles} = \dfrac{mass}{\text{RAM}} = \dfrac{56}{28} = 2.$

b) $\text{Moles} = \dfrac{mass}{\text{RAM}} = \dfrac{3.1}{31} = 0.1.$

c) $\text{Moles} = \dfrac{mass}{\text{RAM}} = \dfrac{72}{12} = 6.$

d) $\text{Moles} = \dfrac{mass}{\text{RMM}} = \dfrac{11}{44} = 0.25.$

e) $\text{Moles} = \dfrac{mass}{\text{RMM}} = \dfrac{68}{34} = 2.$

f) $\text{Moles} = \dfrac{mass}{\text{RFM}} = \dfrac{50}{100} = 0.5.$

Note: Calcium carbonate is an ionic compound and so the correct term is relative formula mass.

A2
a) 48 g.
b) 8 g.
c) 18 g.
d) 4 g.
e) 9.8 g.
f) 120 g.

T2
a) $\text{Mass} = moles \times \text{RAM} = 2 \times 24 = 48\,g.$
b) $\text{Mass} = moles \times \text{RAM} = 0.2 \times 40 = 8\,g.$
c) $\text{Mass} = moles \times \text{RAM} = 1.5 \times 12 = 18\,g.$
d) $\text{Mass} = moles \times \text{RMM} = 2 \times 2 = 4\,g.$
e) $\text{Mass} = moles \times \text{RMM} = 0.1 \times 98 = 9.8\,g.$
f) $\text{Mass} = moles \times \text{RFM} = 3 \times 40 = 120\,g.$

A3 The atomic mass of chlorine must reflect the % of the two isotopes. $(35 \times 75 + 37 \times 25)/100 = 35.5.$

T3 The relative atomic mass is a weighted average of the isotopic masses. No chlorine atom has a relative mass of 35.5.

A4 $TiCl_4$.

T4

	Ti	Cl
$\dfrac{mass}{\text{RAM}}$	$\dfrac{25}{48}$	$\dfrac{75}{35.5}$
moles	0.52	2.1
ratio	$\dfrac{0.52}{0.52} = 1$	$\dfrac{2.1}{0.52} = 4.04$

formula is $TiCl_4$

A5 CuO.

T5

	Cu	O
$\dfrac{mass}{\text{RAM}}$	$\dfrac{6.4}{64}$	$\dfrac{1.6}{16}$
moles	0.1	0.1
ratio	1	1

formula is CuO

EQUATIONS AND REACTING MASSES

One reason for writing chemical equations is that they are a form of short-hand recognised and understood by scientists throughout the world. Another reason is that they allow quantities of **reactants** and **products** to be linked together. In other words, they tell you how much of the products you can expect to make from a fixed amount of reactants.

In a balanced equation the numbers in front of each symbol or formula indicate the number of moles represented. The number of moles can then be converted into a mass in grams.

For example when magnesium (R.A.M. 24) reacts with oxygen (R.A.M. 16):

1. Write down the balanced equation: $2Mg(s) + O_2(g) \rightarrow 2MgO(s)$
2. Write down the number of moles: $2 + 1 \rightarrow 2$
3. Convert moles to masses: $48\,g + 32\,g \rightarrow 80\,g$

So when 48 g of magnesium reacts with 32 g of oxygen, 80 g of magnesium oxide is produced. From this you should be able to work out the mass of magnesium oxide produced from any mass of magnesium. All you need to do is work out a scaling factor.

Worked example

What mass of magnesium oxide can be made from 6 g of magnesium?
(O = 16, Mg = 24.)

1. Equation: $2Mg(s) + O_2(g) \rightarrow 2MgO(s)$
2. Moles: $2 \qquad 1 \qquad 2$
3. Masses: $48\,g \qquad 32\,g \qquad 80\,g$

Instead of 48 g of magnesium, the question asks about 6 g. We can scale down all the quantities by dividing them by 8. Therefore $\frac{48}{8}\,g = 6\,g$ of magnesium would make $\frac{80}{8}\,g = 10\,g$ of magnesium oxide.

Note: In this example, there was no need to work out the mass of oxygen needed. It was assumed that there would be as much as was necessary to convert all the magnesium to magnesium oxide.

Worked example

What mass of ammonia can be made from 56 g of nitrogen? (H = 1, N = 14.)

1. Equation: $N_2(g) + 3H_2(g) \rightarrow 2NH_3(g)$
2. Moles: $1 \qquad 3 \qquad 2$
3. Masses: $28\,g \qquad 6\,g \qquad 34\,g$

In this case we need to multiply the mass of nitrogen by 2 to get the amount asked about in the question. Use the same scaling factor on the other quantities. So $28\,g \times 2 = 56\,g$ of nitrogen makes $34\,g \times 2 = 68\,g$ of ammonia.

Note: In this example there was no need to work out the mass of hydrogen required.

Worked example

How much ethene is needed to make 230 tonnes of ethanol? (H = 1, C = 12, O = 16)

1. Equation: $C_2H_4(g) + H_2O(g) \rightarrow C_2H_5OH(l)$
2. Moles: $1 \qquad 1 \qquad 1$
3. Masses: $28\,g \qquad 18\,g \qquad 46\,g$

In this case we scale up by multiplying the 46 by 5 to get 230 and then scale up again by changing all the units from grams to tonnes. This means that $28\,g \times 5 = 140\,g$ of ethene, so 140 *tonnes* of ethene is needed.

MOLES OF GASES

In reactions involving gases it is often more convenient to measure the volume of a gas rather than its mass. There is a simple relationship between the volume of a gas and the number of moles it contains. One mole of any gas occupies the same volume under the same conditions of temperature and pressure. The conditions chosen are usually room temperature (20 °C) and normal atmospheric pressure.

1 mole of any gas occupies 24 000 cm³ (24 litres) at room temperature and pressure.

The following equation can be used to convert moles and volumes:

$$\text{moles} = \frac{\text{volume in cm}^3}{24\,000}$$

Each of these contains 1 mole (6×10^{23}) of molecules.

Worked example

What volume of hydrogen is formed at room temperature and pressure when 4 g of magnesium is added to excess dilute hydrochloric acid? (H = 1, Mg = 24.)

1. Equation	$Mg(s)$ +	$2HCl(aq)$ →	$MgCl_2(aq)$ +	$H_2(g)$
2. Moles	1	2	1	1
3. Masses/volumes	24 g			24 000 cm³

Now work out the scaling factor needed and use the same scaling factor on the other quantities. Here we divide 24 g by 6 to get 4 g of magnesium. This then produces $\frac{24\,000}{6}$ cm³ = 4000 cm³ of hydrogen gas.

Note: The hydrochloric acid is in excess. This means that there is enough to react with all the magnesium.

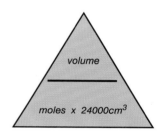

The triangle can be used as before to work out whether to multiply or divide the quantities.

MOLES OF SOLUTIONS

A **solution** is made when a **solute** dissolves in a **solvent**. The concentration of a solution depends on how much solute is dissolved in how much solvent. The concentrations of a solution are defined in terms of moles per litre (1000 cm³) and are referred to as the **molarity** of the solution.

1 mole of solute dissolved to make 1000 cm³ of solution produces a 1 molar (1 M) solution

similarly:

- 2 moles dissolved to make a 1000 cm³ solution produces a 2 M solution

- 0.5 moles dissolved to make a 1000 cm³ solution produces a 0.5 M solution

- 1 mole dissolved to make a 500 cm³ solution produces a 2 M solution

- 1 mole dissolved to make a 250 cm³ solution produces a 4 M solution.

Note: If the same amount of solute is dissolved to make a smaller amount of solution the solution will be more concentrated.

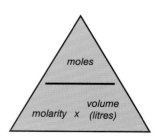

This triangle will help you to calculate concentrations of solutions.

Worked example

How much sodium chloride can be made from reacting $100\,cm^3$ of 1 M hydrochloric acid with excess sodium hydroxide solution? (Na = 23, Cl = 35.5.)

1. Equation		$HCl(aq) + NaOH(aq) \rightarrow NaCl(aq) + H_2O(l)$			
2 Moles		1	1	1	1
3. Masses/volumes		$(1000\,cm^3\ 1\,M)$		$58.5\,g$	

Now work out the scaling factor and use the same scaling factor on the other quantities. In this case we divide by 10 to get $100\,cm^3$ of reactant, which means we get $\frac{58.5}{10}\,g = 5.85\,g$ of product.

Note: $100\,cm^3$ of 1M solution is equal to 0.1 mole.

Check yourself

QUESTIONS

Q1 What mass of sodium hydroxide can be made by reacting 2.3 g of sodium with water? (H = 1, O = 16, Na = 23.)

$2Na(s) + 2H_2O(l) \rightarrow 2NaOH(aq) + H_2(g)$

Q2 What mass of copper(II) oxide can be made by oxidising 32 g of copper? (O = 16, Cu = 64.)

$2Cu(s) + O_2(g) \rightarrow 2CuO(s)$

Q3 What mass of carbon is needed to produce 880 tonnes of carbon dioxide? (C = 12, O = 16.)

$C(s) + O_2(g) \rightarrow CO_2(g)$

Q4 Iron(III) oxide is reduced to iron by carbon monoxide. (C = 12, O = 16, Fe = 56.)

$Fe_2O_3(s) + 3CO(g) \rightarrow 2Fe(s) + 3CO_2(g)$

a) Calculate the mass of iron which could be obtained by the reduction of 800 tonnes of iron(III) oxide.

b) What volume of carbon dioxide would be obtained by the reduction of 320 g of iron(III) oxide?

Q5 What mass of barium sulphate can be produced from $50\,cm^3$ of 0.2 M barium chloride solution and excess sodium sulphate solution? (O = 16, S = 32, Ba = 137.)

$BaCl_2(aq) + Na_2SO_4(aq)$
$\rightarrow BaSO_4(s) + 2NaCl(aq)$

REMEMBER! Cover the answers if you want to.

ANSWERS

A1 4 g.

A2 40 g.

A3 240 tonnes.

A4
a) 560 tonnes.
b) 144 litres at room temp and pressure.

TUTORIAL

T1 $2Na = 2\ moles = 46\,g$. $2NaOH = 2\ moles = 80\,g$. *The scaling factor is $\div 20$.*

T2 $2Cu = 2moles = 128\,g$. $2CuO = 2\ moles = 160\,g$. *The scaling factor is $\div 4$.*

T3 $C = 1\ mole \rightarrow 12\ tonnes$. $CO_2 = 1\ mole \rightarrow 44\ tonnes$. *The scaling factor is $\times 20$.*

T4
a) $Fe_2O_3 = 1\ mole \rightarrow 160\ tonnes$. $2Fe = 2\ moles \rightarrow 112\ tonnes$. *The scaling factor is $\times 5$.*
b) $Fe_2O_3 = 1\ mole \rightarrow 160\,g$. $3CO_2 = 3 \times 24 = 72\ litres$. *The scaling factor is $\times 2$.*

ANSWERS

A5   2.33 g.

TUTORIAL

T5 $BaCl_2 = 1$ *mole* $= 1000\,cm^3$ *(1 M).* $BaSO_4 = 1$ *mole* $= 233\,g$.
First scale by $\div 5$. $1000\,cm^3$ *(0.2 M) and* $46.6\,g$.
Second scale by $\div 20$. $50\,cm^3$ *(0.2 M) and* $2.33\,g$.

EXAMINATION CHECKLIST

The facts and ideas that you should understand by studying this topic.

To be successful at Foundation GCSE Tier you should be able to:

- recall that atomic masses are measured on a relative scale
- recall that a mole is a number equal to 6×10^{23}
- convert moles into mass and vice versa
- work out relative molecular mass and relative formula mass
- recall that 1 mole of any gas occupies 24 litres at room temperature and pressure
- recall that concentrations of solutions are expressed in terms of moles per litre or molarity.

In addition, to be successful at Higher GCSE Tier you should be able to:

- calculate a chemical formula of a compound from its composition by mass
- calculate reacting masses
- calculate the volume of gas involved or produced in a reaction.

Key Words

Tick each word when you are sure of its meaning.

molarity

mole

product

reactant

relative atomic mass

relative formula mass

relative molecular mass

solute

solution

solvent

EXAM PRACTICE

Sample Student's Answers & Examiner's Comments

a)   **(i)**   The subtraction has been done correctly and the correct unit has been given. Units are often missed in questions like this.

   (ii)   The candidate has correctly divided the mass by the relative atomic mass to work out the number of moles. The masses in the question were given to one decimal place so the number of moles should be given to one decimal place.

   (iii)   The ratio has been correctly worked out by dividing by the smallest. One mark is awarded for this. The answer of 1:2.5 cannot be rounded down to 1:2 as the candidate has done. Nor can it be rounded up to 1:3. To obtain the simplest whole number ratio the ratio should be doubled to give 2:5 and a simplest formula of C_2H_5.

b)   Having got the wrong answer in part a) iii) it is still possible for the candidate to score both marks in this part. One mark has been scored for working out the relative formula mass of CH_2. This is what is called a mark for a 'transmitted error'. The idea is to penalise a wrong answer only once. If the candidate had completed the formula as C_4H_8 another transmitted error mark would have been given for using the correct method. The correct working is as follows:

RMM of $C_2H_5 = 29$, $\frac{58}{29} = 2$,

molecular formula is C_4H_{10}

Note: Because marks are awarded for transmitted errors never give up on a question just because the numbers don't seem to be working out.

●   A mark of 5/7 corresponds to a grade B on this higher tier question.

---

1   A gaseous compound contains carbon and hydrogen only. It was found that 100 g of the compound contained 82.8 g of carbon. (H = 1, C = 12)

a)   (i)   Calculate the mass of hydrogen in 100 g of the compound.

    $100 - 82.8 = 17.2$ g   ✓   (1)

  (ii)   Use this information to calculate the number of moles of carbon atoms and of hydrogen atoms in 100 g of the compound.

    Carbon $82.8/12 = 6.9$ ✓   (1)

    Hydrogen $17.2/1 = 17.2$ ✓   (1)

  (iii)   Use these figures to find the simplest formula of the compound.

    Ratio: C   H

    $6.9/6.9$ $17.2/6.9$ CH_2 ✗

    1   2.5 ✓ that is 1:2   (2)

b)   The mass of 1 mole of the gaseous compound is 58 g.

  What is the molecular formula of the gaseous compound?

  $CH_2 = 12 + 2 = 14$ ✓ t.e.

    $58/14 = 4.14$

  Molecular formula =   ⑤/7   (2)

                        (Total 7 marks)

                    Midland Examining Group

Questions to Answer

Answers to questions 2 – 5 can be found on page 369.

2   Laboratory analysis shows that 15.15 g of a mineral called chalcopyrite has the following composition by mass.

copper 5.27 g

iron 4.61 g

Sulphur is the only other element present. (S = 32, Fe = 56, Cu 64)

Use these figures to find the simplest formula of chalcopyrite. (4)

Midland Examining Group

3   What mass of chlorine would be needed to make 150 tonnes of hydrogen chloride?

(H = 1, Cl = 35.5)

$H_2 + Cl_2 \rightarrow 2HCl$ (3)

Midland Examining Group

4   The symbol equation below shows the reduction of iron(III) oxide by carbon monoxide.

$Fe_2O_3 + 3CO \rightarrow 2Fe + 3CO_2$

(Fe = 56, O = 16, C = 12)

a)   Calculate the formula mass of iron(III) oxide. (3)

b)   Calculate the mass of iron that could be obtained from 32 000 tonnes of iron(III) oxide. (3)

Northern Examination and Assessment Board

5   Magnecalc plc is a company that uses limestone (calcium carbonate) to manufacture quicklime (calcium oxide).

The equation for the reaction is:

calcium carbonate → calcium oxide + carbon dioxide

$\qquad\qquad CaCO_3 \rightarrow CaO + CO_2$

(C = 12, Ca = 40, O = 16)

Magnecalc sell 224 tonnes of quicklime a week. Calculate the mass of limestone that has to be quarried to provide 224 tonnes of quicklime. Assume the limestone is pure. (4)

Midland Examining Group

Electricity is extremely useful in our everyday lives. Many devices in the home need electrical circuits to provide them with the energy they need to work. There is also static electricity. Becoming 'charged up' and then discharging when you touch something metallic is quite a common experience for some people – even this has its useful applications. Electricity can also be dangerous, knowing more about it will help you to use it safely.

KEY POINTS

Charge flow and conventional current.

Voltage = joules per coulomb.

Series and parallel circuits.

Direct and alternating current.

ELECTRICAL CIRCUITS

You cannot see electricity but you can see the effects it has. It can:

- make things hot – as in the heating element of an electric fire
- make things magnetic – as in an electromagnet
- produce light – as in a light bulb
- break down certain compounds and solutions – as in electrolysis.

In a nutshell, electricity is very good at transferring energy. To help us understand how it does this it is useful to have a simple model of what happens in an electrical circuit.

A simple model of an electrical circuit

If an electrical circuit is made using a battery and a lamp, the battery can be thought of as pushing electrical charge round the circuit to make a **current**. The battery also transfers energy to the electrical charge. The **voltage** of the battery is a measure of how much 'push' it can provide and how much energy it can transfer to the charge.

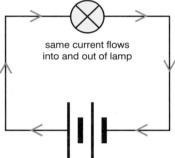

energy transferred from charge to lamp and then to surroundings

same current flows into and out of lamp

battery transfers energy to charge and pushes charge around circuit

In this simple circuit the arrows show the direction of the current.

A simple electrical circuit works rather like this central heating system. Energy is transferred to the water by the boiler (the battery). A pump (the battery) pushes the water particles around the circuit in a current. A radiator (the lamp) transfers energy from the hot water to the surroundings.

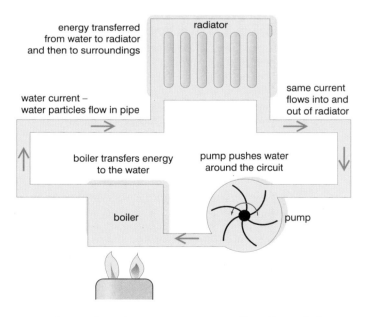

energy transferred from water to radiator and then to surroundings

radiator

same current flows into and out of radiator

water current – water particles flow in pipe

boiler transfers energy to the water

pump pushes water around the circuit

boiler

pump

Scientists now know that electric current is really a flow of electrons. The electrons actually flow around the circuit from negative to positive. Unfortunately, although early scientists knew that there must be a flow of charge in a circuit, they guessed the direction of flow incorrectly. Consequently all diagrams were drawn showing the current flowing from positive to negative. Surprisingly this way of showing the current has not been changed and so the **conventional current** that everyone uses gives the direction that positive charges would flow.

Electric charge is measured in **coulombs** (C). Electric current is measured in **amperes**. The electric current is the amount of charge flowing every second, that is the number of coulombs per second:

$$I = \frac{Q}{t}$$

I = current in amperes (A)

Q = charge in coulombs (C)

t = times in seconds (s)

The charge flowing round a circuit has some potential energy. When the charge flows through a lamp it transfers some of its energy to the lamp.

There are two different ways of connecting two lamps to the same battery. Two very different kinds of circuit can be made. These circuits are called **series** and **parallel** circuits.

MEASURING ELECTRICITY

As charge flows around a circuit it transfers energy to the various components in the circuit. The amount of energy that a unit of charge (a coulomb) transfers between one point and another (the number of joules per coulomb) is called the **potential difference** (p.d.). Potential difference is measured in volts and so it is often referred to as voltage:

$$V = \frac{J}{Q}$$

V = potential difference in volts

J = energy transferred in joules

Q = charge in coulombs

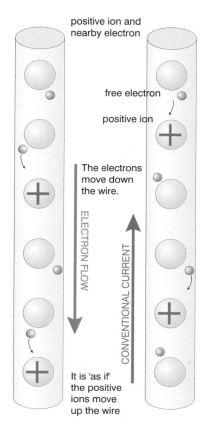

positive ion and nearby electron

free electron

positive ion

The electrons move down the wire.

ELECTRON FLOW

CONVENTIONAL CURRENT

It is 'as if' the positive ions move up the wire

Conventional current is drawn in the opposite direction to electron flow.

	Series	**Parallel**
Circuit diagram		
Appearance of the lamps	Both lamps have the same brightness, both lamps are dim.	Both lamps have the same brightness, both lamps are bright.
Battery	The battery is having a hard time pushing the same charge first through one bulb, then another. This means less charge flows each second, so there is a low current and energy is slowly transferred from the battery.	The battery pushes the charge along two alternative paths. This means more charge can flow around the circuit each second, so energy is quickly transferred from the battery.
Switches	The lamps cannot be switched on and off independently.	The lamps can be switched on and off independently by putting switches in the parallel branches.
Advantages/ disadvantages	A very simple circuit to make. The battery will last longer. If one lamp 'blows' then the circuit is broken so the other one goes out too.	The battery won't last as long. If one lamp 'blows' the other one will keep working.
Examples	Christmas tree lights are often connected in series.	Electric lights in the home are connected in parallel.

Potential difference (p.d.) is the difference in energy of a coulomb of charge between two parts of a circuit.

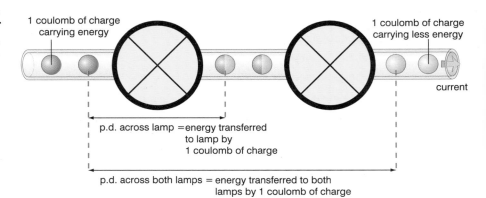

1 coulomb of charge carrying energy

1 coulomb of charge carrying less energy

current

p.d. across lamp = energy transferred to lamp by 1 coulomb of charge

p.d. across both lamps = energy transferred to both lamps by 1 coulomb of charge

Potential difference is measured using a **voltmeter**. If you want to measure the p.d. across a component then the voltmeter must be connected in *parallel* to that component. Testing with a voltmeter does not interfere with the circuit at all.

The voltmeter can be added after the circuit has been made.

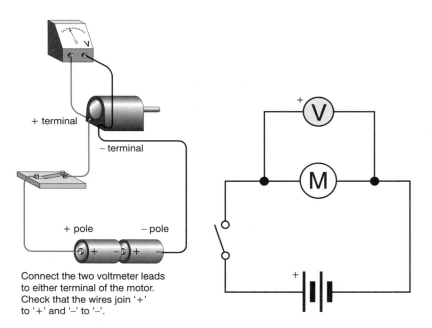

+ terminal

– terminal

+ pole   – pole

Connect the two voltmeter leads to either terminal of the motor. Check that the wires join '+' to '+' and '–' to '–'.

A voltmeter can be used to show how the potential difference varies in different parts of a circuit. In a series circuit you get different values of the voltage depending on where you attach the voltmeter. You can assume that energy is only transferred when the current passes through electrical components such as lamps and motors – the energy transfer as the current flows through copper connecting wire is very small. It is only possible therefore to measure a potential difference or voltage across a component.

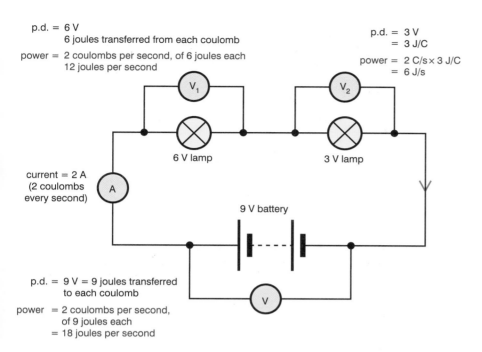

p.d. = 6 V
 6 joules transferred from each coulomb

power = 2 coulombs per second, of 6 joules each
 12 joules per second

V_1

6 V lamp

p.d. = 3 V
 = 3 J/C

power = 2 C/s × 3 J/C
 = 6 J/s

V_2

3 V lamp

current = 2 A
(2 coulombs
every second)

A

9 V battery

V

p.d. = 9 V = 9 joules transferred
 to each coulomb

power = 2 coulombs per second,
 of 9 joules each
 = 18 joules per second

The potential difference across the battery equals the sum of the potential differences across each lamp. That is $V = V_1 + V_2$.

The current flowing in a circuit can be measured using an **ammeter**. If you want to measure the current flowing through a particular component, such as a lamp or motor, the ammeter must be connected in *series* with the component. In a series circuit the current is the same no matter where the ammeter is put. This is not the case with a parallel circuit.

The circuit has to be broken to include the ammeter.

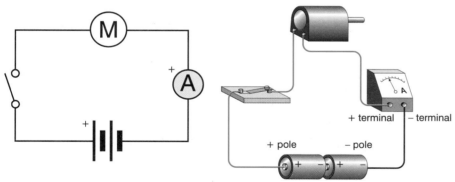

M

+
A

+

+ terminal

− terminal

+ pole

− pole

Make a break anywhere in the circuit and connect the ammeter in the break. Check that the wires join '+' to '+' and '−' to '−'.

In this series circuit the current will be the same throughout the circuit so the readings $A_1 = A_2 = A_3$.

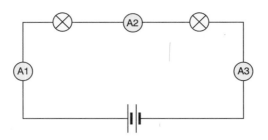

The current flow splits between the two branches of the parallel circuit so the readings $A_1 = A_2 + A_3$.

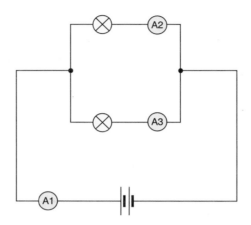

A.C. AND D.C.

A battery produces a steady current. The electrons are constantly flowing from the negative terminal of the battery round the circuit and back to the positive terminal. This produces a **direct current** (d.c.). The mains electricity used in the home is quite different. The electrons in the circuit move backwards and forwards. This kind of current is called **alternating current** (a.c.). Mains electricity moves forwards and backwards 50 times each second, that is, with a frequency of 50 hertz (Hz).

The advantage of using an a.c. source of electricity rather than a d.c. source is that it can be transmitted from power stations to the home at very high voltages which reduces the amount of energy that is lost in the overhead cables (see chapter 15).

Check yourself

QUESTIONS

Q1 Look at the circuit diagram. What happens to lamps A, B and C when switch 2 is closed (pressed)? Explain your answer.

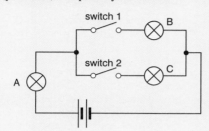

Q2 Look at the following circuit diagrams. They show a number of ammeters and in some cases the readings on these ammeters. All the lamps are identical.

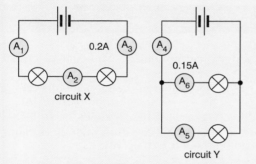

a) For circuit X. What readings would you expect on ammeters A_1 and A_2.

b) For circuit Y. What readings would you expect on ammeters A_4 and A_5.

Q3 Look at the circuit diagram. It shows how three voltmeters have been added to the circuit. What reading would you expect on V_1?

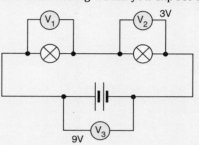

Q4 a) A charge of 10 coulombs flows through a motor in 30 seconds. What is the current flowing through the motor?

b) A heater uses a current of 10 A. How much charge flows through the lamp in: (i) 1 second, (ii) 1 hour?

Q5 Explain what is meant by the following:
a) potential difference
b) a direct current
c) an alternating current.

REMEMBER! Cover the answers if you want to.

ANSWERS

A1 Lamps A and C light but not lamp B. Unless switch 1 is pressed the current cannot flow through lamp B as the circuit is broken.

A2 a) A_1 reading 0.2 A. A_2 reading 0.2 A.
b) A_4 reading 0.30 A. A_5 reading 0.15 A.

TUTORIAL

T1 *With switch 1 open the circuit behaves as a simple circuit with lamps A and C in series. Remember that one of the advantages of parallel circuits is that individual components can have their own switches.*

T2 a) *The current is always the same at any point in a series circuit.*

b) *The ammeter A_6 has been placed on one branch of the parallel circuit. Ammeter A_5 is on the other branch. As the lamps are identical the current flowing through them must be the same. Ammeter A_4 gives the current before it 'splits' in half as it flows through the two parallel branches.*

ANSWERS

A3 Reading on $V_1 = 6\,V$.

A4
a) 0.33 A.
b) i) 10 C, ii) 36 000 C.

A5
a) It is the amount of energy transferred per unit of charge.
b) A direct current is one formed when electrons move round a circuit flowing from negative to positive.
c) An alternating current is formed when electrons change direction constantly as they flow in a circuit.

TUTORIAL

T3 *The potential difference across the battery is 9 V. This must equal the total p.d. in the circuit. Assuming there is no loss along the copper wiring, the potential difference across the lamp must be 9 − 3 = 6 V.*

T4
a) *Use the equation, $I = \frac{Q}{t}$, $I = \frac{10}{30} = 0.33\,A$*
b *Rearranging the formula, $Q = It$, $Q = 10 \times 1 = 10\,C$, $Q = 10 \times 60 \times 60 = 36\,000\,C$*

T5
a) *In fact the potential difference is the energy transferred from one coulomb of charge.*
b) *Remember that conventional current flow is in the opposite direction to electron flow, i.e. from positive to negative.*
c) *The electrons in a mains circuit change direction and back 50 times each second.*

KEY POINTS

Factors affecting resistance:
● **material**
● **length**
● **cross-sectional area**
● **temperature**

cover *I* to find $I = \dfrac{V}{R}$

RESISTANCE

The relationship between voltage, current and resistance in electrical circuits is given by **Ohm's law**:

$$V = I\,R$$

V is the voltage in volts (V)
I is the current in amps (A)
R is the resistance in ohms (Ω)

It is important to be able to rearrange this equation when performing calculations. Use the triangle on the left to help you.

Worked example

Calculate the resistance of a heater element if the current is 10 A when it is connected to a 230 V supply.

Write down the formula in terms of R: $R = \dfrac{V}{I}$

Substitute the values for V and I: $R = \dfrac{230}{10}$

Work out the answer and write down the unit: $R = 23\,\Omega$

Worked example

A 6 V supply is applied to 1000 Ω resistor. What current will flow?

Write down the formula in terms of I: $I = \dfrac{V}{R}$

Substitute the values for V and R: $I = \dfrac{6}{1000}$

Work out the answer and write down the unit: $I = 0.006\,A$

EFFECT OF MATERIAL

Substances which allow an electric current to flow through them are called **conductors**. Those which do not are called **insulators**. Metals behave as conductors because of their structure. In a metal structure the metal atoms release their outermost electrons to form an 'electron cloud' throughout the whole structure. In other words, the atoms in a metal exist as ions surrounded by an electron cloud. If a potential difference is applied to the metal the electrons in this cloud are able to move.

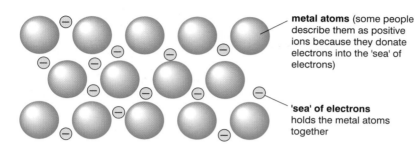

metal atoms (some people describe them as positive ions because they donate electrons into the 'sea' of electrons)

'sea' of electrons holds the metal atoms together

In a metal structure metal ions are surrounded by a cloud or 'sea' of electrons.

Some conductors are better than others. Copper is a better conductor than iron. When the electrons are moving through the metal structure they bump into the metal ions and this causes **resistance** to the electron flow or current. In different conductors the ease of flow of the electrons is different and so the conductors have different resistance.

For a particular conductor the resistance will depend on its length and cross-sectional area. The longer the conductor, the further the electrons have to travel, the more likely they are to have collisions with the metal ions and so the greater the resistance. Resistance is **proportional** to length. The greater the cross-sectional area of the conductor, the more electrons available to carry the charge along the conductor's length and so the lower the resistance is. Resistance is **inversely proportional** to cross-sectional area.

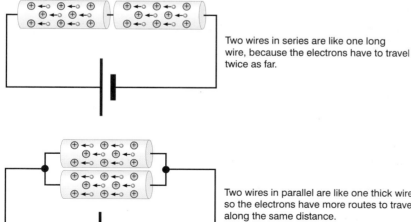

Two wires in series are like one long wire, because the electrons have to travel twice as far.

Two wires in parallel are like one thick wire, so the electrons have more routes to travel along the same distance.

The amount of current flowing through a circuit can be controlled by changing the resistance of the circuit. This can be done with a **variable resistor** or rheostat. Adjusting the rheostat changes the length of the wire the current has to flow through.

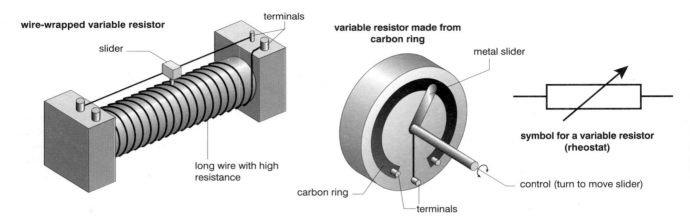

wire-wrapped variable resistor

slider

terminals

long wire with high resistance

variable resistor made from carbon ring

metal slider

carbon ring

terminals

symbol for a variable resistor (rheostat)

control (turn to move slider)

Variable resistors are commonly used in electrical equipment for example in the speed controls of model racing cars or in volume controls on radios and hi-fi systems.

EFFECT OF TEMPERATURE

If the resistance of a conductor remains constant then a graph showing voltage plotted against current will give a straight line. The gradient of the line will be the resistance of the conductor. However, the resistance of most conductors changes if the temperature of the conductor changes. This can be explained using the simple model of a conductor. As the temperature increases the metal ions vibrate more and therefore provide greater resistance to the flow of the electrons. The resistance of a filament lamp becomes greater as the voltage is increased and the lamp gets hotter.

In an 'ohmic' resistor, such as carbon, Ohm's law applies and the voltage is directly proportional to the current – a straight line is obtained. In a filament lamp, Ohm's law is not obeyed because the heating of the lamp changes its resistance.

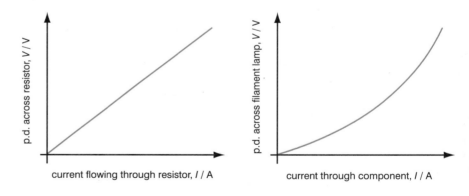

p.d. across resistor, V / V

current flowing through resistor, I / A

p.d. across filament lamp, V / V

current through component, I / A

In some substances, increasing the temperature actually lowers the resistance. This is the case with **semiconductors** such as silicon. Silicon has few free electrons and so behaves more like an insulator than a conductor. But if silicon is heated, more electrons are removed from the outer electron shells of the atoms producing an increased electron cloud. The released electrons can move throughout the structure creating an electric current. This effect is large enough to outweigh the increase in resistance that might be expected from the increased movement of the silicon ions in the structure as the temperature increases.

Semiconducting silicon is used to make **thermistors**, which are used as temperature sensors, and **light dependent resistors** (LDRs), which are used as light sensors. In LDRs it is light energy that removes electrons from the silicon atoms, increasing the electron cloud. Silicon **diodes** also use resistance as a means of controlling current flow in a circuit. In one direction the resistance is very high, effectively preventing current flow, in the other direction the resistance is relatively low and current can flow. Diodes are used to protect sensitive electronic equipment that would be damaged if a current flowed in the wrong direction.

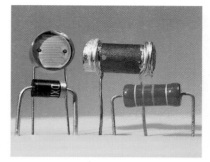

A light-dependent resistor, LDR (top left) conducts better when light shines on it. A thermistor (top right) conducts better when it is hot. A diode (bottom left) only conducts in one direction. An ordinary (ohmic) resistor is shown bottom right.

Check yourself

QUESTIONS

Q1 Explain why metals are good conductors of electricity.

Q2 Explain how the resistances of the following change as the temperature is increased:
a) a nichrome wire (the wire used as the element in an electric toaster)
b) a semiconductor such as silicon.

Q3 a) Draw a circuit diagram to show how you could measure the resistance of a piece of nichrome wire? Explain how you would calculate the resistance of the wire.
b) How would the resistance of the wire change if (i) its length was doubled, (ii) its cross-sectional area was doubled?

Q4 Use Ohm's law to calculate the following:
a) The voltage required to produce a current of 2 A in a 12 Ω resistor.
b) The voltage required to produce a current of 0.1 A in a 200 Ω resistor.
c) The current produced when a voltage of 12 V is applied to a 100 Ω resistor.
d) The current produced when a voltage of 230 V is applied to 10 Ω resistor.
e) The resistance of a wire which under a potential difference of 6 V allows a current of 0.1 A to flow.
f) The resistance of a heater which under a potential difference of 230 V allows a current of 10 A to flow.

Q5 A graph of current against voltage is plotted for a piece of wire. The graph is shown below.

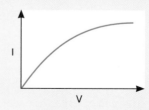

a) Describe how the resistance of the wire changes as the voltage is increased.
b) Suggest an explanation for this change.

QUESTIONS

Q6 The graph shows how the current through a
diode varies for different values of the
potential difference across it.

a) Describe the change in resistance of
the diode when the potential
difference across it increases from
zero to 0.6 V.

b) What does the part of the curve from 0
to −1.0 V tell you about the diode?

c) A resistor has a constant value of
625 Ω. Calculate the current through the
resistor when the p.d. across it is 1.0 V.

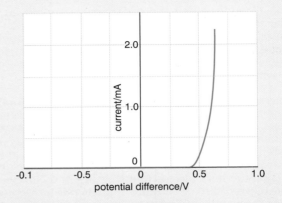

current/mA — *potential difference/V*

REMEMBER! Cover the answers if you want to.

ANSWERS

A1 The atoms in the metal have released their
outermost electrons into an electron cloud.
When a potential difference is applied these
electrons are free to move through the
structure, creating an electric current.

A2 a) As the temperature increases, the
metal ions vibrate more and more and
become a greater barrier for the
moving electrons. Hence the
resistance increases with temperature.

b) As the temperature increases,
electrons are 'knocked off' the silicon
atoms and a much greater cloud of
electrons is created. Consequently the
resistance of the silicon decreases as
the temperature increases.

A3 a)

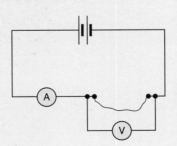

The resistance is calculated using
Ohm's law, $R = \frac{V}{I}$.

b) (i) The resistance would double.
(ii) The resistance would be halved.

TUTORIAL

T1 *These 'free' electrons are sometimes referred to as a
'sea' of electrons or more correctly as a delocalised
cloud. Metals are much more ready to lose electrons
than non-metals.*

T2 *a) Remember that the electrons must be free to move
through the structure. They will bump into the
ions even at very low temperatures. However, as
the ions vibrate more about their fixed positions
they will restrict the electron movement even more.*

*b) Semiconductors are affected by temperature in the
opposite sense to conductors. The restriction in the
movement of electrons by the increasingly
vibrating ions is overcome by the fact that many
more electrons are released into the electron cloud.*

T3 *a) Remember that the ammeter must be in series
with the nichrome wire, but the voltmeter must be
in parallel.*

*b) Resistance is directly proportional to length.
Resistance is inversely proportional to cross-
sectional area. If necessary look back at the
models shown on page 253.*

ANSWERS

A4
a) 24 V
b) 20 V
c) 0.12 A
d) 23 A
e) 60 Ω
f) 23 Ω

A5
a) The resistance increases as the voltage increases.
b) As the voltage increases the temperature of the wire must also be increasing.

A6
a) The resistance at 0 V is very high. It starts to decrease at about 0.4 V and becomes very low at 0.6 V.
b) The diode is not conducting – it has very high resistance.
c) 0.0016 A.

TUTORIAL

T4
a) $V = IR = 2 \times 12 = 24\,V$
b) $V = IR = 0.1 \times 200 = 20\,V$
c) $I = \dfrac{V}{R} = \dfrac{12}{100} = 0.12\,A$
d) $I = \dfrac{V}{R} = \dfrac{230}{10} = 23\,A$
e) $R = \dfrac{V}{I} = \dfrac{6}{0.1} = 60\,Ω$
f) $R = \dfrac{V}{I} = \dfrac{230}{10} = 23\,Ω$

T5
a) If the resistance were constant the line would be straight with a constant gradient. It curves in such a way that increasing voltage produces a smaller increase in the current, so resistance must be increasing.
b) This is a common characteristic of non-ohmic resistors – that is resistors that do not obey Ohm's law.

T6
a) The key word here was 'describe'. It is important to describe in words the key features of the graph.
b) You should remember that diodes only allow current to pass in one direction.
c) $I = \dfrac{V}{R} = \dfrac{1.0}{625} = 0.0016\,A$ or $1.6\,mA$.

POWER IN ELECTRICAL CIRCUITS

All electrical equipment has a **power** rating. The power indicates how many joules of energy are supplied each second. The unit of power used is the **watt**. Light bulbs often have power ratings of 60 W or 100 W. Electric kettles have ratings of about 2 kilowatts (2 kW = 2000 W). A 2 kW kettle supplies 2000 J of energy each second.

The power of a piece of electrical equipment depends on the voltage and the current.

$P = V\,I$

P = power in watts (W)
V = voltage in volts (V)
I = current in amps (A)

Worked example

What is the power of an electric toaster if a current of 7 A is obtained from a 230 V supply?

Write down the formula in terms of P: $\quad P = V\,I$
Substitute the values: $\quad P = 230 \times 7$
Work out the answer and write down the unit: $\quad P = 1610\,W$

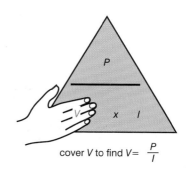

cover *V* to find $V = \dfrac{P}{I}$

Worked example

An electric oven has a power rating of 2 kW. What current will flow when the oven is used with a 230 V supply?

Write down the formula in terms of I:	$I = \dfrac{P}{V}$
Substitute the values:	$I = \dfrac{2000}{230}$
Work out the answer and write down the unit:	$I = 8.7\,A$

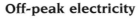

PAYING FOR ELECTRICITY

Electricity meters in the home and in industry measure the amount of energy used. They don't measure the energy in joules or even kilojoules. The unit used is a larger unit of energy, the kilowatt-hour (kWh).

> 1 kilowatt-hour (1 kWh) is the amount of energy transferred by a 1 kW device in 1 hour.
> 1 kWh = 3 600 000 J.

Worked example

Calculate the energy transferred by a 3 kW electrical immersion heater which is used for 30 minutes.

Write down the formula:	Energy = power (kW) × time (h)
Substitute the values:	Energy = 3 × 0.5
Work out the answer and write down the unit:	Energy = 1.5 kWh

Off-peak electricity

The electricity companies charge customers for the units of electricity they use. They cannot simply switch power stations off at night when the demand for electricity is low, so instead they try and balance the demand by encouraging people to use electricity at night. They do this by selling electricity at two different rates. Standard rate applies to electricity used during the day whereas an 'off-peak' rate can be applied to electricity used at night. Off-peak rate electricity is less than half the cost of standard rate electricity. To take advantage of this cheaper electricity a special meter has to be fitted. The cheaper off-peak electricity is usually available from midnight until 7 a.m. – hence it is sometimes called 'Economy 7' electricity. Electrical heating systems are often programmed to switch on at night and make use of this cheaper electricity. In this way the consumers benefit from cheaper electricity and the companies are able to smooth out the demand for electricity.

An off-peak white meter has two displays. The lower display shows the number of kWh units used at the standard rate. The upper display shows the units used at off-peak rate.

Worked example

A 3 kW night storage heater is switched on full power for 7 hours one night. The rate on the Economy 7 tariff is 4p per unit. Calculate the cost of using the heater.

Write down the formula:	Units = power (kW) × time (h)
Substitute the values:	Units = 3 × 7 = 21
Include cost of each unit:	Cost = units × 4p = 21 × 4 = 84p

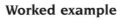

USING ELECTRICITY SAFELY

Electrical appliances can be damaged if too high a current flows through them. They need to be protected by a **fuse** which is usually fitted into the plug of the appliance. The plug is designed so that the electric current travelling to the appliance along the **live wire** has to first pass through the fuse. If a fault occurs in the appliance and this causes a sudden surge in the current the wire in the fuse will heat up and melt. The fuse 'blows'. This breaks the circuit and stops any further current flowing.

It is very important that an appliance is fitted with the correct fuse. The fuse must have a value above the normal current that the appliance needs but should be as small as possible. The most common fuses are rated at 3 A, 5 A and 13 A. As a rough rule of thumb any electrical appliance with a heating element in it should be fitted with a 13 A fuse.

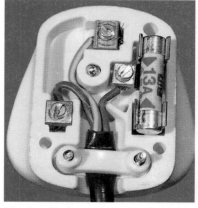

The fuse fits between the live brown wire and the pin. The brown live wire and the blue neutral wire carry the current. The green and yellow striped earth wire is needed to make metal appliances safer.

Worked example

What fuse should be fitted in the plug of a 2.2 kW electric kettle used with a supply voltage of 230 V?

Calculate the normal current: $I = \dfrac{P}{V} = \dfrac{2200}{230} = 9.6\,A$. Choose the fuse with the smallest rating bigger than the normal current. The fuse must be 13 A.

Worked example

What fuse should be fitted to the plug of a reading lamp which has a 60 W lamp and a supply of 230 V?

Calculate the normal current: $I = \dfrac{P}{V} = \dfrac{60}{230} = 0.26A$. Choose the fuse with the smallest rating bigger than the normal current. The fuse must be 3 A.

Other safety measures

In newer houses the traditional fuse box has been replaced by a box containing **circuit breakers**. These spring open ('trip') a switch if there is an increase in current in the circuit. Their advantage is that they can be reset easily once the fault in the circuit has been corrected.

Metal-cased appliances, as well as having a fuse, must also have an **earth wire**. In these appliances the fuse has two functions: firstly it protects the appliance from too high a current and secondly, along with the earth wire, it protects the user from being electrocuted. If the live wire of a cooker became loose and came into contact with the metal casing, the casing would become live and very dangerous. The earth wire prevents this from happening by providing a very low resistance route to the 0 V earth – usually water pipes buried deep underground. This low resistance means that a large current passes from the live wire to earth causing the fuse to melt and break the circuit. The earth wire and the fuse work together to protect the user.

Appliances that are made with plastic casing do not need an earth wire. The plastic is an insulator and so can never become live. Appliances like this are said to be **double insulated**.

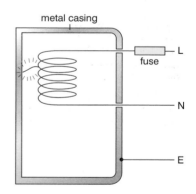

The earth wire and fuse work together to make sure that metal outer casing of this appliance can never become live and electrocute someone.

Check yourself

QUESTIONS

Q1 A hairdryer works on mains electricity of 230 V and takes a current of 4 A. Calculate the power of the hairdryer.

Q2
a) An iron has a power rating of 1000 W. Calculate the cost of using the iron for 3 hours if electricity costs 8p for a kilowatt-hour.
b) The power of a television set is 200 W. Calculate how much it costs to watch the television for 5 hours. A unit (kWh) of electricity costs 8p.

Q3 The diagram shows the inside of an electrical fan heater.

a) What is the job of the earth wire?
b) The plug contains a fuse. How does the fuse make the heater safer to use?
c) Explain why some electrical devices do not need an earth wire.

Q4 A lawn mower operates at 230 V and has a current of 7 A.
a) What size fuse should be fitted in the plug?
b) Calculate the power of the lawn mover.

Q5 A 3 kW immersion heater transfers energy to water for 10 minutes. How much energy does the heater transfer in 10 minutes?

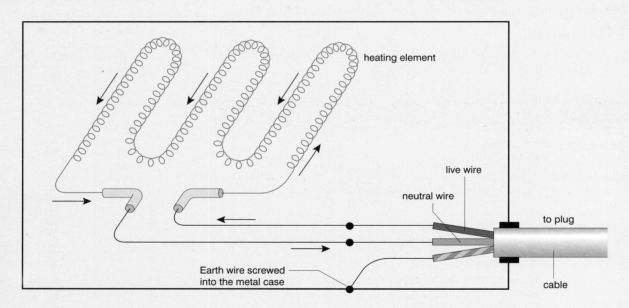

heating element

live wire

neutral wire

to plug

Earth wire screwed into the metal case

cable

REMEMBER! Cover the answers if you want to.

ANSWERS

A1 920 W.

A2
a) 24p.
b) 8p.

TUTORIAL

T1 *Power* = V × I = 230 × 4 = 920 W. *Don't forget the unit. Power is measured in watts.*

T2
a) *Cost = units × 8 = 1 × 3 × 8 = 24p. Remember that a unit is a kilowatt-hour so the 1000 W has to be converted into kilowatts, that is 1 kW.*
b) *200 W = 0.2 kW. Cost = units × 8 = 0.2 × 5 × 8 = 8p.*

ANSWERS

A3 a) If there is a fault in the heater and the metal case becomes live, the earth wire provides a low resistance route for the current to flow to earth.

b) If there is a large increase in the current the wire inside the fuse will melt and break the circuit. This stops the heater overheating and possibly catching fire.

c) Devices which have a plastic outer casing (where no metal can be touched) do not need to be earthed. The plastic is an insulator and so cannot become live.

A4 a) 13 A.

b) 1610 W.

A5 1800 kJ.

TUTORIAL

T3 a) *The cable may become frayed or the live wire may become loose and touch the side of the metal casing. As the high current flows to earth the fuse would blow, protecting the fan heater from any possible damage.*

b) *The fuse must have a rating as small as possible but above the size of current that usually flows through the heater.*

c) *These devices are said to be double insulated. A hairdryer is a good example of a double-insulated appliance.*

T4 a) *The fuse must have a rating of more than 7 A. Most common fuses are 3 A, 5 A or 13 A.*

b) *Power = V × I = 230 × 7 = 1610 W.*

T5 *Energy = Power × time(s) = 3000 × 10 × 60 = 1 800 000 J or 1800 kJ. Remember that 1 watt is 1 joule/second.*

STATIC ELECTRICITY

Electric charge doesn't flow in an insulator. However, you can put an electric charge on the surface of an insulator simply by rubbing the insulator with another insulator. For example, you can charge up a polythene ruler by rubbing it with a duster. The charge is fixed in place, which is why it is called **static electricity**.

Charging up a plastic ruler.

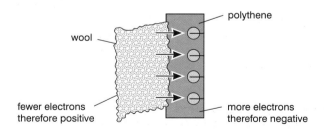

wool
polythene
fewer electrons therefore positive
more electrons therefore negative

When two insulators are rubbed together electrons from the atoms on the surface of one insulator are transferred to the surface of the other insulator. This leaves an excess of electrons – a negative charge – on one insulator and a shortage of electrons – a positive charge – on the other insulator.

Static charges interact in a simple way:

like charges repel – unlike charges attract.

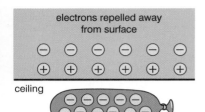

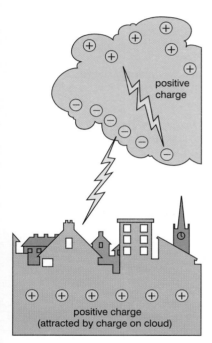

The balloon induces a charge on the ceiling's surface.

Thunder clouds have a high concentration of negative charge at the bottom. They have very high voltages. They ionise the air so that it conducts electricity. Lightning is caused by a burst of electric current through the air.

In a photocopier, charged particles attract the toner. Light is used to remove charge from parts that are not to be printed.

When a balloon is rubbed against clothing it will then 'stick' to a wall or ceiling. This is because of **electrostatic induction**. Rubbing the balloon transfers electrons to the balloon. It becomes negatively charged. When this is placed near the ceiling the electrons on the ceiling surface are repelled leaving the surface positively charged. The balloon 'sticks' because of the attraction between the negative charges on the balloon and the induced positive charges on the ceiling.

Disadvantages of static electricity

The disadvantages of static electricity are more familiar than the advantages. Air currents in thunder clouds move up and down very quickly rubbing past water molecules. This charges up the water molecules leaving the bottom of the thunder cloud with a strong negative charge. This induces a strong positive charge in buildings and trees on the ground. A strong electric field is created between the cloud and the ground. This tends to ionise the air until eventually there is a short burst of electric current through the air.

The sudden discharge of electric charge can cause shocks in everyday situations. For example:

- combing your hair
- pulling clothes over your head
- walking on synthetic carpets
- getting out of a car.

In all cases the charge originates from friction between two insulators. Sparks can also cause explosions and so great care has to be taken when emptying fuel tankers at service stations and airports or in grain silos or flour factories.

Advantages of static electricity

The properties of static electricity are put to good effect in paint spraying, ink-jet printers and photocopiers.

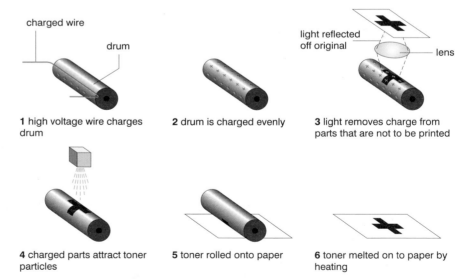

1 high voltage wire charges drum

2 drum is charged evenly

3 light removes charge from parts that are not to be printed

4 charged parts attract toner particles

5 toner rolled onto paper

6 toner melted on to paper by heating

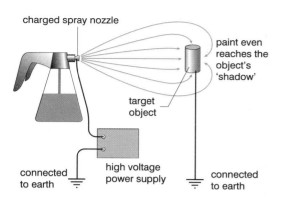

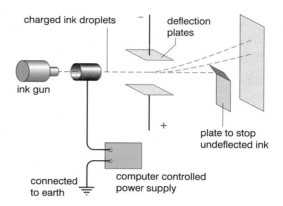

In electrostatic paint spraying, ions in the air collide with paint droplets and give them charge. Even parts of the object pointing away from the spray get painted.

In inkjet printers, uncharged ink droplets do not reach the paper. This is how the spaces between words are made.

Check yourself

QUESTIONS

Q1 A plastic rod is rubbed with a cloth.
a) How does the plastic become positively charged?
b) The charged plastic rod attracts small pieces of paper. Explain why this attraction occurs.

Q2 a) A car stops and one of the passengers gets out. When she touches a metal post she feels an electric shock. Explain why she feels a shock.
b) Write down two other situations where people might get this type of shock.

Q3 Give two examples of where static electricity can be dangerous.

Q4 The diagram shows one way of painting using a charged paint-gun.

The paint gun has a positive charge.
a) What kind of charge should the tube have? Explain your answer.
b) Suggest a reason why a charged paint gun produces a finer spray of paint than an uncharged gun.

REMEMBER! Cover the answers if you want to.

ANSWERS

A1 a) Electrons are rubbed off the surface of the plastic onto the cloth.
b) The positively charged rod induces a negative charge on the surface of the paper near to the rod. Opposite charges then attract. (A positive charge is therefore induced in the paper well away from the charged rod.)

TUTORIAL

T1 a) Static electricity is produced by removing electrons from one insulator to another. An excess of electrons leads to a negative charge.
b) Electrostatic induction is a common phenomenon. Remember, this is how you can make a balloon stick to the ceiling.

ANSWERS

A2

a) The passenger is charged by friction as her clothes rub against the seat covers. The car is also charged by friction with the road and the air. Touching metal allows the charge to flow to earth.

b) Touching a door handle after walking on a synthetic carpet, or removing clothing.

A3 Lightning, fuelling aircraft, in grain silos.

A4

a) Negative. To attract the paint.

b) The charged paint particles repel each other.

TUTORIAL

T2

a) *Remember static electricity is produced by friction.*

b) *Synthetic fibres are more likely to cause this effect than natural fibres. This must be due to the different atom arrangements.*

T3 *Static electricity can cause explosions in any situation when there are fuels in the gaseous form or where there are very fine particles of solid material (dust).*

T4 *In electrostatics you need to remember that unlike charges attract each other and like charges repel each other.*

EXAMINATION CHECKLIST

The facts and ideas that you should understand by studying this topic.

To be successful at Foundation GCSE Tier you should be able to:

- recall that an ammeter measures current and is connected in series
- use switches to control devices in series and parallel circuits
- recall the uses of the three wires in a mains cable
- explain how variable resistors can be used to change the current in a circuit
- recall that current involves a flow of charge
- explain the behaviour of simple circuits in terms of charge flow
- recall that batteries produce direct current (d.c.) and that mains electricity is supplied as alternating current (a.c.)
- recall that a voltmeter measures potential difference and is connected in parallel
- use the equation $V = I \times R$
- compare devices such as cookers and water heaters in terms of power
- recall that the unit of consumption of electrical power is the kilowatt-hour and be able to use the equation, units $= kW \times h$
- calculate the costs of using common electrical appliances
- use the equation for power, $P = V \times I$
- recall that there are two kinds of electrostatic charge, positive and negative
- recall that static electricity is caused by the movement of electrons
- recall examples of when static electricity is useful and when it is dangerous.

*In addition, to be successful at **Higher** GCSE Tier you should be able to:*

- explain how the fuse and earth wire work together to make some electrical equipment safe

- recall that current is the rate of flow of charge, $I = Q/t$

- recall that charge carriers in a circuit are either electrons or ions

- recall that resistors are heated when charge flows through them and the use of this in fuses

- recall the advantages of off-peak electricity for both producers and consumers

- recall that the kilowatt-hour is a measure of the energy supplied

- recall that voltage measures the energy transferred per unit charge flowing

- recall how current varies with voltage in an ohmic resistor, a filament bulb and a diode

- recall that diodes, LDRs and thermistors can control electric currents

- recall how light and temperature affect the resistance of LDRs and thermistors

- explain static electricity in terms of the movement of electrons, positive charges as a lack of electrons and negative charges as an excess of electrons.

Key Words

Tick each word when you are sure of its meaning.

alternating current	parallel
conductors	potential difference
current	power
diode	semiconductors
direct current	series
double insulated	static electricity
electrostatic induction	thermistor
insulators	voltage
light dependent resistor	

EXAM PRACTICE

Sample Student's Answers & Examiner's Comments

EXAMINER'S COMMENTS

a) (i) *All of the points have been plotted correctly (2 marks) and a line through the points has been drawn accurately. As the first set of points are clearly in a straight line, this part of the curve can be drawn with a ruler.*
(ii) *The straight line part of the graph has been correctly used to align a current of 1.8 A with 9 volts. The student has sensibly marked on the graph how this alignment has been done. The mark would not have been awarded if the unit had been missed.*

b) *Full marks have been awarded. The equation has been given (1 mark), the substitution has been done correctly (1 mark) and the answer worked out correctly. The final mark is only given because the correct units have been given also.*

c) *The candidate has only scored 1 mark here for stating that according to Ohm's law current is directly proportional to voltage. Insufficient explanation has been given to score further marks. Marks were available for any **two** of the following points: stating that as the voltage increases the temperature increases (1 mark); this increases the resistance of the resistor (1 mark) causing non-proportionality between voltage and current at the higher voltages (1 mark). This idea is commonly tested on Higher Tier papers.*

● *A mark of 8/10 corresponds to a grade A on this higher tier question.*

---

1 a) A student wanted to know what happened to the current flowing in a resistor when the voltage changed. She used the circuit shown in the diagram below. The readings she took are shown in the table.

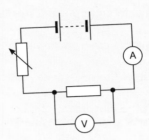

Voltage in volts	Current in amps
2.0	0.4
4.0	0.8
6.0	1.2
8.0	1.6
10.0	2.0
12.0	2.2
14.0	2.3

(i) Plot a graph of her results.

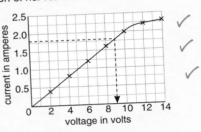

(3)

(ii) What voltage is needed to get a current of 1.8 A?

 9 volts ✓ (1)

b) What is the resistance of the resistor she is using? Include in your answer the equation you are going to use. Show clearly how you are going to get your final answer and give the unit.

 V = IR ✓ 9 = 1.8 x R R = 9/1.8 ✓

 R = 5 ohms ✓ (3)

c) Explain why the results she obtained did **not** result in a straight line graph.

 Ohm's law (V = IR) applies only to certain resistors, in this case current is not directly proportional to voltage and so a curve is formed. ✓

(3)

(Total 10 marks)

Southern Examining Group

Questions to Answer

Answers to questions 2 – 4 can be found on pages 369–370.

2 This question is about electric circuits.

The diagram shows an electric circuit. It has three switches, 1, 2 and 3. There are three identical lamps (bulbs) **A**, **B** and **C**.

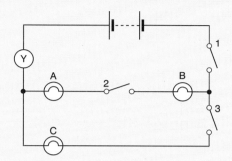

a) All the switches are open. Explain why none of the lamps are lit. (1)

b) Switches 1 and 2 are pressed at the same time (they are both closed). Which lamp(s) light? Choose from **A**, **B** and **C**. (1)

c) Switches 1 and 3 are pressed (closed) at the same time. Switch 2 is open. Which lamp(s) light? Choose from **A**, **B** and **C**. (1)

d) All three switches are pressed (closed) at the same time. Which lamp(s) will be the brightest? Choose from **A**, **B** and **C**. (1)

e) Two lamps are in series with each other. Which two? Choose from **A**, **B** and **C**. (1)

f) The part marked Y in the diagram measures the current in the circuit.

   (i)   What is the name we use for the part marked Y? (1)

   (ii)   What unit do we use for measuring current? (1)

g) Look at the diagram.

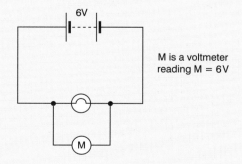

M is a voltmeter
reading M = 6V

(i) Look at the diagram below. The lamps and meter are identical.

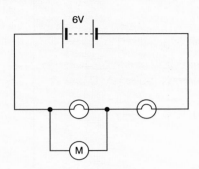

What will the meter read now? (1)

(ii) Look at the diagram below. The lamps and meter are identical.

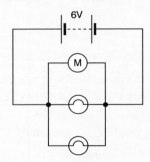

What will the meter read now? (1)

Midland Examining Group

3 This shows part of the mains wiring in a house.

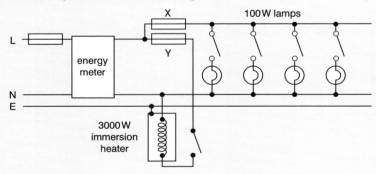

a) Why is fuse X rated at 5 A but fuse Y at 30 A? (2)

b) Explain what might cause the fuse labelled **X** to melt. (2)

c) Why are the lamps connected in parallel rather than in series? (2)

d) The cable to the immersion heater has an Earth wire in it.

Explain why this Earth wire improves safety. (2)

Midland Examining Group

4 The circuit below was used to measure the resistance of a lamp.

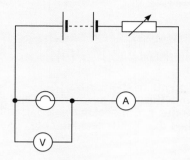

a) The first readings were:

voltmeter   4.0 V

ammeter   1.25 A

   (i)   Calculate the resistance of the lamp at this current.   (2)

   (ii)   If the resistance of the lamp did not change, what current
       would a voltage of 12 V drive through this lamp?   (2)

b) Further sets of readings were taken. The resistance of the
lamp was calculated for each set of readings.

The graph shows how the resistance of the lamp changes
with the current passing through it.

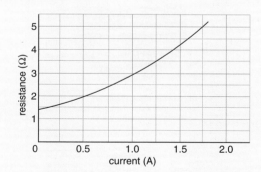

   (i)   Explain why the resistance changed with increasing current.   (2)

   (ii)   Use information from the graph to calculate the potential
       difference across the lamp when the current is 0.5 A.   (3)

   (iii)   Calculate the power of this lamp when the current passing
       through it is 0.5 A.   (2)

Northern Examination and Assessment Board

The attraction of opposite poles and the repulsion of like poles of a magnet is a very familiar characteristic of simple magnetism. In many everyday devices that depend on magnetism, such as electric motors and relays, the magnetic effect is created by an electric current. In devices such as dynamos and generators, however, magnetism is used to create electricity.

USING MAGNETISM

A magnetic compass always points to the Earth's Magnetic North Pole. The compass needle is made from a thin steel magnet. The end that points to the Magnetic North Pole is called the 'north-seeking pole', usually shortened to **N-pole**. The other end is called the 'south-seeking pole', usually shortened to **S-pole**.

The N-pole of a freely moving magnet will always point to the Magnetic North Pole.
Like poles repel. Unlike poles attract.

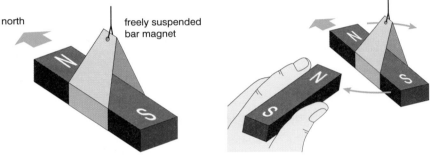

A magnet's N-pole is 'north-seeking' so it points to the Earth's Magnetic North Pole. This means that the magnetic field near the Earth's North Pole is similar to the field near a bar magnet's S-pole.

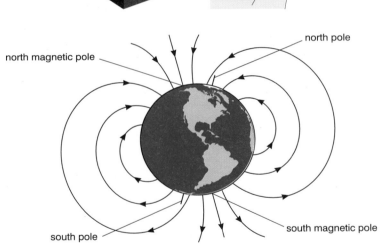

Certain materials are attracted to magnets. A common mistake is to think that all metals are magnetic. In fact, very few metals are magnetic. Only iron, cobalt, nickel and their alloys are magnetic.

The region round a magnet where its magnetic effect can be detected is called a **magnetic field**. The shape of a magnetic field can be determined by using iron filings or small plotting compasses.

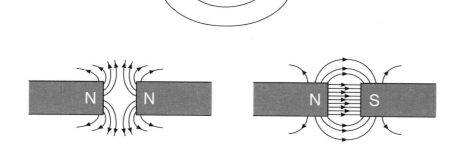

small compasses

Lines of magnetic field can be traced with a plotting compass. Field lines point from the north-seeking pole to the south-seeking pole.

The magnetic fields from two bar magnets interact with each other. Notice the difference between the field lines when the magnets are attracting and when they are repelling.

ELECTROMAGNETISM

An electric current flowing through a wire creates a magnetic field in the region of the wire. Magnetism created in this way is known as **electromagnetism**. The magnetic field is stronger if the wire is made into a coil. It is even stronger if the coil is wrapped around a piece of magnetic material such as iron. It is very easy, therefore, to make a simple electromagnet. The strength of the electromagnet can be increased by:

- increasing the number of coils in the wire
- increasing the current flowing through the wire
- placing a soft iron core inside the coils.

Relay switches

Electromagnets produce a magnetic field that can be turned on and off at will. A **relay** is an electromagnetic switch. It has the advantage of using a small current from a low-voltage circuit to switch on a higher current.

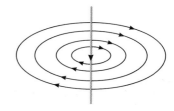

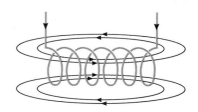

Magnetic fields are created around any wire that carries a current.

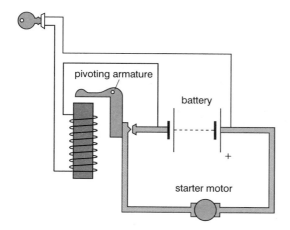

pivoting armature

battery

starter motor

A relay circuit used to switch on a starter motor.

A relay switch operates the starter motor of a car. Thin low-current wires are used in the circuit that contains the ignition switch operated by the driver. Much thicker wires are used in the circuit containing the battery and the high-current starter-motor. Turning the ignition key causes a current to flow through the relay coil. This creates a magnetic field which attracts the armature to the core. The armature pivots and connects the starter motor to the car battery.

D.C. MOTORS

An **electric motor** is a device that transfers electrical energy to kinetic energy. It is made from a coil of wire positioned between the poles of two permanent magnets. When a current flows through the coil of wire, it creates a magnetic field. This magnetic field interacts with the magnetic field produced by the two permanent magnets. The two fields exert a force that pushes the wire at right angles to the permanent magnetic field. This is the basis of the electric motor.

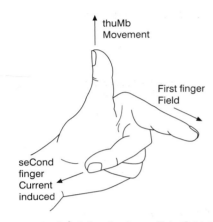

Fleming's left-hand rule predicts the direction of the force on a current carrying wire.

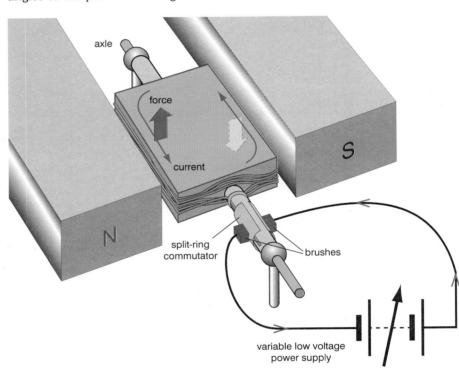

A motor coil as set up in the diagram will be forced round as indicated by the arrows (1 and 2 below). The split-ring commutator ensures that the motor continues to spin. Without the commutator, the coil would rotate 90° and then stop. This would not make a very useful motor! The commutator reverses the direction of the current at just the right point (3) so that the forces on the coil flip around and continue the rotating motion (4).

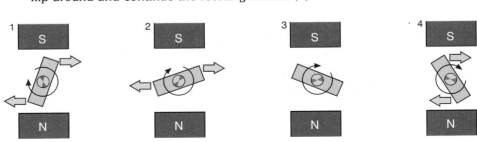

Check yourself

QUESTIONS

Q1

The circles in the diagram represent small magnetic compasses placed near a bar magnet. What would be the directions of the needles in the two compasses where the direction is not shown?

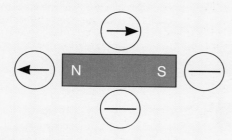

Q2

In a recycling plant an electromagnet separates scrap metal from household rubbish.

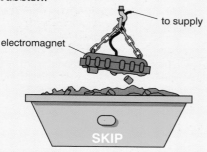

a) Can this method be used to separate aluminium drink cans from other household rubbish? Explain your answer.
b) How can the operator drop the scrap metal into the skip?

Q3

The diagram shows a simple electromagnet made by a student.

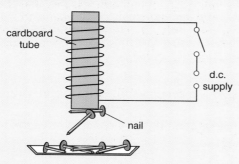

Suggest two ways in which the electromagnet can be made to pick up more nails.

Q4

Rachel placed a small magnetic compass close to the end of a coil of wire.

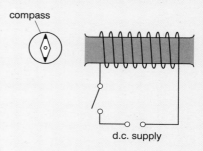

a) When the switch is closed what happens to the compass needle?
b) What happens to the compass needle when the current is switched off?
c) If the connections to the power supply were reversed and then the current switched on, what would happen to the compass needle now?

Q5

The diagram shows an electric bell.

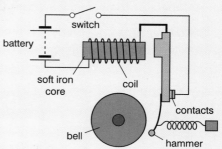

Explain how the bell works when the switch is closed.

ANSWERS

A1 The right hand compass should point to the S-pole. The lower compass should point to the right.

A2
a)   No. Aluminium is not magnetic.
b)   Switch off the current to the electromagnet.

A3 Increasing the current flowing through the coil. Increase the number of coils. Add a soft iron core inside the cardboard tube.

A4
a)   The compass needle will point towards or away from the coil.
b)   The compass needle will return to its original position.
c)   The compass needle would point in the opposite direction to that in a).

A5 When the switch is pressed the electromagnet attracts the hammer support and the hammer hits the bell. The movement of the hammer support breaks the circuit and so the electromagnet ceases to operate. The hammer support then returns to its original position forming the circuit again so the process is repeated.

TUTORIAL

T1 *The lines of the magnetic field around a bar magnet are shown on page 271. The field lines always travel from N to S.*

T2
a)   Remember that very few metals are magnetic, e.g. iron, cobalt and nickel. The electromagnet would separate steel cans as steel is an alloy containing iron.
b)   The magnetic field is only present when the current is flowing. An electromagnet is a temporary magnet.

T3 *These will produce a stronger magnetic field.*

T4
a)   Insufficient information is given for you to know whether the coil closest to the compass is an N-pole or an S-pole.
b)   Remember, an electromagnet is a temporary magnet.
c)   Reversing the direction of the current will change the direction of the magnetic field. Therefore the compass will point in the opposite direction to the way it did originally.

T5 *The electromagnet is constantly activated and then deactivated. This means that the hammer will continually hit the bell and then retract.*

ELECTROMAGNETIC INDUCTION

We have seen that an electric current can create a magnetic field. It was the famous physicist Michael Faraday who was the first person to do the reverse, that is, generate electricity from a magnetic field. This involves a process known as **electromagnetic induction**. The large **generators** in power stations generate the electricity we need using this process.
Current is created in a wire when:

- the wire is moved through a magnetic field ('cutting' the field lines)
- the magnetic field is moved past the wire (again 'cutting' the field lines)
- the magnetic field around the wire changes strength.

The faster these changes, the larger the current. Current created in this way is said to be **induced**.

A simple example of a current generator is a **dynamo**. A dynamo looks very like an electric motor. Turning the permanent magnet near to the coil induces or creates a current in the wires. This time, the split-ring commutator ensures that the current generated only flows in one direction. Power station generators don't have a commutator, so they produce alternating current. Power stations use electromagnets rather than permanent magnets.

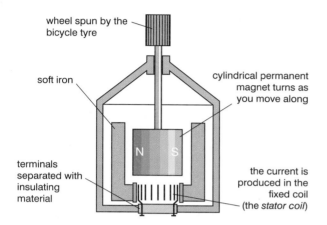

In a bicycle dynamo the magnet rotates and the coil is fixed.

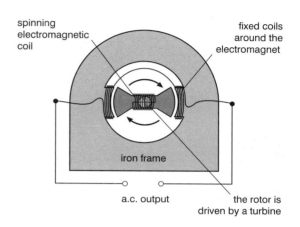

The generator rotates at 50 times per second producing a.c. at 50 hertz.

TRANSFORMERS

A **transformer** is a device consisting of two coils of insulated wire wound on a piece of iron. If an alternating voltage is applied to the first (primary) coil the alternating current produces a changing magnetic field in the core. This changing magnetic field induces an alternating current in the second coil. Hence an alternating voltage is created across the secondary coil.

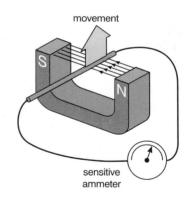

If there are more turns on the secondary coil than on the primary coil, then the voltage in the secondary coil will be greater than the voltage in the primary coil. The exact relationship between turns and voltage is:

$$\frac{\text{primary coil voltage } (V_p)}{\text{secondary coil voltage } (V_s)} = \frac{\text{number of primary turns } (N_p)}{\text{number of secondary turns } (N_s)}$$

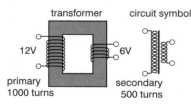

step-down transformer
ratio of number of turns is 2:1
voltage ratio is 2:1

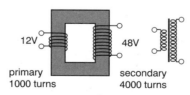

step-up transformer
ratio of number of turns is 1:4
voltage ratio is 1:4

Transformers are widely used to change voltages. They are frequently used in the home to step down the mains voltage of 230 V to 6 V or 12 V.

So when the secondary coil has more turns than the primary coil, the voltage increases in the same proportion. Such a transformer is referred to as a **step-up** transformer. A transformer with fewer turns on the secondary coil than on the primary coil is a **step-down** transformer which produces a smaller voltage in the secondary coil.

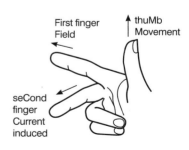

Fleming's right-hand rule predicts the direction of the current induced in a moving wire.

Worked example

Calculate the output voltage from a transformer when the input voltage is 230 V and the number of turns on the primary coil is 2000 and the number of turns on the secondary coil is 100.

Write down the formula:

$$\frac{V_p}{V_s} = \frac{N_p}{N_s}$$

Substitute the values known:

$$\frac{230}{V_s} = \frac{2000}{100} = 20$$

Rewrite this so that V_s is the subject:

$$V_s = \frac{230}{20}$$

Work out the answer and write down the unit:

$$V_s = 11.5\,V$$

TRANSMITTING ELECTRICITY

Most power stations use a fuel to change water into high pressure steam which is then used to drive a turbine. The turbine turns a generator which produces the electricity.

| Fuel burnt in furnace to heat water into steam | → | turbine is turned by the high pressure steam | → | the turbine turns an a.c. generator which produces the electricity |

The most common fuels used in power stations are still coal, oil and gas.

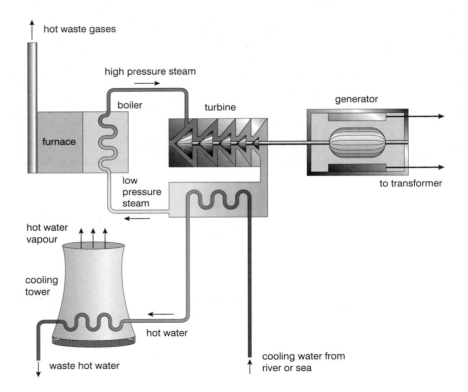

All the electricity that is generated is fed into the **National Grid** which links all the power stations to all parts of the country. To minimise the power loss in transmitting the electricity around the grid the current has to be kept as low as possible. The higher the current the more the transmission wires will be heated by the current. So the higher the current, the more energy is wasted as heat.

This is where transformers come in useful. This is also the reason that mains electricity is generated as alternating current – because transformers only work with an alternating current. When a transformer steps up a voltage it also steps down the current (current multiplied by voltage, $I \times V$, is identical in each coil) and vice versa.

Power stations generate electricity with a voltage of 25 000 V. Before this is transmitted on the grid it is converted by a step up transformer to 400 000 V. This is then reduced by a series of step down transformers to 230 V before it is supplied to homes.

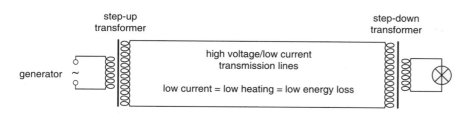

Mains electricity is a.c. so that it can be easily stepped up and down. High voltage/low current transmission lines lose less energy than low voltage/high current lines.

This familiar sight of an electricity sub-station is really a step down transformer.

Check yourself

QUESTIONS

Q1 Two students are using the equipment shown in the diagram.

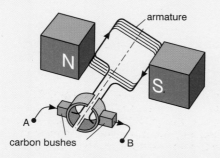

The students cannot decide whether it is an electric motor or a generator. Explain how you would know which it is.

QUESTIONS

Q2 The diagram shows a transformer.

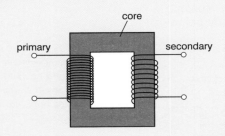

a)   What material is used for the transformer core?
b)   What happens in the core when the primary coil is switched on?
c)   What happens in the secondary coil when the primary coil is switched on?
d)   If the primary coil has 12 turns and the secondary coil has 7 turns, what will the primary voltage be if the secondary voltage is 14 V.

Q3 Explain why electricity is transmitted on the national grid at very high voltages.

REMEMBER! Cover the answers if you want to.

ANSWERS

A1 If the rotation of the armature is producing electricity which is being withdrawn via A and B, then it is a generator. If alternatively, electricity is being supplied via A and B to produce rotation then it is a motor.

A2
a)   Iron.
b)   A magnetic field is produced.
c)   A current is induced.
d)   24 V.

A3 An electric current produces a heating effect in a wire thus transferring energy as heat. To reduce energy loss electricity is therefore transmitted at as low a current as possible which means transmitting at very high voltage.

TUTORIAL

T1 *The same device can be used as either a motor or a generator. A motor needs an electrical supply in order to produce movement. In a generator the movement is used to generate electricity.*

T2
a)   *The core must be a magnetic material. It concentrates the magnetic effect.*
b)   *Remember that a current flowing in a wire will produce a magnetic field around it.*
c)   *The varying magnetic field in the core induces a current in the secondary coil.*
d) $\dfrac{V_p}{V_s} = \dfrac{N_p}{N_s}; \dfrac{V_p}{14} = \dfrac{12}{7}; V_p = \dfrac{12}{7} \times 14 = 24\,V.$

T3 *The heating effect of an electric current is considered in chapter 14.*

EXAMINATION CHECKLIST

The facts and ideas that you should understand by studying this topic.

To be successful at Foundation GCSE Tier you should be able to:

- recall that iron and steel can be attracted by magnets
- recall that like magnetic poles repel and unlike magnetic poles attract
- recall that electric motors use magnets and are found in a variety of everyday applications
- recall that a force is exerted on a current-carrying wire in a magnetic field
- recall that electricity can be generated by moving a wire in a magnetic field
- recall how simple a.c. generators work
- recall that transformers can step-up and step-down voltages
- recall that domestic electricity is generated by burning fuel in a power station and is transported to the consumer via the National Grid.

In addition, to be successful at Higher GCSE Tier you should be able to:

- recall the application of electromagnets in simple electric motors and relays
- recall that a voltage is induced in a coil when a nearby magnetic field changes
- use Fleming's left-hand rule to work out the direction of the movement of a current carrying wire in a magnetic field
- use Fleming's right-hand rule to work out the direction of the induced current when a wire 'cuts through' a magnetic field
- use the quantitative relationship between voltage across the coils in a transformer and the number of turns
- explain how electricity is generated at a power station
- recall that when electricity is distributed via the National Grid energy loss is reduced by transmitting at very high voltage.

Key Words

Tick each word when you are sure of its meaning.

dynamo	N-pole
electric motor	magnetic field
electromagnetic induction	relay
electromagnetism	S-pole
generator	transformer
National Grid	

EXAM PRACTICE

Sample Student's Answers & Examiner's Comments

a) The correct magnetic field pattern has been drawn and scores one mark. However, the direction of the field lines has been shown from S to N which is incorrect. This is a common error because in many text books the poles of a bar magnet are usually written with the N-pole on the left.

b) This has been answered correctly. For the N-pole of the compass needle to point north the Earth must behave as if it has an S-pole at that point.

c) The correct answer has been given. The magnetic field strength decreases the further you move from a magnet.

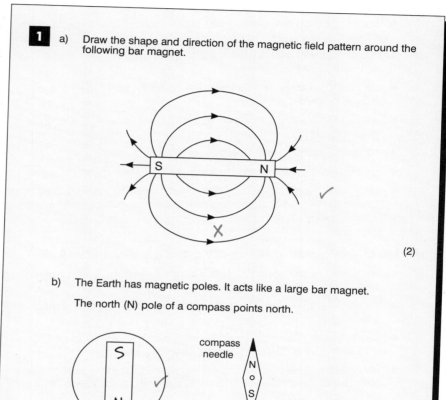

1 a) Draw the shape and direction of the magnetic field pattern around the following bar magnet.

(2)

b) The Earth has magnetic poles. It acts like a large bar magnet.

The north (N) pole of a compass points north.

compass needle

Label the poles of the magnet in the diagram of the Earth. (1)

c) What happens to the strength of the Earth's magnetic field as the distance from the Earth increases?

it decreases ✓

(1)

d) The following diagram shows an electromagnet.

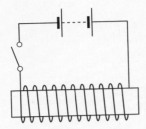

(i) Why is electromagnetism sometimes called temporary magnetism?

the core is only magnetic when the circuit is complete ✓ (1)

(ii) How could the strength of this electromagnet be increased?

increase the current ✓ .... (1)

e) The diagram below shows an electric bell.

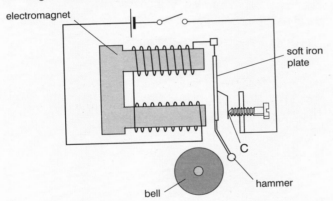

electromagnet

soft iron plate

C

hammer

bell

Explain how the electric bell can produce a continuous ringing sound.

When the switch is pressed the soft iron plate is attracted ✓
to the electromagnet and the hammer hits the bell
✓   ✓
....
.... (5)
....

(Total 11 marks)

London Examinations

d) (**i**) *The mark has been given. An alternative, and probably better, answer would be to say 'the core is only magnetic when a current is flowing through the coil'.*
(**ii**) *The mark has been given. Increasing the number of coils would be an alternative answer.*

e) *The candidate has correctly explained how the hammer hits the bell. What has not been explained is how the bell 'can produce a continuous ringing sound'. The two remaining marks would be given for saying 'as the hammer moves to the bell the contact at C is broken switching the electromagnet off' (1 mark) and 'the soft iron plate then springs back and the contact is made again' (1 mark). In these extended answer questions with a large number of marks available it is absolutely vital that you read the question carefully.*

● *A mark of 8/11 corresponds to a very good grade C on this foundation tier question.*

Questions to Answer

Answers to this question can be found on page 371.

2 a) (i) The diagram shows a transformer.

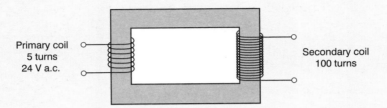

Primary coil
5 turns
24 V a.c.

Secondary coil
100 turns

Calculate the output voltage. Include in your answer the equation you are going to use. Show clearly how you get your final answer.

(2)

(ii) Electrical cables, connected to a transformer, supply a factory with its electrical energy. Explain why it is necessary for the cables to operate at as high a voltage as possible.

(3)

b) The relay circuit shown is used to switch on a car starter motor.

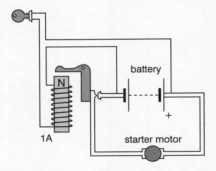

battery

N

1A   starter motor

Explain how this relay works when the key is turned.

(2)

c) When the switch in the circuit below is closed, in which direction will the bare wire AB move? Explain why.

(5)

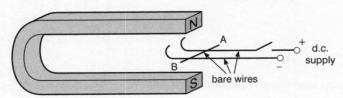

N

A

B

bare wires

+
−

d.c.
supply

S

Southern Examination Group

It is very unusual for a single force to be acting on an object. Usually there will be two or more. The size and direction of these forces determine whether the object will move and the direction it will move in. An understanding of the different types of forces and how they act has resulted in considerable advances in car safety. The way that forces act in fluids (liquids and gases), and the link between force and pressure, forms the basis of hydraulics.

THE EFFECT OF FORCES

Forces take many forms and have many effects including pushing, pulling, bending, stretching, squeezing and tearing. In short, forces can:

- change the speed of an object
- change the direction of movement of an object
- change the shape of an object.

Forces are measured in **newtons**.

Friction is a very common force. It is the force that tries to stop movement between touching surfaces. Friction will oppose the force that causes the movement. In many situations friction can be a disadvantage, e.g. friction in the bearings of a bicycle wheel, and lubricants are used to reduce the frictional force. In other situations friction can be an advantage, e.g. between brake pads and a bicycle wheel, or between the bicycle wheel and the ground.

Weight is another common force and because of this, scientists insist it is measured in newtons. The weight of an object depends on two things: its mass (the amount of stuff in it) and **gravity**. Any mass near the Earth has weight due to the Earth's gravitational pull. If you travel to the Moon, your **mass** remains constant but your weight drops! The force of attraction of an object to the Moon is about one sixth of that on the Earth. This is because the gravity on the Moon is one sixth of that on the Earth. An astronaut weighing 900 N on the Earth weighs 150 N on the Moon. Far away from the force of attraction of a planet, an object can be weightless. That is, the gravity is zero and the object has no weight.

The blades of a bobsleigh are polished to reduce friction between the track and the sled. But the runners initially have to push the sleigh so it important to have a lot of friction between their shoes and the ice. Their shoes have dozens of tiny spikes in them to maintain a firm grip on the ice.

Who cares how much you weigh? It's your mass that people really care about.

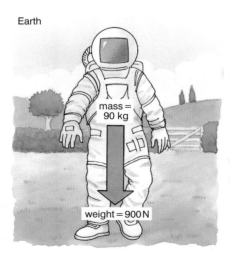

Earth

mass = 90 kg

weight = 900 N

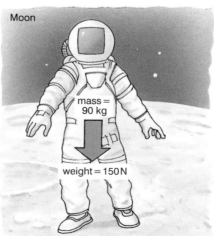

Moon

mass = 90 kg

weight = 150 N

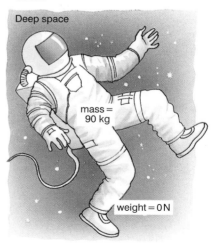

Deep space

mass = 90 kg

weight = 0 N

STRETCHING FORCES

Certain materials are **elastic**. An elastic material will return to its original length after stretching. Hooke's law shows the relationship between the force applied to an elastic material and the amount of stretching produced. The force applied is proportional to the extension it creates (force ∝ extension), which is often written as:

force = constant × extension
$F = k x$
F = force, k = constant, x = extension

As the force gets larger, the effect stops being proportional at a point known as the 'limit of proportionality'. If the force gets larger still, the material can be stretched so much that it will not return to its original length. This limit of its elastic properties is known as the **elastic limit**. Seat belts in cars should always be replaced after a crash as the belts are likely to have stretched beyond their elastic limit.

When a material reaches the elastic limit, it will no longer return to its original length – it has stopped being elastic.

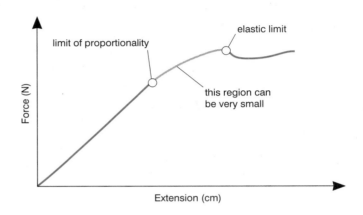

BALANCED FORCES

Usually there will be at least two forces acting on an object. If these two forces are **balanced** then the object will either be stationary or moving at a constant speed.

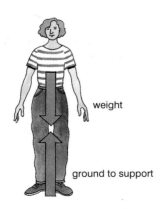

This person is stationary because her weight, acting towards the centre of the Earth, is balanced by the ground's supporting force pushing upwards.

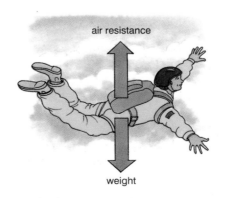

This skydiver is moving at a constant speed because the weight (the force due to gravity) is balanced by the force created by air resistance.

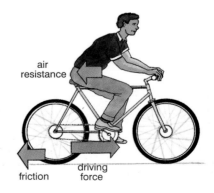

This cyclist is travelling at a constant speed because the driving force is equally balanced by the force caused by air resistance and friction.

A space craft in deep space, such as Voyager, will have no forces acting on it. There will be no air resistance, no force of gravity, and no need to produce a forward force from its rockets – it will travel at a constant speed.

UNBALANCED FORCES

To change the speed or direction of movement of an object the forces acting on it must be **unbalanced**.

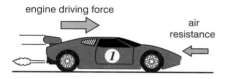

Acceleration. The force provided by the engine is greater than the force provided by air resistance and so the dragster increases its speed.

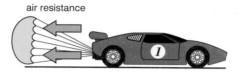

Deceleration. The engine is now providing no forward force. The 'drag' force provided by the parachute will slow the dragster down.

Voyager usually travels at constant speed in deep space without using its rockets. There are no forces acting on it.

As a gymnast first steps on to a trampoline, his weight is much greater than the supporting force of the trampoline, so he starts to move downwards, stretching the trampoline. As the trampoline stretches, its supporting force increases. Only when the trampoline's supporting force is equal to the gymnast's weight, but acting upwards, does the trampoline stop stretching. If an elephant stood on the trampoline it would stretch and eventually break because it could never produce a support force equal to the elephant's weight.

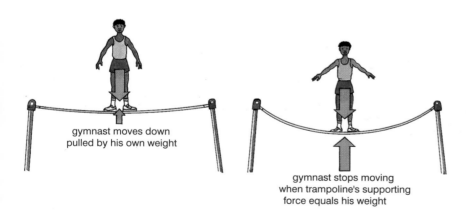

gymnast moves down
pulled by his own weight

gymnast stops moving
when trampoline's supporting
force equals his weight

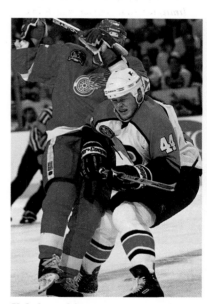

Unbalanced forces can cause a change in direction!

A trampoline stretches until it supports the weight on it.

As a skydiver jumps from a plane, the force of gravity will be much greater than the opposing force caused by air resistance. The skydiver's speed will increase rapidly. The force caused by the air resistance will increase as the skydiver's speed increases. Eventually the force due to the air resistance will exactly match the force of gravity. At this point the forces will be balanced and the speed of the skydiver will remain constant. This speed is known as the **terminal speed**.

If the skydiver spreads him or herself so that a greater surface is in contact with the air, the resistive force will be greater and the terminal speed will be lower than if he or she adopted a more compact shape. A parachute produces a very large resistive force and so the terminal speed of a parachutist is quite low. This means that they can land with relative safety.

Check yourself

QUESTIONS

Q1 Look at the diagrams A, B and C. In each case describe the effect the forces would have on the object.

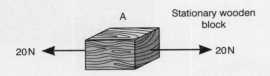

A
Stationary wooden block
20N ← → 20N

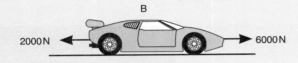

B
2000N ← → 6000N

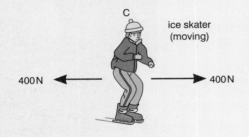

C
ice skater (moving)
400N ← → 400N

Q2 Skiers can travel at very high speeds. A friction force acts on the skis.

a) Use an arrow on the diagram to show the direction of the friction force acting on the skier.

b) Explain why it is important for the friction forces to be as small as possible.

c) Describe one way the skier could reduce the friction force.

Q3 The diagram shows the stages in the descent of a sky diver.

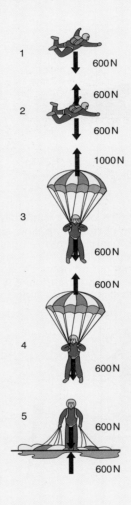

1 600N

2 600N
   600N

3 1000N
   600N

4 600N
   600N

5 600N
   600N

a) Describe and explain the motion of the skydiver in each stage.

b) In stage 5 explain why the parachutist doesn't sink into the ground.

Q4 A spring is 80 cm long when a force of 60 N is applied. It is 120 cm long when 120 N is applied. Calculate:

a) the length of the spring when no force is applied,

b) the extension of the spring when 90 N is applied.

ANSWERS

A1 A – the wooden block would not move. B – the car would accelerate. C – the ice skater would maintain a constant speed.

A2 a)

b) The frictional forces slow the skier down.

c) Using narrower skis, or using wax on the skis.

A3 a) Stage 1 – The skydiver is accelerating. The downward force of gravity is greater than the upward force caused by air resistance. Stage 2 – The sky diver is travelling at constant speed. The forces of gravity and air resistance are balanced. Stage 3 – The sky diver is slowing down. The force caused by the air in the parachute is greater than the force of gravity. Stage 4 – The sky diver is travelling at a constant speed. The forces are balanced again.

b) The force of gravity is balanced by the upward force of the ground on the sky diver.

A4 a) 40 cm.
b) 60 cm.

TUTORIAL

T1 A – The forces are balanced and so the stationary wooden block would not move. B – The resultant force would be 4000 N. This unbalanced force will cause the car to accelerate. (As the car moves faster the resistive force caused by air resistance will increase and eventually reach 6000 N.) C – The forces are balanced. As the ice skater is moving she will continue to do so at the same speed.

T2 a) Friction will always oppose motion and so the arrow points in the opposite direction to the direction of motion.

b) As the skier travels downhill the friction force opposes the motion.

c) Lubricant reduces frictional forces.

T3 a) In questions of this sort the first thing to do is to decide which forces will be acting on the object. The next thing is to decide whether the forces are balanced or unbalanced. If they are unbalanced then the object will be either accelerating or decelerating. If they are balanced the skydiver will either be travelling at constant speed or not moving. In Stage 2 the sky diver will have reached the terminal speed and because the forces are balanced will not accelerate or decelerate.

b) The upward force from the ground is equal to the sky diver's weight.

T4 a) Increasing the force by 60 N (from 60 N to 120 N) has increased the length of the spring by 40 cm. Therefore the original force of 60 N must have also caused an extension of 40 cm. Hence the original length of the spring must be 80 − 40 = 40 cm.

b) A force of 90 N would produce an extension mid-way between 60 cm and 120 cm. It is a common error for candidates to write down the length instead of the extension and vice versa. Always read the question carefully and give what is asked for.

VELOCITY AND ACCELERATION

The **speed** of an object can be calculated using the following formula:

$$\text{speed} = \frac{\text{distance}}{\text{time}}$$

$$v = \frac{s}{t}$$

v = speed in m/s
s = distance in m
t = time in s

Velocity is almost the same as speed. It has a size (called speed) and a direction.

Worked example

Calculate the average speed of a motor car that travels 500 m in 20 seconds.

Write down the formula: $\quad v = \frac{s}{t}$

Substitute the values for s and t: $\quad v = \frac{500}{20}$

Work out the answer and write down the unit: $\quad v = 25$ m/s

Worked example

A horse canters at an average speed of 5 m/s for 2 minutes. Calculate the distance it travels.

Write down the formula in terms of s: $\quad s = v \times t$
Substitute the values for v and t: $\quad s = 5 \times 2 \times 60$
Work out the answer and write down the unit: $\quad s = 600$ m

ACCELERATION

Acceleration is a measure of how much the speed changes in a certain time. Acceleration also occurs when an object changes direction, i.e. when the velocity changes. It can be calculated using the following formula:

$$\text{acceleration} = \frac{\text{change in speed}}{\text{time taken}}$$

$$a = \frac{(v - u)}{t}$$

a = acceleration in m/s/s
v = final speed in m/s
u = starting speed in m/s
t = time in s

Note that the units of acceleration 'm/s/s' (metres per second per second) is sometimes written as 'm/s^2' (metres per second squared).

Worked example

Calculate the acceleration of a car that travels from 0 to 28 m/s in 10 seconds.

Write down the formula: $\quad a = \frac{(v - u)}{t}$

Substitute the values for v, u and t: $\quad a = \frac{(28 - 0)}{10}$

Work out the answer and write down the unit:. $\quad a = 2.8$ m/s/s

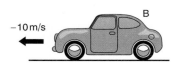

Both cars have the same speed. Car A has a velocity of +10 m/s, car B has a velocity of –10 m/s.

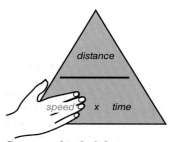

Cover speed to find that
$$speed = \frac{distance}{time}$$

GRAPHING MOTION

Distance–time graphs

Journeys can be summarised using graphs. The simplest type is a distance–time graph where the distance travelled is plotted against the time of the journey.

Steady speed is shown by a straight line. Steady acceleration is shown by a smooth curve.

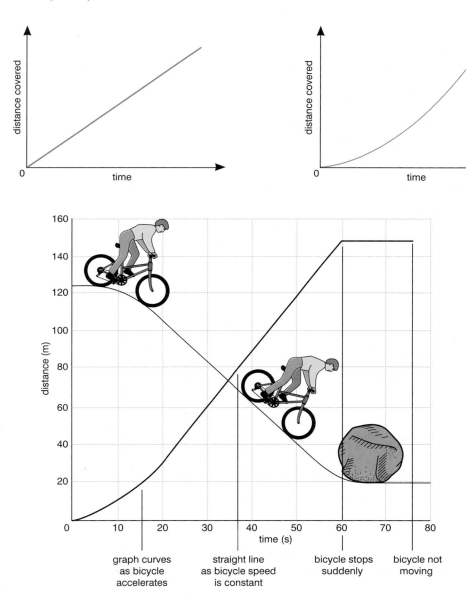

graph curves as bicycle accelerates

straight line as bicycle speed is constant

bicycle stops suddenly

bicycle not moving

A distance–time graph for a bicycle travelling down a hill. The graph slopes when the bicycle is moving. The slope gets steeper when the bicycle goes faster. The slope is straight (has a constant gradient) when the bicycle's speed is constant. The line is horizontal when the bicycle is at rest.

Speed–time graphs

Steady speed is shown by a horizontal line. Steady acceleration is shown by a line sloping up.

A speed–time graph provides information on speed, acceleration and distance travelled.

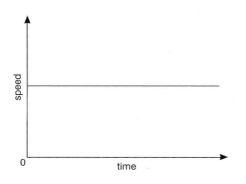

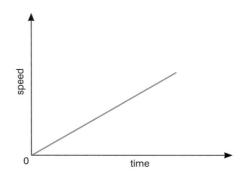

A speed–time graph for a car travelling between two sets of traffic lights.

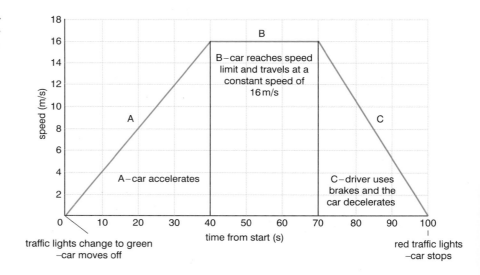

B–car reaches speed limit and travels at a constant speed of 16 m/s

A–car accelerates

C–driver uses brakes and the car decelerates

traffic lights change to green –car moves off

red traffic lights –car stops

The graph shows a car travelling between two sets of traffic lights. It can be divided into three regions. In region A, the car is accelerating at a constant rate (the line has a constant positive gradient). The distance travelled by the car can be calculated:

> average velocity = (16 + 0)/2 = 8 m/s
> time = 40 s
> so, distance = $v \times t$ = 8 × 40 = 320 m

This can also be calculated from the area under the line ($\frac{1}{2}$ base × height = $\frac{1}{2}$ × 40 × 16 = 320 m). In region B, the car is travelling at a constant speed (the line has a gradient of zero). The distance travelled by the car can be calculated:

> velocity = 16 m/s
> time = 30 s
> so, distance = $v \times t$ = 16 × 30 = 480 m

This can also be calculated from the area under the line (base × height = 30 × 16 = 480 m). In region C, the car is decelerating at a constant rate (the line has a constant negative gradient). The distance travelled by the car can be calculated:

> average velocity = (16 + 0)/2 = 8 m/s
> time = 30 s
> so, distance = $v \times t$ = 8 × 30 = 240 m

This can also be calculated from the area under the line ($\frac{1}{2}$ base $\times$ height = $\frac{1}{2} \times 30 \times 16 = 240$ m).

THINKING, BRAKING AND STOPPING DISTANCES

When a car driver has to brake it takes time for him or her to react. During this time the car will be travelling at its normal speed. The distance it travels in this time is called the **thinking distance**. The driver then puts on the brakes. The distance the car travels while it is braking is called the **braking distance**. The overall stopping distance is made up from the thinking distance and the braking distance. The thinking distance can vary from person to person and from situation to situation – the braking distance can vary from car to car. Other factors include:

Factors affecting thinking distance	Factors affecting braking distance
• speed	• speed
• tiredness	• condition of tyres (amount of tread)
• alcohol	• condition of brakes
• medication, drugs	• road conditions (dry, wet, icy, gravel, etc.)
• level of concentration and distraction.	• mass of the car.

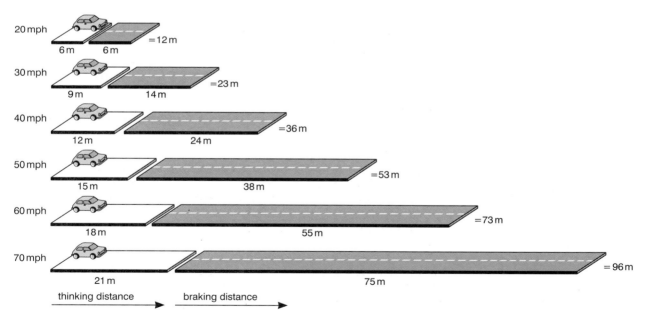

	thinking distance	braking distance	
20 mph	6 m	6 m	=12 m
30 mph	9 m	14 m	=23 m
40 mph	12 m	24 m	=36 m
50 mph	15 m	38 m	=53 m
60 mph	18 m	55 m	=73 m
70 mph	21 m	75 m	=96 m

Shortest stopping distances for a car on a good, dry road.

MASS, FORCE AND ACCELERATION

The acceleration of an object depends on its mass and the force that is applied to it. The relationship between these three factors is given by the formula:

force = mass × acceleration   F = force in newtons
$F = m\,a$ m = mass in kg
                                           a = acceleration in m/s/s

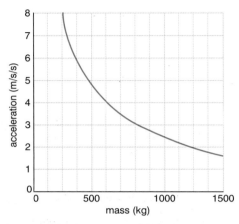

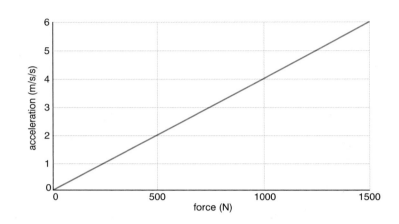

Acceleration is directly proportional to force. Acceleration is inversely proportional to mass.

Worked example

What force would be required to give a mass of 5 kg an acceleration of 10 m/s/s?

Write down the formula: $F = m\,a$
Substitute the values for m and a: $F = 5 \times 10$
Work out the answer and write down the unit: $F = 50\,N$

Worked example

A car has a resultant driving force of 6000 N and a mass of 1200 kg. Calculate the car's initial acceleration.

Write down the formula in terms of a: $a = \dfrac{F}{m}$

Substitute the values for F and m: $a = \dfrac{6000}{1200}$

Work out the answer and write down the unit: $a = 5\,m/s/s$

Check yourself

QUESTIONS

Q1 The graph shows a distance–time graph for a journey.

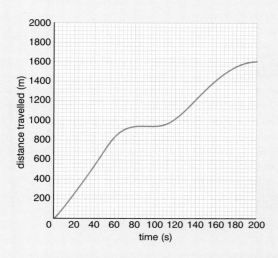

a) What does the graph tell us about the speed of the car between 20 and 60 seconds?

b) How far did the car travel between 20 and 60 seconds?

c) Calculate the speed of the car between 20 and 60 seconds.

d) What happened to the car between 80 and 100 seconds?

Q2 Look at the velocity–time graph for a toy tractor.

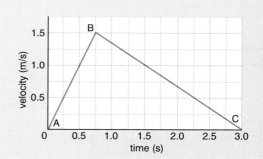

a) Calculate the acceleration of the tractor from A to B.

b) Calculate the total distance travelled by the tractor from A to C.

Q3 a) Explain the terms: (i) thinking distance, (ii) braking distance.

b) List three factors that can affect the braking distance of a lorry.

Q4 The manufacturer of a car gave the following information:

Mass of car 1000 kg. The car will accelerate from 0 to 30 m/s in 12 seconds.

a) Calculate the average acceleration of the car during the 12 seconds.

b) Calculate the force needed to produce this acceleration.

······ **REMEMBER! Cover the answers if you want to.** ······

ANSWERS

A1 a) The speed remained constant.

b) 600 m.

c) 15 m/s

d) It stopped.

TUTORIAL

T1 a) *If the line on a distance–time graph has a constant gradient then the speed is constant.*

b) *This can be read off the graph: after 20 s the car had travelled 200 m, after 60 s it had travelled 800 m. So the distance travelled is 800 − 200 = 600 m.*

c) *Speed = $\dfrac{distance}{time}$, $v = \dfrac{600}{40} = 15$ m/s.*
Don't forget the units.

d) *The line is horizontal during this time indicating that the car was not moving (no distance was travelled).*

ANSWERS

A2 a) 2.0 m/s/s.
b) 2.25 m.

A3 a) (i) The distance travelled during the reaction time of the driver.
(ii) The distance travelled from the point where the brakes are applied to the point where the vehicle stops.
b) Speed, tyre conditions, road conditions, brake conditions, mass. (Any three.)

A4 a) 2.5 m/s/s
b) 2500 N.

TUTORIAL

T2 a) Acceleration $= \dfrac{change\ in\ velocity}{time}$, $a = \dfrac{(1.5-0)}{0.75}$,
$a = 2$ m/s/s.
b) Total distance $=$ area under the line
$= (\frac{1}{2} \times 0.75 \times 1.5) + (\frac{1}{2} \times 2.25 \times 1.5)$
$= 2.25$ m.

Note: the tractor showed constant acceleration from A to B and then constant deceleration from B to C.

T3 a) Try not to confuse thinking distance with thinking time. Factors affecting thinking distance are shown on page 291. Remember that it is a distance not a time. Note: the overall stopping distance = thinking distance + braking distance.
b) If you are still not clear about the difference between thinking distance and braking distance look back at the section on page 291.

T4 a) Acceleration $= \dfrac{change\ in\ speed}{time}$, $a = \dfrac{(30-0)}{12}$
$= 2.5$ m/s/s. Don't forget the units!
b) Force $=$ mass $\times$ acceleration, F $= 1000 \times 2.5$
$= 2500$ N.

PRESSURE AND HYDRAULICS

Car brakes use a large force to stop a car but the driver uses a much smaller force on the brake pedal. The braking system in a car depends on a hydraulic pressure system which multiplies the force exerted by the driver.

You can push a drawing pin into some wood but it is much more difficult to push a nail into the same piece of wood, even if you are able to exert the same force. The reason is that the tip of the drawing pin and the nail have different surface areas. The same force is acting over different areas, creating different pressures. **Pressure** is defined as the force per unit area:

pressure $= \dfrac{force}{area}$

$P = \dfrac{F}{A}$

P $=$ pressure in N/m^2
F $=$ force in N
A $=$ area in m^2.

Worked example

A box weighs 500 N and has a base of area 0.5 m^2. Calculate the pressure the box exerts on the ground.

Write down the formula: $P = \dfrac{F}{A}$

Substitute the values for F and A: $P = \dfrac{500}{0.5}$

Work out the answer and write down the unit: $P = 1000$ N/m^2

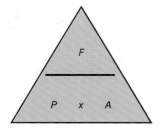

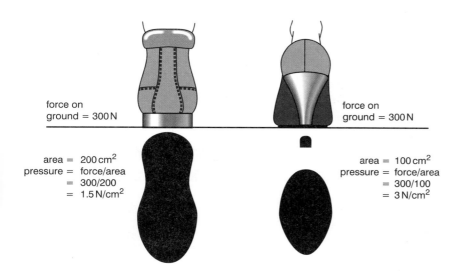

The same force is acting on the heel of each shoe. The heel on the right has a smaller area and so exerts a greater pressure on the ground.

force on ground = 300 N

force on ground = 300 N

area = 200 cm²
pressure = force/area
= 300/200
= 1.5 N/cm²

area = 100 cm²
pressure = force/area
= 300/100
= 3 N/cm²

Worked example

The heel of a shoe exerts a pressure of 20 N/cm² on the ground. If the area of the heel is 25 cm² calculate the force being applied.

Write down the formula with F as the subject: $F = P \times A$
Substitute the values for P and A: $F = 20 \times 25$
Work out the answer and write down the unit: $F = 500 N$

HYDRAULICS

Solids exert a downward pressure due to their weight. Liquids and gases exert pressure in all directions. Unlike gases, however, liquids cannot be compressed. The particles in a liquid are close together and cannot be pushed any closer. This means that pressure can be transmitted through a liquid. This is the basis of **hydraulics**.

Two different sized syringes are filled with liquid and connected together by a thin tube as shown in the diagram. If a force of 20 N is applied to the plunger of the smaller syringe (area of 2 cm²) the force exerted by the plunger of the larger syringe (area = 12 cm²) is 120 N.

This can be explained in the following way:

The force applied to the smaller syringe produces a pressure in the liquid:

$$P = \frac{F}{A} = \frac{20}{2} = 10 \text{ N/cm}^2$$

This pressure is transmitted equally throughout the liquid:

P = 10 N/cm² in both syringes.

This pressure then creates a force on the plunger of the larger syringe:

F = P × A = 10 × 12 = 120 N

So, a small force acting on a small area can create a big force acting over a big area. Earth diggers use hydraulic systems to generate the very large forces they need.

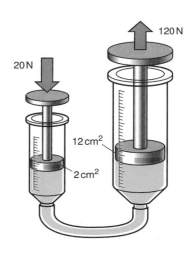

120 N

20 N

12 cm²

2 cm²

Hydraulic disc brakes. The force from pushing the brake pedal acts on a piston in a cylinder with small surface area. The pressure is transmitted to larger cylinders and so a greater force is exerted by the pistons in these cylinders on the disc brake pads.

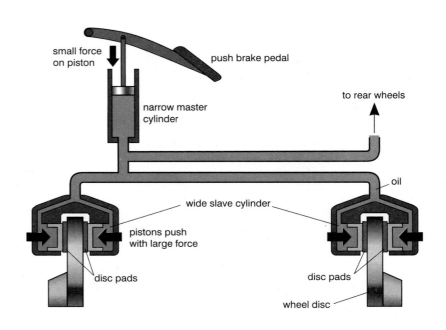

COMPRESSING GASES

Gases squash when pressure is applied. The compressibility of gases is used in car air bags. When a crash occurs the air bag quickly fills with gas. The inflated air bag allows the person's head to decelerate gently – instead of very quickly when it hits the windscreen or dashboard. A slower deceleration reduces the force exerted and the risk of injury.

The relationship between the volume of a gas and its pressure at constant temperature is given by Boyle's law. The law states that pressure is inversely proportional to volume (pressure $\propto$ 1/volume). This is more usually written as:

$$\text{pressure} = \frac{\text{constant}}{\text{volume}}$$

$$P V = k \ \text{ or } \ P_1V_1 = P_2V_2$$

P = pressure in N/m^2
V = volume in dm^3 (litres)
k = constant.

Worked example

The air in a bicycle pump has a pressure of 8 N/m^2 and a volume of 100 cm^3. It is compressed to a volume of 50 cm^3 before it is released into the tyre. What would be the pressure of the air in the pump now?

Write down the formula in terms of P_2:

$$P_2 = \frac{P_1V_1}{V_2}$$

Substitute the values for P_1, V_1 and V_2:

$$P_2 = \frac{(8 \times 100)}{50}$$

Work out the answer and write down the unit: $P_2 = 16\,N/m^2$

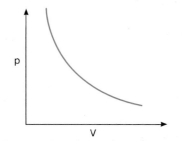

Pressure is inversely proportional to volume. If the pressure is doubled the volume is halved.

Check yourself

QUESTIONS

Q1 What is the pressure exerted by a girl weighing 450 N standing on the heel of one shoe with an area of 2 cm²?

Q2 A machine has been constructed that will lift a 200 N weight using a smaller force.

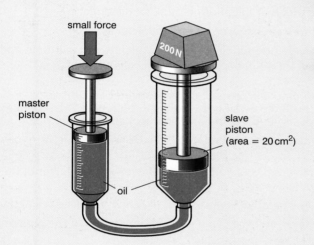

small force

200 N

master piston

slave piston (area = 20 cm²)

oil

a) Calculate the pressure of the slave piston on the oil.

b) The area of the master piston is 3 cm². What force must be applied to the master piston to lift the 200 N weight?

Q3 A volume of gas occupies 5 litres at a pressure of 100 000 N/m². What volume would the gas occupy if the pressure was increased to 250 000 N/m²?

Q4 An air bag is filled with air. The driver's head hits it hard.
a) What happens to the volume of gas in the air bag?
b) What happens to the pressure of the gas in the air bag?
c) What happens to the movement of the head? Why does this help to reduce injury?

REMEMBER! Cover the answers if you want to.

ANSWERS

A1 225 N/cm².

A2
a) 10 N/cm².
b) 30 N.

A3 2 l.

A4
a) The volume decreases.
b) The pressure increases.
c) The head decelerates gently. Smaller forces are involved in the smaller acceleration, so less injuries occur.

TUTORIAL

T1 $P = \dfrac{F}{A} = \dfrac{450}{2} = 225$ N/cm²

T2
a) $P = \dfrac{F}{A} = \dfrac{200}{20} = 10$ N/cm²

b) $F = P \times A = 10 \times 3 = 30$ N
Remember that in a liquid the pressure is transmitted equally throughout the liquid.

T3 $P_1V_1 = P_2V_2; \ V_2 = P_1V_1/P_2$
$V_2 = \dfrac{(100\,000 \times 5)}{250\,000} = 2\,l.$
You don't need to worry about the units of volume and pressure as long as they are the same on both sides of the equation.

T4 *Boyle's law states that the volume of a certain amount of gas is inversely proportional to the pressure providing the temperature remains constant.*

Key Words

Tick each word when you are sure of its meaning.

acceleration
balanced forces
braking distance
elastic
elastic limit
friction
gravity
hydraulics
mass
pressure
speed
terminal speed
thinking distance
unbalanced forces
velocity
weight

EXAMINATION CHECKLIST

The facts and ideas that you should understand by studying this topic.

To be successful at Foundation GCSE Tier you should be able to:

- recall and use the formula $v = s/t$ to calculate speed
- draw simple graphs of distance against time and speed against time
- explain what the steepness of distance–time and speed–time graphs represents
- recall the factors that might increase thinking distance
- recall the factors that might affect braking distance
- recall the difference between speed and velocity
- recall that acceleration is the change in velocity per unit time and be able to use the formula, acceleration = change in speed/time taken
- recall that an object keeps still or moves with constant velocity when either no force acts on it or when acting forces are equal in size and opposite in direction and act through the same point
- explain how acceleration is related to force and mass in everyday situations
- recall that falling objects go faster and faster as they fall until they reach a terminal speed
- explain why falling objects accelerate in terms of the forces acting
- recall that the greater the force exerted on a material the greater the extension
- recall that an elastic material will not return to its original size if a very large force is applied and removed
- recall that the gas in an air bag is compressed during a collision
- know that the transmission of pressure in a liquid is used in hydraulics
- use the formula, pressure = force/area.

In addition, to be successful at Higher GCSE Tier you should be able to:

- calculate speed from the gradient of a distance–time graph
- calculate distance travelled and acceleration from a speed–time graph for uniform acceleration
- explain everyday situations where braking distance is increased
- recall that acceleration involves a change in speed or a change in direction
- use the formula $F = m\,a$
- recall that when two bodies interact, the forces they exert on each other are equal and opposite
- explain why a falling object reaches a terminal speed
- recall that the extension of certain materials is directly proportional to the force applied
- explain how air bags reduce the force that acts on a person in a collision
- use the equation $P\,V = k$ for gases at constant temperature
- explain how a small force exerted over a small area can be transmitted through a liquid to produce a large force over a large area.

EXAM PRACTICE

Sample Student's Answers & Examiner's Comments

1 The graph shows how the braking distance of a car in good condition depends on the speed at which it is travelling.

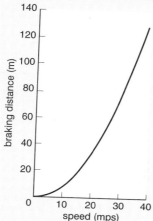

speed (mps)

a) (i) Is it true to say that 'if speed is doubled the braking distance is also doubled'? Use data from the graph to support your answer.

Speed	Braking distance
10	8
20	32 ✓
40	128

No, when the speed is doubled the braking distances increases by 4x. ✓ (2)

(ii) Explain why the braking distance at a speed of 60 m/s would be 288 m.

At 30 m/s braking distance = 72m ✓
At 60 m/s braking distance = 4 x 72 ✓
= 288 m. ✓ (3)

(iii) The driver of a car travelling at 25 m/s suddenly notices a person stepping into the road, and applies the brakes.

Explain why the total stopping distance is greater than 50 m.

The stopping distance = thinking ditance + braking distance ✓
The car will continue to travel as the driver reacts
before starting to brake. ✓ (2)

b) A car travelling at 20 m/s takes 32 m to stop.

Its average speed during braking is 10 m/s.

Use this information to calculate the time it takes to stop a car from a speed of 20 m/s.

speed = distance / time ✓   time = (distance x speed) ✗
= 32 x 20
= 640 s (3)

EXAMINER'S COMMENTS

a) (i) One mark has been obtained by selecting data from the graph. Notice it is important to choose speeds that progressively double. The second mark has been given for recognising the relationship, i.e. doubling the speed quadruples the braking distance.
(ii) The relationship found in part (i) needs to be used here. One mark is given for using the graph to read off the braking distance at 30 m/s. The second mark is for appreciating that at 60 m/s the braking distance will be 4 times that amount. The final mark is gained for showing that $4 \times 72 = 288$.
(iii) The candidate has correctly referred to reaction time for 1 mark and shown that the overall stopping distance includes thinking distance and braking distance for the second mark.

b) 1 mark has been scored for writing down the correct equation but this has been rearranged incorrectly. Time = distance/speed = 32/10 = 3.2 s. Notice that the candidate has also used the incorrect value for speed.

c) *(i)* Full marks have been obtained for the calculation with the correct formula being used and values substituted correctly. A mark has been lost for choosing the wrong units for acceleration. The unit should have been m/s/s.

(ii) This is a very good answer. The first mark has been obtained for stating that the mass will be greater. Reference is then made to the equation F = ma which scores the second mark. The final mark is given for linking the reduced rate of deceleration with the increased distance travelled.

- A mark of 14/17 corresponds to a grade A on this higher tier question.

c)

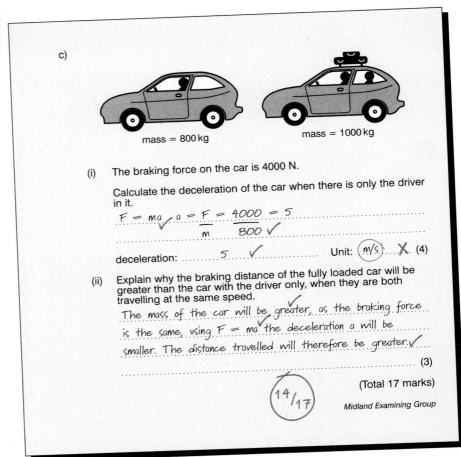

mass = 800 kg   mass = 1000 kg

(i) The braking force on the car is 4000 N.

Calculate the deceleration of the car when there is only the driver in it.

$$F = ma \quad a = \frac{F}{m} = \frac{4000}{800} = 5$$

deceleration: ....5........   Unit: ...(m/s)... ✗ (4)

(ii) Explain why the braking distance of the fully loaded car will be greater than the car with the driver only, when they are both travelling at the same speed.

The mass of the car will be greater, as the braking force is the same, using F = ma the deceleration a will be smaller. The distance travelled will therefore be greater. ✓

(3)

14/17   (Total 17 marks)

Midland Examining Group

Questions to Answer

Answers to questions 2 and 3 can be found on page 371.

2 a) Peter cycles from home to school. The following graph represents the journey.

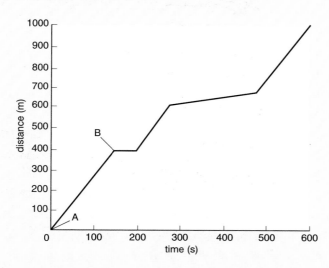

(i) During the journey Peter has to stop at traffic lights and to cycle up a steep hill. Apart from this he travels at a constant speed. On the graph:
   • write a letter 'L' where Peter is stopped at the traffic lights;
   • write a letter 'H' where Peter is cycling up the steep hill. (2)

(ii) How far is it from Peter's home to school? (1)

(iii) Calculate Peter's speed between points A and B. (2)

b) The combined weight of Peter and his bicycle is 600 N.
This weight is divided equally over each of the two wheels.
Each tyre is in contact with the road through a strip 10 cm by 2 cm.

(i) Calculate the area of each tyre in contact with the road. (1)

(ii) Calculate the pressure, in N/cm^2, which each tyre exerts on the road. (2)

London Examinations

3 a) An athlete is training. He runs 240 m in 30 s.

(i) Calculate the athlete's average speed. (2)

(ii) The maximum speed during the race is higher than this average speed. Explain why. (2)

301

b) The athlete has a friendly race with a cyclist.

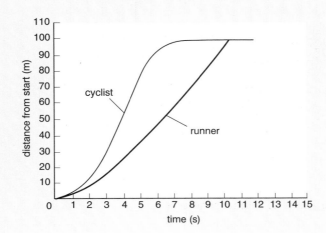

(i) Mark on the graph with an **X** where the cyclist's speed started to decrease. (1)

(ii) Mark on the graph with a **Y** where the runner's speed stopped increasing. (1)

(iii) What is the difference in the competitors' times for this race? (1)

c) (i) Look at the graph for the runner. The runner has a steady speed for some of the race. Between what times does the runner have a steady speed? (1)

(ii) The graph is not straight between 0 s and 3.5 s. Explain why. (2)

d) As soon as the race begins, the starter walks behind the athlete and the cyclist at a steady speed of 2 m/s.

(i) Draw a line on the graph to show the starter's movement for 15 s. (1)

(ii) After 10 s of walking, how far was the starter from the finishing line? (2)

Midland Examining Group

Energy is a fundamental requirement of an increasingly technological world. It is obtained from a large number of sources, but mainly from a dwindling supply of fossil fuels. Just as important as looking for new sources of energy is ensuring that we find ways of using available energy supplies more efficiently.

SOURCES OF ENERGY

At the moment most of the energy we use is obtained from fossil fuels. The fossil fuels of coal, oil and natural gas are still the most common sources of energy used in power stations to generate electricity (see chapter 15). These fuels are **non-renewable**. They were formed over millions of years with intense heat and pressure and once supplies have been used up they cannot be replaced. It has been estimated that, at current levels of use, oil and gas supplies will last for about another 40 years, and coal supplies for about a further 300 years. The development of other sources of energy is therefore becoming increasingly important. Many of the new sources being developed are **renewable**, that is, they can be regenerated relatively quickly.

Solar power

Solar panels are used to 'trap' the energy from the Sun. The Sun's energy is transferred into electrical energy or, as with domestic solar panels, is used to heat water. The high cost of installing solar panels is a disadvantage, and the weather limits the time when the panels are effective.

These solar panels have transparent tops and black bases. This lets the infrared radiation in and absorbs it. Water flows over the hot, black base, providing cheap hot water.

Wind power

The wind is used to turn windmill-like turbines which generate electricity directly from the rotating motion of their blades. Modern wind turbines are very efficient but several thousand would be required to equal the generating capacity of a modern fossil-fuel power station.

Wave and tidal power

Large floats move up and down with the waves and their movement is used to generate electricity. A very large number of floats are needed to produce a significant amount of electricity.

Dams on tidal estuaries trap the water at high tide. When the water is allowed to flow back at low tide, electricity can be generated. This technique obviously limits the use of the estuary by ships.

You need a large number of wind-powered generators to produce the same amount of electricity as a small fuel-powered station. There are no poisonous waste products and the wind is free. Unfortunately, the wind does not blow all the time.

Wave and tidal energy also work by turning a generator. The up-and-down motion of the float is converted to a turning motion by the ratchet. At low tide, the trapped high-tide water runs out past a turbine.

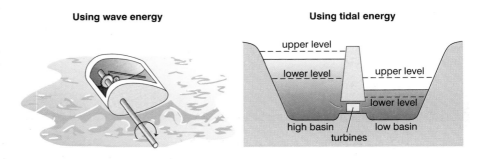

Using wave energy

Using tidal energy

upper level

lower level

upper level

lower level

high basin   low basin

turbines

Hydroelectric power

Dams are used to store water which is allowed to fall in a controlled way that generates electricity. This is a particularly useful method in mountainous regions. When demand for electricity is low, electricity can be used to pump water back up into the high dam for use in times of high electricity demand.

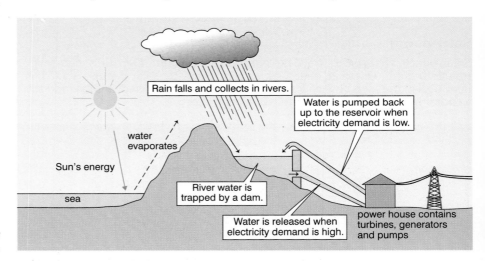

Rain falls and collects in rivers.

Water is pumped back up to the reservoir when electricity demand is low.

water evaporates

Sun's energy

River water is trapped by a dam.

sea

Water is released when electricity demand is high.

power house contains turbines, generators and pumps

A 'pumped storage' hydroelectric power station.

A geothermal power station surrounded by hot springs.

Biomass power

Plants use energy from the Sun in photosynthesis. Plant material can then be used as a fuel. Burning wood is a good example of the direct use of biomass fuel. A good example of indirect use is when sugar cane is fermented to make ethanol – the ethanol is then used as an alternative to petrol. Waste plant material can be used in 'biodigesters' to produce methane gas. The methane is then used as a fuel.

Geothermal power

This source makes use of the heat of the Earth. In certain parts of the world, water forms hot springs which can be used directly for heating. Water can also be pumped deep into the ground to be heated.

Check yourself

QUESTIONS

Q1
a) What is meant by a *non-renewable* energy source?
b) Name three non-renewable energy sources.
c) Which non-renewable energy source is likely to last the longest?

Q2 Look at the graph which shows the amount of energy from different sources used in the UK in 1955, 1965, 1975 and 1985.

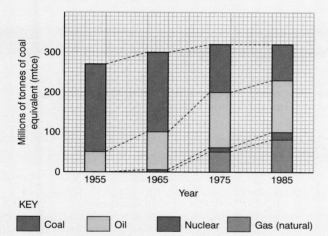

KEY

■ Coal   □ Oil   ■ Nuclear   ■ Gas (natural)

a) Describe the trend in the total energy used in this period.
b) Describe the main changes in the sources of this energy.

Q3 A site has been chosen for a wind farm (a series of windmill-like turbines).
a) Give two important factors in choosing the site.
b) Give one advantage and one disadvantage of using wind farms to generate electricity.

REMEMBER! Cover the answers if you want to.

ANSWERS

A1
a) A source that cannot be regenerated – it takes millions of years to form.
b) Coal, oil and natural gas.
c) Coal.

A2
a) The total energy increased from 1955 to 1975 and then levelled out.
b) The amount of coal used has significantly reduced. The amounts of oil, gas and nuclear have significantly increased.

TUTORIAL

T1
a) *A common mistake is to say that it is a source that 'cannot be used again'. Many energy sources cannot be used again but they can be regenerated, e.g. wood.*
b) *These are the fossil fuels. Substances obtained from fossil fuels such as petrol and diesel are not strictly speaking fossil fuels.*
c) *Coal is becoming increasingly more difficult to mine as more inaccessible coal seams are tackled. The 300 year estimate could be very optimistic.*

T2
a) *Always describe a graphical trend carefully. Be as precise as you can.*
b) *It is important to refer to each of the energy sources given on the graph.*

ANSWERS

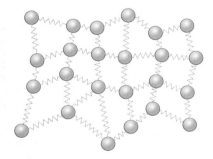

A3   a)   Any two from: strength of the wind,
high ground, constant supply of wind,
open ground.

      b)   Advantage: renewable energy source,
no air pollution. Disadvantage:
unsightly, takes up too much space.

TUTORIAL

T3   a)   *Higher ground tends to be more windy than
lower ground. However, the wind farm cannot be
built too far away from the centres of population
otherwise costs will be incurred in connecting to
the National Grid.*

      b)   *Wind turbines can be very efficient. However,
they need to be reasonably small and so a large
number are needed to generate significant
amounts of electricity. Environmentally, whilst
they produce no air pollution they do take up a
lot of land.*

TRANSFERRING ENERGY

Energy flows from high temperatures to low temperatures – this is called
thermal transfer. Thermal energy can be transferred in four main ways:
conduction, **convection**, **radiation** and **evaporation**.

CONDUCTION

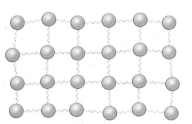

*Conduction in a solid. Particles in a
hot part of a solid (top) vibrate further
and faster than particles in a cold part
(bottom). The vibrations are passed on
through the bonds from particle to
particle.*

Materials that allow thermal energy to transfer through them quickly are
called **conductors**. Those that do not are called **insulators**. If one end of a
conductor is heated the atoms that make up its structure start to vibrate more
vigorously. As the atoms in a solid are linked together by chemical bonds the
increased vibration can be passed on to other atoms. The energy of
movement (kinetic energy) passes through the whole material.

Metals are particularly good conductors because their structure contains
freely moving electrons which transfer energy very rapidly. Air is a good
insulator and reduces thermal transfer by conduction. As air is a gas there are
no bonds between the particles and so energy can only be transferred by the
particles colliding with each other. To obtain the best results the air needs to
be trapped so that energy cannot be transferred by convection (see later).
Conduction cannot occur when there are no particles present, so a vacuum is
a perfect insulator.

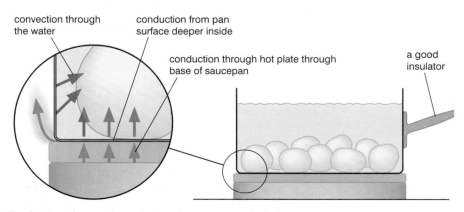

Conduction plays an important part when you cook food.

CONVECTION

Convection occurs in liquids and gases because these materials flow (which is why they are described as 'fluids'). Convection occurs when hot fluids rise and cold fluids sink. The circulation of fluids that this causes is called a **convection current**. The particles in a fluid move all the time. When a fluid is heated, energy is transferred to the particles causing them to move faster and further apart. This makes the heated fluid less dense than the unheated fluid. Consequently the less dense warm fluid will rise above the more dense colder fluid. This is how the thermal energy is transferred.

If a fluid's movement is restricted, then energy cannot be transferred. That is why many insulators, such as ceiling tiles, have trapped air pockets. Wall cavities in houses are filled with fibre to prevent air from circulating and transferring thermal energy by convection.

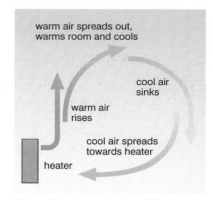

Convection currents caused by a room heater.

RADIATION

Radiation, unlike conduction and convection, does not need particles at all. Radiation can travel through a vacuum. All objects take in and give out infrared radiation all the time. Hot objects radiate more infrared than cold objects. The amount of radiation given out by an object depends on its temperature and on the object's surface. The surface of an object also determines how good it is at absorbing infrared radiation.

Type of surface	As an emitter of radiation	As an absorber of radiation	Examples
Dull black	good	good	Cooling fans on the back of a refrigerator are dull black to radiate away more energy.
Bright shiny	poor	poor	Marathon runners wrap themselves in shiny blankets to prevent thermal transfer by radiation. Fuel storage tanks are sprayed with shiny silver paint to reflect radiation from the Sun.

EVAPORATION

When particles break away from the surface of a liquid and form a vapour, the process is known as evaporation. The particles that escape are those with the greatest energy and so the average energy of those remaining is reduced. Evaporation therefore causes cooling. The evaporation of sweat helps to keep a body cool in hot weather. The cooling obtained in a refrigerator (right) is also due to evaporation.

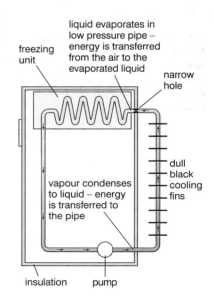

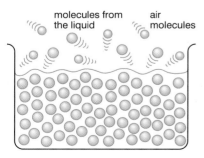

The more energetic molecules of the liquid escape from the surface. This reduces the average energy of the molecules remaining in the liquid and so the liquid cools down.

SAVING ENERGY COSTS

Heating a house can account for over 60% of a family's total energy bill. Reducing thermal energy transfer from the house to the outside can create very large savings and greatly reduce the amount of energy that is being used. Ways of reducing wasteful energy transfers are shown in the table.

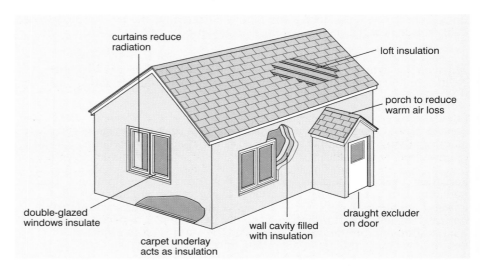

There are a number of ways of reducing wasteful energy transfer.

Source of energy wastage	% of energy wasted	Insulation technique
Walls	35	*Cavity wall insulation*. Modern houses have cavity walls, that is, two single walls separated by an air cavity. The air reduces energy transfer by conduction but not by convection as the air is free to move within the cavity. Fibre insulation is inserted into the cavity to prevent the air from moving and so reduces convection.
Roof	25	*Loft insulation*. Fibre insulation is placed on top of the ceiling and between the wooden joists. Air is trapped between the fibres thus reducing energy transfer by conduction and convection.
Floors	15	*Carpets*. Carpets and underlay prevent energy loss by conduction and convection. In some modern houses foam blocks are placed under the floors.
Draughts	15	*Draught excluders*. Cold air can get into the home through gaps between windows and doors and their frames. Draught excluder tape can be used to block these gaps.
Windows	10	*Double glazing*. Energy is transferred through glass by conduction and radiation. Double glazing has two panes of glass with a layer of air between the panes. Double glazing reduces energy transfer by conduction but not by radiation. Radiation can be reduced by drawing the curtains.

One thing to consider when insulating a house is the balance of the cost of the insulation against the potential saving in energy costs. The **pay-back time** is the time it takes for the savings to repay the costs of installation. There are very different pay-back times associated with the different methods of insulation. However, the pay-back time is not the only consideration. Double glazing reduces noise and condensation inside the home, and carpets provide increased comfort.

Means of insulation	Approximate pay-back time (years)
Cavity wall	5
Loft	2
Carpets	10
Draught excluders	1
Double glazing	20

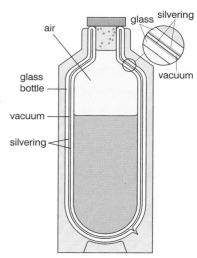

A vacuum flask has a hollow glass lining. The absence of air in the lining prevents thermal transfer by conduction. The insides of the lining are silvered to reduce thermal transfer by radiation. The flask will therefore keep hot drinks hot, or cold drinks cold, for hours.

EFFICIENCY

Energy transfers can be summarised using simple energy transfer diagrams or Sankey diagrams. The thickness of each arrow is drawn to scale to show the amount of energy. Energy is always conserved, that is, the total amount of energy after the transfer must be the same as the total amount of energy before the transfer. Unfortunately, in nearly all energy transfers some of the energy will end up as 'useless' heat.

Power stations are not as efficient as you might think. Only some of the energy originally produced from the fuel is transferred to useful electrical output. Energy **efficiency** can be calculated from the following formula:

$$\text{efficiency} = \frac{\text{useful energy output}}{\text{energy input}} \times 100\%$$

For example, in the energy transfer diagram for a power station shown below, only 30% of the energy from the fuel ends up as useful electricity. In other words the power station is only 30% efficient.

In this power station 70% of the energy transfers do not produce useful energy. Many power stations are now trying to make use of the large amounts of energy 'lost' in the cooling towers.

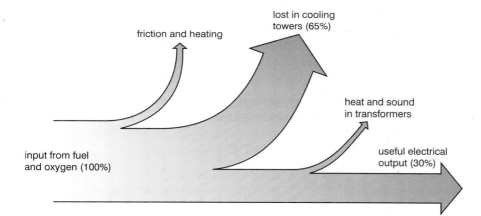

friction and heating

lost in cooling towers (65%)

heat and sound in transformers

input from fuel and oxygen (100%)

useful electrical output (30%)

Check yourself

QUESTIONS

Q1 Why are several thin layers of clothing more likely to reduce thermal transfer than one thick layer of clothing?

Q2 Explain how sweating helps to keep a body cool in hot weather.

Q3 Hot water in an open container transfers energy by evaporation. Explain how the loss of molecules from the surface of the liquid causes the liquid to cool.

Q4 The diagram shows a cross-section of a steel radiator positioned in a room next to a wall.

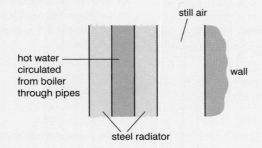

Describe how energy from the hot water reaches the wall behind the radiator.

Q5 Look at the information in the table.

Way of saving energy	Cost of fitting (£)	Money saved on energy each year	Pay-back time in years
double glazing	4000	200	20
cavity wall insulation	1000	250	
loft insulation	300		2

a) Calculate the pay-back time for cavity wall insulation.
b) Calculate the money saved each year after fitting loft insulation.
c) Loft insulation, cavity wall insulation and double glazing all have air in them. Explain why the air is important in helping to reduce energy transfer.

REMEMBER! Cover the answers if you want to.

ANSWERS

A1 The thin layers will trap air between them. The air is an insulator and so will reduce thermal transfer from the body.

A2 The sweat evaporates from the surface of the skin. As evaporation requires energy, energy is removed from the surface of the skin keeping the skin cool.

A3 The molecules that are able to leave the surface of the liquid are those that are moving the fastest, that is, the molecules with the most energy. The average energy of the remaining molecules is therefore reduced and so the liquid is cooler.

TUTORIAL

T1 *Remember that air is a very good insulator. Trapping it between layers of clothing means that convection is inhibited as well as conduction.*

T2 *Vasodilation is another mechanism the body uses to maintain its body temperature. See chapter 3.*

T3 *The temperature of a liquid is related to the average speed of its molecules. If the temperature is reduced, on average, the molecules will move more slowly.*

ANSWERS

A4 Energy from the hot water is transferred from the inner to the outer surface of the metal by conduction. Energy is transferred through the still air by radiation.

A5
a) 4 years.
b) £150
c) Air is an insulator and so reduces thermal transfer by conduction. When the air is trapped it reduces thermal transfer by convection.

TUTORIAL

T4 *In a question like this it is important to take the energy transfer stage by stage. Thermal transfer through a solid involves conduction. Thermal transfer through still air must involve radiation. In fact convection would be occurring and it is very likely that the air behind the radiator would be moving (convection currents).*

T5
a) *Pay-back time* $= \dfrac{cost}{money\ saved} = \dfrac{1000}{250}$
   $= 4\ years.$
b) *Money saved* $= \dfrac{cost}{pay\text{-}back\ time} = \dfrac{300}{2}$
   $= £150.$
c) *You should be getting the message that air is a very useful insulator! In both the cavity wall insulation and loft insulation the air is trapped in the fibres of the insulating material.*

WORK, POWER AND ENERGY

Work and power are familiar words in everyday use. In science they have very precise meanings which do not always match those in common usage.

WORK

Work is done when the application of a force results in movement. Work can only be done if the object or system has energy. When work is done energy is transferred. Work done can be calculated using the following formula:

work done = force × distance moved
W = F s

W = work done in joules (J)
F = force in newtons (N)
s = distance moved in the direction of the force in metres (m)

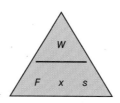

In this position the gymnast is not doing any work against his body weight – he is not moving (he will be doing work pumping blood around his body though!).

Worked example

A cyclist pedals along a flat road. She exerts a force of 60 N and travels 150 m. Calculate the work done by the cyclist.

Write down the formula:	W = F s
Substitute the values for F and *s*:	W = 60 × 150
Work out the answer and write down the unit:	W = 9000 J

Worked example

A person does 3000 J of work in pushing a supermarket trolley 50 m across a level car park. What force was the person exerting on the trolley?

Write down the formula with F as the subject:
$$F = \frac{W}{s}$$

Substitute the values for W and *s*:
$$F = \frac{3000}{50}$$

Work out the answer and write down the unit:
$$F = 60\ N$$

The gymnast is doing work. He is moving upwards against the force of gravity. Energy is being transferred as he does the work.

POWER

Power is defined as the rate of doing work or the rate of transferring energy. The more powerful a machine is the quicker it does a fixed amount of work or transfers a fixed amount of energy. Electrical power has already been considered in chapter 14. Power can be calculated using the following formula:

power = work done/time taken = energy transfer/time taken

$$P = \frac{W}{t}$$

P = power in joules per second or watts (W)
W = work done in joules (J)
t = time taken in seconds (s)

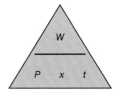

Worked example

A crane does 20 000 J of work in 40 seconds. Calculate its power over this time.

Write down the formula:	$P = \dfrac{W}{t}$
Substitute the values for W and t:	$P = \dfrac{20\,000}{40}$
Work out the answer and write down the unit:	$P = 500\,W$

Worked example

A student with a weight of 600 N runs up the flight of stairs shown in the diagram (left) in four seconds. Calculate the student's power.

5 m

Write down the formula for work done:	$W = F\,s$
Substitute the values for F and s:	$W = 600 \times 5 = 3000\,J$
Write down the formula for power:	$P = \dfrac{W}{t}$
Substitute the values for W and t:	$P = \dfrac{3000}{4} = 750\,W$

The student is lifting his body against the force of gravity, which acts in a vertical direction. The distance measured must be in the direction of the force (that is, the vertical height).

POTENTIAL ENERGY

Stored, or hidden, energy is called **potential energy** (P. E.). If a spring is stretched the spring will have potential energy. If a load is raised above the ground it will have **gravitational potential energy**. If the spring is released or the load moves back to the ground the stored potential energy is transferred to movement energy, which is called **kinetic energy** (K. E.).

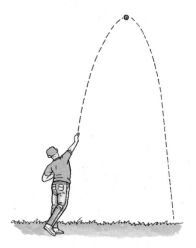

The kinetic energy given to the stone when it is thrown is transferred to potential energy as it gains height and slows down. At the top of its flight practically all the kinetic energy will have been converted into gravitational potential energy. A small amount of energy will have been lost due to friction between the stone and the air.

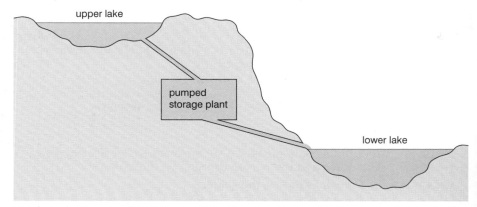

upper lake

pumped storage plant

lower lake

The water in the upper lake has a considerable amount of gravitational potential energy. As the water falls to the lower lake its potential energy is transferred into kinetic energy which is used to turn turbines and generate electricity.

Gravitational potential energy can be calculated using the following formula:

gravitational potential energy = mass × gravitational field strength × height
P.E. = $m\,g\,h$
P.E. = gravitational potential energy in joules (J)
m = mass in kilograms (kg)
g = gravitational field strength of 10 N/kg
h = height in metres (m)

Worked example

A skier has a mass of 70 kg and travels up a ski lift a vertical height of 300 m. Calculate the change in the skier's gravitational potential energy.

Write down the formula: P.E. = $m\,g\,h$
Substitute values for m, g and h: P.E. = $70 \times 10 \times 300$
Work out the answer and write down the unit: P.E. = 210 000 J or 210 kJ

KINETIC ENERGY

The kinetic energy of an object depends on its mass and its velocity. The kinetic energy can be calculated using the following formula:

kinetic energy = $\frac{1}{2}$ × mass × velocity^2

K.E. = $\frac{1}{2}\,m v^2$

K.E. = kinetic energy in joules (J)
m = mass in kilograms (kg)
v = velocity in m/s

The skier's potential energy is transferred to kinetic energy. Her potential energy depends on her weight and her height on the mountain. Her kinetic energy depends on her mass and velocity.

Worked example

An ice skater has a mass of 50 kg and travels at a velocity of 5 m/s. Calculate the ice-skater's kinetic energy.

Write down the formula: K.E. = $\frac{1}{2}\,m v^2$

Substitute the values for m and v: K.E. = $\frac{1}{2} \times 50 \times 5 \times 5$

Work out the answer and write down the unit: K.E. = 625 J

Check yourself

QUESTIONS

Q1   Look at the diagrams.

a)   Write down the letters of two pictures where work is not being done.
b)   Write down the letter of a picture where kinetic energy is increasing.
c)   Write down the letter of a picture where potential energy is increasing.

Q2   50 000 J of work are done as a crane lifts a load of 400 kg. How far did the crane lift the load? (Gravitational field strength, g, is 10 N/kg.)

Check yourself

QUESTIONS

Q3 A student is carrying out a personal fitness test.

She steps on and off the 'step' 200 times. She transfers 30 J of energy each time she steps up.
a) Calculate the energy transferred during the test.
b) She takes 3 minutes to do the test. Calculate her average power.

Q4 A lift on a delivery van raises a dishwasher weighing 80 kg through 90 cm. Calculate the gain in gravitational potential energy of the dishwasher. (Gravitational field strength, g, is 10 N/kg.)

Q5 An athlete with a mass of 65 kg runs at a steady speed of 6 m/s. Calculate his kinetic energy.

Q6 A child of mass 35 kg climbed a 30 m high snow-covered hill.
a) Calculate the change in the child's gravitational potential energy.
b) The child then climbed on to a light-weight sledge and slid down the hill. Calculate the child's maximum speed at the bottom of the hill. (Ignore the mass of the sledge.)
c) Explain why the actual speed at the bottom of the hill is likely to be less than the value calculated in part b).

REMEMBER! Cover the answers if you want to.

ANSWERS

A1
a) B and D.
b) E.
c) C.

A2 12.5 m.

A3
a) 6000 J.
b) 33.3 W.

A4 720 J.

A5 1170 J.

TUTORIAL

T1
a) *Work is not being done unless there is movement. Both B and D are static situations.*
b) *For kinetic energy to increase either the mass or the velocity must increase.*
c) *For the gravitational potential energy to increase either the mass or the height above the ground must increase.*

T2
Work done = F s
$F = 400 \times 10 = 4000 \, \text{N}$
$s = \dfrac{W}{F} = \dfrac{50\,000}{4000} = 12.5 \, m$

T3
a) *Energy transferred = $30 \times 200 = 6000 \, J$*
b) *Power = $\dfrac{work\ done}{time\ taken} = \dfrac{6000}{180} = 33.3 \, W.$*
Remember that a watt is 1 joule/sec so the time must be in seconds.

T4 *P.E. = $m\,g\,h = 80 \times 10 \times 0.9 = 720 \, J$. Remember that the distance must be in metres.*

T5 *K.E. = $\frac{1}{2} \, m \, v^2 = \frac{1}{2} \times 65 \times 6 \times 6 = 1170 \, J$*

ANSWERS

A6
a) 10 500 J.
b) 24.5 m/s.
c) Some of the gravitational potential energy will have been transferred to thermal energy due to the friction between the sledge and the snow.

TUTORIAL

T6
a) P.E. $= m g h = 35 \times 10 \times 30 = 10\,500\,J$
b) Assuming all the potential energy is transferred into kinetic energy:
K.E. $= \frac{1}{2} m v^2$, so $v^2 = 2$ K.E./m
$= 2 \times \dfrac{10\,500}{35} = 600$, and $v = 24.5$ m/s.
c) Energy must be conserved but friction is a very common cause of energy being wasted, that is, transferred into less useful forms.

EXAMINATION CHECKLIST

The facts and ideas that you should understand by studying this topic.

To be successful at Foundation GCSE Tier you should be able to:

- recall that objects hotter than their surroundings transfer energy to the surroundings
- recall how energy is transferred by conduction, convection and radiation
- describe everyday examples of energy saving methods in the home
- recall that many insulating materials contain air and that air is a good insulator
- describe everyday examples in which work is done
- describe the relationship between force, distance and work and use the formula $W = F\,s$
- describe the relationship between work, power and time and use the formula $P = W/t$
- describe examples in which objects have kinetic and/or potential energy
- describe situations in which energy is transformed between gravitational potential energy and kinetic energy.

In addition, to be successful at Higher GCSE Tier you should be able to:

- explain how energy is transferred by the movement of particles in evaporation
- describe the relationship between gravitational potential energy, mass and height and use the formula, potential energy $= m g h$
- explain the link between work done and gravitational potential energy
- describe the relationship between kinetic energy, mass and velocity and use the formula, kinetic energy $= \frac{1}{2} m v^2$.

Key Words

Tick each word when you are sure of its meaning.

conduction

conductors

convection

convection current

evaporation

gravitational potential energy

insulators

kinetic energy

non-renewable

pay-back time

power

radiation

renewable

work

EXAM PRACTICE

Sample Student's Answers & Examiner's Comments

EXAMINER'S COMMENTS

a)   *(i)*   *One mark has been given for using the correct formula. However, the incorrect distance has been substituted. The distance must be in a vertical direction as this is the direction the force must act to overcome the gravity. The calculation should have been: work = F s = 550 × 40 = 22 000 J*
  (ii)   *The correct method has been used to calculate the power and the candidate has not been penalised again for the incorrect answer in part (i). The correct units for power have been given. The correct calculation should have been: power = 22000/50 = 440 W.*

b)   *Both marks have been scored. 1 mark for appreciating that energy will be transferred due to friction in the pulleys and the second mark for appreciating that friction will also occur between the skis and the snow.*

●   *A mark of 6/8 corresponds to a good grade C on this common question between foundation and higher tiers.*

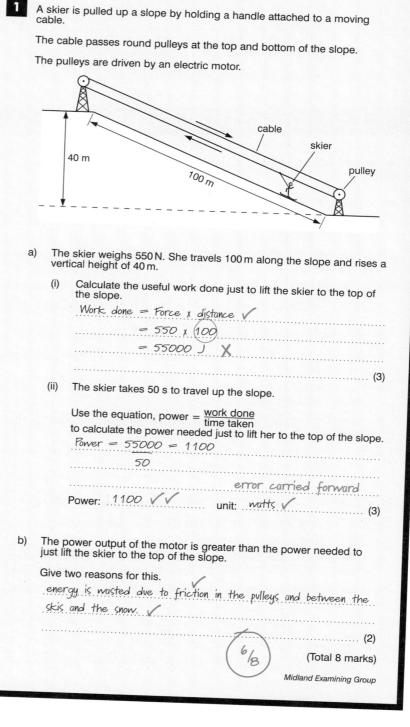

1   A skier is pulled up a slope by holding a handle attached to a moving cable.

The cable passes round pulleys at the top and bottom of the slope.

The pulleys are driven by an electric motor.

a)   The skier weighs 550 N. She travels 100 m along the slope and rises a vertical height of 40 m.

  (i)   Calculate the useful work done just to lift the skier to the top of the slope.

Work done = Force × distance ✓
= 550 × (100)
= 55000 J ✗

(3)

  (ii)   The skier takes 50 s to travel up the slope.

Use the equation, power = $\frac{\text{work done}}{\text{time taken}}$
to calculate the power needed just to lift her to the top of the slope.

Power = $\frac{55000}{50}$ = 1100

error carried forward

Power: 1100 ✓✓   unit: watts ✓   (3)

b)   The power output of the motor is greater than the power needed to just lift the skier to the top of the slope.

Give two reasons for this.

energy is wasted due to friction in the pulleys and between the ✓
skis and the snow. ✓

(2)

6/8

(Total 8 marks)

Midland Examining Group

Questions to Answer

Answers to questions 2–4 can be found on page 372.

2 The following diagram shows a solar panel used to heat water.

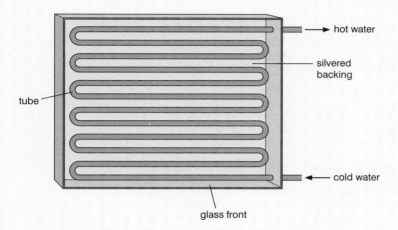

a)   Name the process by which energy is transferred:

   (i)   from the Sun to the tube in the panel

   (ii)   from the tube to the water. (2)

b)   (i)   Which of the following would be the most suitable material
         for the tube containing the water?

         black plastic   frosted glass   copper painted black   polished stainless steel (1)

   (ii)   Give TWO reasons for your choice. (2)

c)   Discuss ONE advantage and ONE disadvantage of heating
     water with a solar panel rather than a gas boiler. (2)

London Examinations

3 The following diagram shows details of the hydroelectric power
station at Cruachan in Scotland.

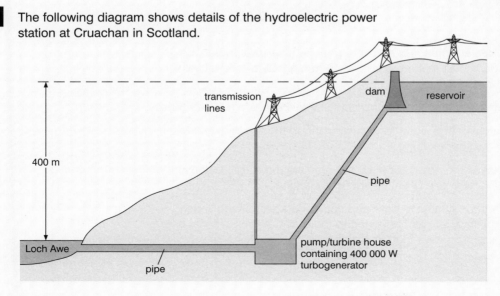

Water flows from the reservoir to Loch Awe driving a turbo-generator to produce electricity.

a)   How many joules of gravitational potential energy does 1 kg of water lose when it flows from the reservoir to Loch Awe? (Gravitational field strength is 10 N/kg).   (2)

b)   How much electrical energy is generated when 1 kg of water flows from the reservoir to Loch Awe? Explain your answer.   (1)

The energy transfer during electricity generation at Cruachan is:

gravitational potential energy → electrical energy

Cruachan is also used for another process called pumped storage. Electricity from the National Grid can be used to pump water from Loch Awe back up to the reservoir.

c)   State the energy transfer during pumped storage.   (1)

d)   The hydroelectric power station at Cruachan uses more electricity for pumped storage than it generates. Suggest why this is so.   (2)

e)   Suggest why pumped storage is useful.   (1)

London Examinations

4

10 m

Hydroslides are great fun and very popular in leisure pools. The hydroslide shown in the diagram is 10 m high. Gravitational field strength is 10 N/kg.

a)   The boy climbs to the top of the slide. He has a mass of 50 kg.

(i)   Calculate the work done in climbing to the top of the slide. Show your working.   (3)

(ii)   It takes the boy 6 seconds to complete the climb. Calculate his power during the climb. Show your working.   (3)

b)   He goes down the slide and comes to a stop in the water. Explain what has happened to the gravitational potential energy which he had at the top of the slide.   (3)

Northern Examination and Assessment Board

Light and sound exist as waves. Light is part of a family of waves known as the electromagnetic spectrum. Gaining a full understanding about waves has been fundamental to the development of modern communication systems and advances that have been made in the field of medicine.

THE PROPERTIES OF WAVES

There are two types of waves. These are known as **longitudinal** and **transverse** waves. Sound is an example of a longitudinal wave. If you hit a piece of wood the particles in the wood vibrate backwards and forwards in the same direction that the waves travel. This type of wave can be shown by pushing and pulling a spring. The spring stretches in places and squashes in others. The stretching produces regions of **rarefaction**, whilst the squashing produces regions of **compression**.

Light, radio and other electromagnetic waves are examples of transverse waves. In a transverse wave the vibrations are at right angles to the direction of motion. Water waves are transverse waves, the water in a particular place moves up and down whilst the wave shape moves along. Water waves are often used to demonstrate the properties of waves because the **wavefront** of a water wave is easy to see. A wavefront is the moving line that joins all the points on the crest of a wave.

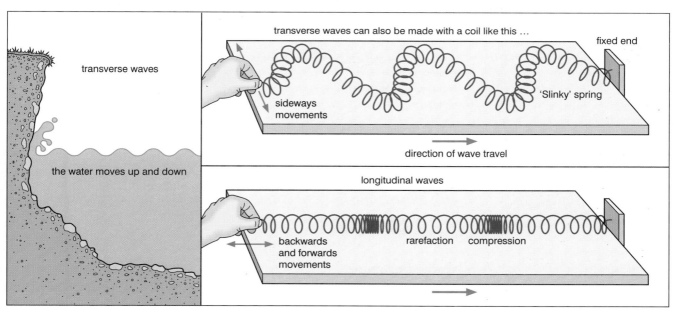

Transverse and longitudinal waves are made by vibrations.

There are many different types of waves, but they all have certain features in common:

- Waves travel at a speed that can vary depending on the substance or **medium** they are passing through.
- Waves have a repeating shape or pattern.
- Waves have a **frequency**, **wavelength** and **amplitude**.
- Waves carry energy without moving material along.

transverse

longitudinal

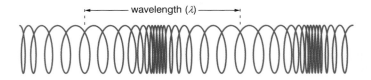

- Wavelength – the length of the repeating pattern.
- Frequency – the number of repeated patterns that go past any point each second.
- Amplitude – the maximum displacement of the medium's vibration. In transverse waves, this is half the crest-to-trough height.

SPEED, FREQUENCY AND WAVELENGTH

The speed of a wave in a given medium is constant. This means that you if you change the wavelength the frequency must change also. The relationship between the speed, frequency and wavelength of a wave is given by the equation:

wave speed $=$ frequency $\times$ wavelength
$v = f \times \lambda$
v = wave speed, usually measured in metres/second (m/s)
f = frequency, measured in cycles per second or hertz (Hz)
λ = wavelength, usually measured in metres (m)

Worked example

A loudspeaker makes sound waves with a frequency of 300 Hz. The waves have a wavelength of 1.13 m. Calculate the speed of the sound waves.

Write down the formula:	$v = f \times \lambda$
Substitute the values for f and λ:	$v = 300 \times 1.13$
Work out the answer and write down the unit:	$v = 339\,\text{m/s}$

Worked example

A radio station broadcasts on a wavelength of 250 m. The speed of the radio waves is 3×10^8 m/s. Calculate the frequency.

Write down the formula with f as the subject:	$f = \dfrac{v}{\lambda}$
Substitute the values for v and λ:	$f = \dfrac{3 \times 10^8}{250}$
Work out the answer and write down the unit:	$f = 12\,000\,000\,\text{Hz or}$ $12\,000\,\text{kHz}$

WAVES REFLECT

When a wave hits a barrier the wave will be reflected. If it hits the barrier at an angle then the angle of reflection will be equal to the angle of incidence. Echoes are a common consequence of the **reflection** of sound waves. The reflection of light is covered later in this chapter.

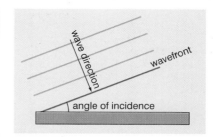

Waves hit a barrier. The angle between a wavefront and the barrier is the angle of incidence.

WAVES REFRACT

When a wave moves from one medium into another, it will either speed up or slow down. When a wave slows down, the wavefronts crowd together – the wavelength gets smaller. When a wave speeds up, the wavefronts spread out – the wavelength gets larger.

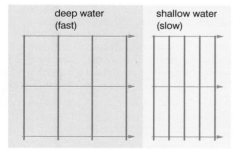

When waves slow down, their wavelength gets shorter.

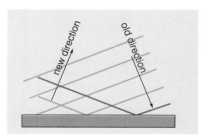

The waves bounce off.

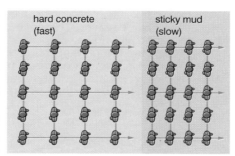

Rows of marching soldiers get closer together when they walk from concrete into mud.

If a wave enters a new medium at an angle then the wavefronts also change *direction*. This is known as **refraction**. Refraction happens whenever there is a change in wave speed. Water waves are faster in shallower water than in deep water, so water waves will refract when the depth changes.

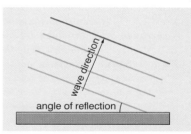

The angle of reflection is the same as the angle of incidence.

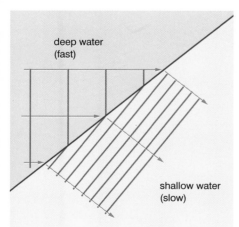

If waves cross into a new medium at an angle, their wavelength and direction changes.

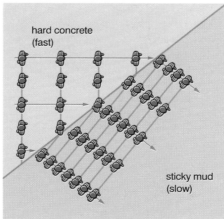

When rows of marching soldiers meet another surface at an angle their direction changes because, one by one, they start to move more slowly.

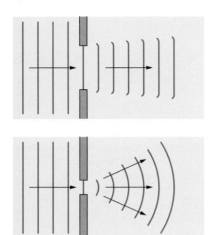

Diffraction is most noticeable when the size of the gap equals the wavelength of the waves.

WAVES DIFFRACT

Wavefronts change shape when they pass the edge of an obstacle or go through a gap. This process is known as **diffraction**. Diffraction is strong when the width of the gap is similar in size to the wavelength of the waves.

Diffraction is a problem in communications when radio and television signals are transmitted through the air in a narrow beam. Diffraction of the wavefront means that not all the energy transmitted with the wavefront reaches the receiving dishes. On the other hand, diffraction allows long-wave radio waves to spread out and diffract around buildings and hills.

SEISMIC WAVES

Earthquakes make waves which travel right through the Earth. These waves are called **seismic waves**. There are longitudinal waves called P-waves (Primary) and transverse waves called S-waves (Secondary). The waves travel through rock and are partially reflected at the boundaries between different types of rock. Instruments called seismometers detect the waves. Monitoring these seismic waves after earthquakes has provided geologists with evidence that the Earth is made up of layers and that part of the core is made of liquid.

S-waves can travel through solid rock but not through liquid. P-waves can travel through solid and liquid rock. Only the P-waves from an earthquake, not the S-waves, are received at seismic stations on the opposite side of the world.

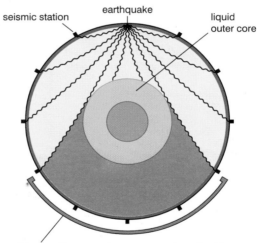

sideways (transverse) waves are not received here

Check yourself

QUESTIONS

Q1 The diagram shows a trace of a sound wave obtained on a cathode ray oscilloscope.

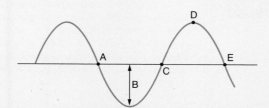

a) Which letter shows the crest of the wave?
b) The wavelength is the distance between which two letters?
c) Which letter shows the amplitude?
d) The frequency of the wave is 512 Hz. How many waves are produced each second?

Q2 Radio waves of frequency 900 MHz are used to send information to and from a portable phone. The speed of the waves is 3×10^8 m/s. Calculate the wavelength of the waves.
(1 MHz = 1 000 000 Hz, 3×10^8 = 300 000 000)

Q3 The speed of sound is approximately 340 m/s. Calculate the wavelength of middle C, which has a frequency of 256 Hz.

Q4 a) The diagram shows an experiment with water in a ripple tank.

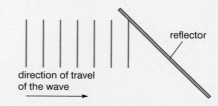

(i) What is the wavelength of the water wave.
(ii) Copy the diagram and draw on the reflected rays.

b) The diagram below shows another experiment with water waves in a ripple tank.

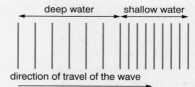

(i) What property of waves is shown in this experiment?
(ii) Explain why the wavelength decreases as the waves enter the shallow region.

REMEMBER! Cover the answers if you want to.

ANSWERS

A1
a) D.
b) A and E.
c) B.
d) 512.

A2 0.33 m.

TUTORIAL

T1
a) *The crest is the top of the wave.*
b) *The wavelength is the length of the repeating pattern.*
c) *The amplitude is half the distance between the crest and the trough. If you had difficulty with parts a to c, look back at the diagram on page 320.*
d) *Frequency is measured in hertz and 1 Hz = 1 cycle (or wave) per second.*

T2 $v = f \times \lambda$ *or* $\lambda = v/f$
$$\lambda = \frac{3 \times 10^8}{9 \times 10^8} = 0.33\,m$$

ANSWERS

A3 1.33 m.

A4

a) (i) 0.4 cm or 0.5 cm.
    (ii)

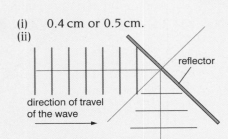

reflector

direction of travel
of the wave

b) (i) Wavelength change due to speed change. N*ot* refraction.
    (ii) The speed of the waves changes.

TUTORIAL

T3 $v = f \times \lambda$ or $\lambda = \dfrac{v}{f}$

$\lambda = \dfrac{340}{256} = 1.33\,m$

T4

a) *The wavelength is the distance between two wavefronts. Remember that the angle of reflection equals the angle of incidence.*

b) *When a wave moves into a different medium, in this case shallow water, the speed changes causing a change in the wavelength of the wave. If the wave's angle of incidence was not zero, then it would have changed direction too. Refraction is the direction change due to the change in speed.*

THE ELECTROMAGNETIC SPECTRUM

The prism splits the white light into the colourful spectrum of visible light.

White light is a mixture of different colours and can be split by a prism into its component colours known as the **visible spectrum**. Each of the different colours of light travels at the same speed in a vacuum but they have different frequencies and wavelengths. When they enter glass or perspex they all slow down but by different amounts. The different colours are therefore refracted through different angles. Violet is refracted the most, red the least.

frequency / 10^{14} Hz	4.3	6.0	7.5
wavelength / 10^{-6} m	0.7	0.5	0.4

All the colours in the visible spectrum travel at the same speed but they have different wavelengths and frequencies.

The visible spectrum is only a small part of the full **electromagnetic spectrum**. The electromagnetic spectrum is a family of different kinds of waves. Electromagnetic waves all travel at the same speed in a vacuum, i.e. the speed of light, 300 000 000 m/s.

The complete electromagnetic spectrum.

Type of wave	gamma rays	X rays	ultraviolet	visible	infrared	microwaves	TV and radio waves
frequency	high						low
wavelength	low						high
use	killing cancer cells	to look at bones	sun tan beds	photography	TV remote controls	cooking	transmission of TV and radio

X-rays and gamma rays

X-rays are produced when high energy electrons are fired at a metal target. They are commonly used for photographing the body. Bones absorb more X-rays than other body tissue. If a person is placed between the X-ray source and a photographic plate, the bones appear to be white on the developed photographic plate compared to the rest of the body. This is because the bones have absorbed more of the X-rays and fewer are able to pass through to the photographic plate. Great care needs to be taken when using X-rays. They have very high energy and can damage or destroy parts of body cells. There is evidence to suggest that X-rays can cause cancer. X-rays are also used to treat cancer. Cancer cells absorb X-rays more readily than normal healthy cells do.

Gamma rays have even higher energy than X-rays and so have to be used with even greater care. Gamma rays are frequently used in radiotherapy to kill cancer cells. Also radioactive substances that emit gamma rays are used as tracers (see chapter 20).

Ultraviolet radiation

Ultraviolet radiation (UV) is the component of the Sun's rays that produces a suntan. UV is also created in fluorescent light tubes by exciting the atoms in a mercury vapour. The UV radiation is then absorbed by the coating on the inside of the fluorescent tube and re-emitted as visible light. Fluorescent tubes are more efficient than light bulbs because they don't depend on heating. Consequently, more energy is available to produce light.

Infrared radiation

All objects give out **infrared radiation** (IR). The hotter the object is, the more radiation it gives out. Thermograms are photographs taken to show the infrared radiation given out from objects. Infrared radiation grills and cooks our food in an ordinary oven and is used in remote controls to operate televisions and videos.

Microwaves

Microwaves are high frequency radio waves. They are used in radar in finding the position of aeroplanes and ships. Metal objects reflect the microwaves back to the transmitter enabling the distance between the object and the transmitter to be calculated.
Microwaves are also used for cooking. Water particles in food absorb the energy carried by microwaves. They vibrate more and get much hotter. Microwaves penetrate several centimetres into the food and so speed up the cooking process.

Radio and TV

Radio waves are the electromagnetic waves with the longest wavelengths and lowest frequencies. UHF (Ultra High Frequency) waves are used to transmit TV programmes to homes. VHF (Very High Frequency) waves are used to transmit local radio programmes. Medium wave and long wave radio are used to transmit over longer distances because their longer wavelengths allow them to diffract around obstacles such as buildings and hills. Communication satellites have been positioned above the Earth in order to receive signals carried by high-frequency (short-wave) radio waves. These signals are then amplified and re-transmitted to other parts of the world.

Check yourself

QUESTIONS

Q1 This is a list of types of wave:

gamma   *infrared*   *light*   *microwaves*
radio   *ultraviolet*   *X-rays*

Choose from the list the type of wave which best fits each of these descriptions:
a) Stimulates the sensitive cells at the back of the eye.
b) Necessary for a suntan.
c) Used for *rapid* cooking in an oven.
d) Used to take a photograph of the bones in a broken arm.
e) Emitted by a video remote control unit.

Q2 Gamma rays are part of the electromagnetic spectrum. Gamma rays are useful to us but can also be very dangerous.
a) Explain how the properties of gamma rays make them useful to us.
b) Explain why gamma rays can cause damage to people.
c) Give one difference between microwaves and gamma rays.
d) Microwaves travel at 300 000 000 m/s. What speed do gamma rays travel at?

REMEMBER! Cover the answers if you want to.

ANSWERS

A1
a) Light.
b) Ultraviolet.
c) Microwaves.
d) X-rays.
e) Infrared.

A2
a) They can pass through soft tissue and kill cancer cells.
b) They can damage healthy cells and cause cancer because of their very high energy.
c) They have different frequencies and wavelengths.
d) 300 000 000 m/s

TUTORIAL

T1
a) *The sensitive cells form a part of the eye called the retina.*
b) *Sun-tan lotions contain chemicals that absorb some of the UV radiation from the Sun before it can act on the skin.*
c) *The key word here was 'rapid'. Electric cookers make use of infrared radiation for cooking but microwaves produce much more rapid cooking.*
d) *A gamma camera would not be as good for this purpose as the gamma rays penetrate the bone as well as the flesh.*
e) *Remote car locking sometimes uses infrared beams.*

T2
a) *Gamma rays are useful because of their great penetrating power. This is also their disadvantage. Consequently they have to be used extremely carefully. They are the highest energy waves in the electromagnetic spectrum.*
b) *All waves in the electromagnetic spectrum have different frequencies and wavelengths.*
c) *All waves in the electromagnetic spectrum travel at the same speed in a vacuum – the speed of light.*
d) *As you might have guessed this is an important point and one that is often tested in exams!*

LIGHT

As with all types of electromagnetic waves, light waves can be reflected and refracted.

LIGHT REFLECTS

Light rays are reflected from mirrors in such a way that:

angle of incidence (i) = angle of reflection (r)

The angles are measured to an imaginary line at 90° to the surface of the mirror. This line is called the **normal**. The normal line is used because with a curved mirror it is difficult to measure the angle between the ray and the mirror.

When you look in a plane mirror you see an **image** of yourself. The image is said to be **laterally inverted** because if you raise your right hand the image raises its left hand. The image is formed as far behind the mirror as you are in front of it and is the same size as you. The image is not a **real image**, it cannot be projected onto a screen, it is known as a **virtual image**.

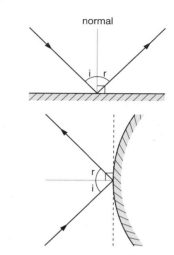

The angles of incidence and reflection are the same when a mirror reflects light. This type of curved mirror is known as a convex mirror.

Uses of mirrors

- Plane – The image is always the same size as the object.

  Household 'dressing' mirror. Dental mirror for examining teeth. Security mirror for checking under vehicles. Periscope.

- Concave – If close to a concave mirror the image is larger than the object.

  In torches and car headlamps to produce a beam of light. Make-up and shaving mirrors. Satellite dishes.

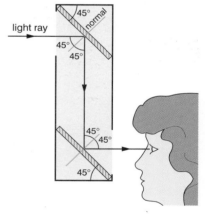

- Convex – The image is always smaller than the object.

  Car driving mirror. Shop security mirror.

LIGHT REFRACTS

Light waves slow down when they travel from air into glass. If they are at an angle to the glass they bend towards the normal. When the light rays travel out of the glass into the air, their speed increases and they bend away from the normal.

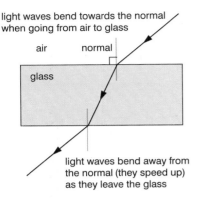

light waves bend towards the normal
when going from air to glass

air   normal

glass

light waves bend away from
the normal (they speed up)
as they leave the glass

Refraction is observed when a triangular prism (a glass or plastic block) is used to obtain the visible spectrum (see page 324). Lenses also refract light rays.

Uses of lenses

- Convex (Converging, positive) – Parallel rays of light converge to the principal focus.

  Magnifying glass. Camera lens. Telescope lenses. Binocular lenses. Microscope lenses. Film projector lenses. Spectacles.

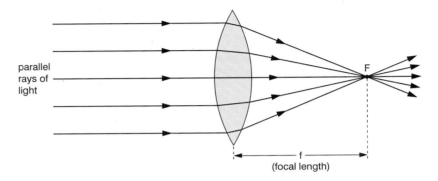

parallel
rays of
light

F

f
(focal length)

- Concave (Diverging, negative) – Parallel rays of light diverge as if they were coming from the principal focus.

  Spectacles. Telescopes. Correcting lenses in all optical equipment.

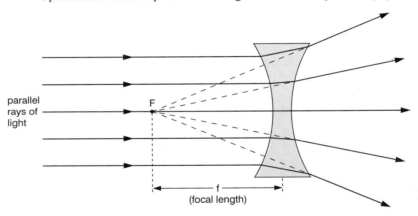

parallel
rays of
light

F

f
(focal length)

How a convex lens makes an image.

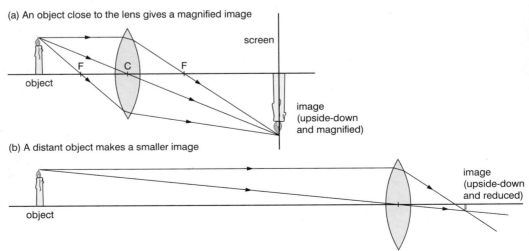

(a) An object close to the lens gives a magnified image

screen

F   C   F

object

image
(upside-down
and magnified)

(b) A distant object makes a smaller image

object

image
(upside-down
and reduced)

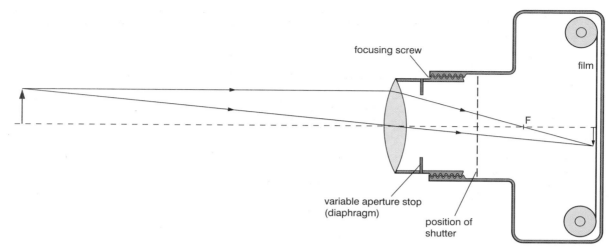

focusing screw

film

F

variable aperture stop
(diaphragm)

position of
shutter

*A simple camera. The image on the
film is smaller than the object and
upside down.*

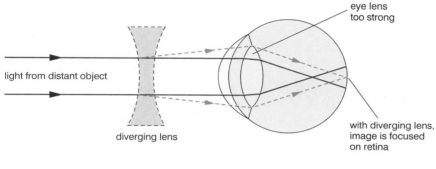

eye lens
too strong

light from distant object

diverging lens

with diverging lens,
image is focused
on retina

*Short sight. The eye lens focuses the
image in front of the retina. This is
corrected using a concave or
diverging lens.*

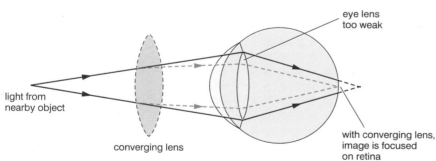

eye lens
too weak

light from
nearby object

converging lens

with converging lens,
image is focused
on retina

*Long sight. The eye lens is not strong
enough to focus the image on the
retina. This is corrected using a
convex or converging lens.*

Refraction of light can also produce some odd effects. When a pencil is put into
water it appears to be bent. Water in a swimming pool appears to be shallower
than it really is. These effects are due to the bending of the rays of light.

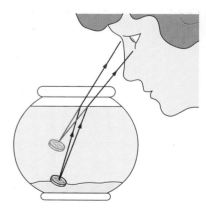

*The rays of light appear to come from
an imaginary coin higher in the water.*

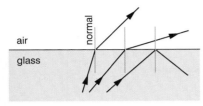

Total internal reflection occurs when a ray of light tries to leave the glass. If the angle of incidence equals or is greater than the critical angle the ray will be totally internally reflected.

Light does not escape from the fibre because it always hits it at an angle greater than the critical angle and is internally reflected.

TOTAL INTERNAL REFLECTION

When rays of light pass from a slow medium to a faster medium they move away from the normal. As the angle of incidence is progressively increased, an angle is reached at which the light rays will have to leave with an angle of refraction greater than 90°! These rays cannot refract, so they are entirely reflected back inside the medium. This process is known as **total internal reflection**. The angle of incidence at which all refraction stops is known as the **critical angle** for the material. The critical angle of glass is 42°, the critical angle of water is 49°.

Total internal reflection is used in fibre-optic cables. These are made up of large numbers of very thin, glass fibres. The cables can bend around corners, but the light continues along the fibres by being constantly internally reflected. Fibre-optic cables in the form of medical endoscopes can be used for internal examination of the body.

Telephone and TV communications systems are increasingly relying on fibre optics instead of the more traditional copper cables. Fibre-optic cables do not use electricity and the signals are carried by infrared rays. The signals are very clear as they don't suffer from electrical interference. Other advantages are that they are cheaper than the copper cables and can carry thousands of different signals down the same fibre at the same time.

Check yourself

QUESTIONS

Q1
a) Rays of light can be reflected and refracted. State one difference between reflection and refraction.
b) The diagram shows a glass block and two rays of light.

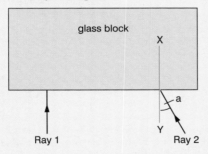

(i) Complete the paths of the two rays as they pass into and then out of the glass block.
(ii) What name is given to the angle marked a?
(iii) What name is given to the line marked XY?

Q2
a) Tom looks into the mirror. E is his eye and X an object.

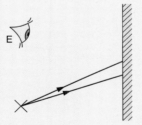

Complete the ray diagram to show clearly how Tom sees an image of X.
b) Jane uses two mirrors to look at the back of her head. Explain how this works.

QUESTIONS

Q3 The diagram shows light entering a prism. Total internal reflection takes place at the inner surfaces of the prism.

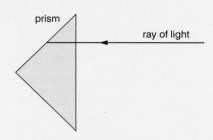

a) Complete the path of the ray.
b) Suggest one use for a prism like this.
c) Complete the table about total internal reflection.

Use T for total internal reflection or R for refraction.

angle of incidence / degrees	total internal reflection (T) or refraction (R)
36	
42 (critical angle)	
46	

Q4 Give one use of each of the following:
a) A convex mirror.
b) A concave mirror.
c) A convex lens.
d) A concave lens.

REMEMBER! Cover the answers if you want to.

ANSWERS

A1 a) In reflection light changes direction when it bounces off a surface. In refraction the light changes direction when it passes from one medium to another.
b) i)

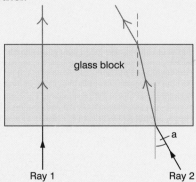

Ray 1   Ray 2

(ii) The angle of incidence. (iii) The normal line

TUTORIAL

T1 a) *In reflection the angle of incidence and the angle of reflection are the same. In refraction the angle of incidence does not equal the angle of refraction, as the ray of light bends towards or away from the normal.*
b) *The ray of light will only be bent if it hits the block at an angle. In both cases the speed of the light in the block will be slower than in air.*

ANSWERS

TUTORIAL

A2 a)

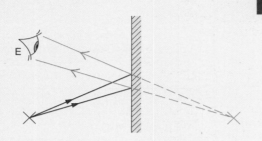

b) Light will be reflected off the hair onto the first mirror. If the angle is correct the reflected ray will then travel to the second mirror. If the angle is correct it will then be reflected into the eye.

A3 a)

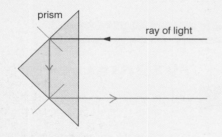

b) Binoculars, bicycle reflector, cat's eyes.

c) R, T, T.

A4 a) Car driving mirror, shop security mirror.

b) Make-up and shaving mirrors, torches, car headlamps.

c) Magnifying glass, camera, telescope, microscope, projector, spectacles.

d) Spectacles (correcting short sight).

T2 a) *Remember that with a plane mirror the image is the same size as the object and will be the same distance behind the mirror as the object is in front. When drawing the ray diagram check that the angles of incidence and reflection are the same.*

b) *Don't forget that the light travels from the object to the eye! In an examination you might find it easier to draw a diagram showing the passage of the ray of light. In this way you can show the correct angles on the mirrors.*

T3 a) *You should draw a normal line to ensure that for each reflection the angles of incidence and reflection are the same.*

b) *Periscopes are also often made using prisms rather than mirrors.*

c) *Total internal reflection only occurs when the angle of incidence is equal to or greater than the critical angle.*

T4 a) *Remember that a convex mirror gives a wider field of view than a plane mirror.*

b) *If close to a concave mirror a magnified image is formed. In torches and car headlamps the bulb is placed at the principal focus of the mirror thus forming a sharp beam of light.*

c) *This is the more commonly used lens. The eye contains a convex lens.*

d) *Concave lenses are diverging lenses. In short sightedness, without the concave lens the image is focused to a point in front of the retina. The image on the retina will therefore be blurred.*

SOUND

Sound waves are very different from electromagnetic waves. They travel much slower than the speed of light. A sound wave travels at about 340 m/s in the air. This is in contrast to the speed of light which is about 300 000 000 m/s. This explains why the flash of lightning is almost always seen before the crash of the thunder is heard. Sound is caused by vibrations and travels as longitudinal waves.

The compressions and rarefactions of sound waves result in small differences in air pressure. High **pitch** sounds have a high frequency whereas low pitch sounds have a low frequency. The human ear can detect sounds with pitches ranging from 20 Hz to 20 000 Hz. Sound with frequencies above this range is known as ultrasound. Loud sounds have a high amplitude whereas quiet sounds have a low amplitude. Sound amplitude is measured in **decibels**.

USING ULTRASOUND

Ultrasound is used in echo sounding or SONAR (SOund NAvigation and Ranging). An echo sounder on a ship sends out ultrasonic waves which reflect off the bottom of the sea. Knowing the time taken for the echo to be received and the speed of sound in water enables the depth of the water to be calculated.

Worked example

A ship sends out an ultrasound wave and receives an echo in 2 seconds. If the speed of sound in water is 1500 m/s, how deep is the water?

Write down the formula: $\text{speed} = \dfrac{\text{distance}}{\text{time}}$ or $v = \dfrac{s}{t}$

Rearrange to make s the subject: $s = v \times t$
Substitute the values for v and t: $s = 1500 \times 2 = 3000$ m

This is the distance travelled by the sound wave. Therefore the depth of the water must be half this. Depth $= 1500$ m.

Ultra sound is used in medicine for scanning patients. Ultrasonic waves will pass through material but dense material, such as bone, reflects more waves than less dense material, such as skin and tissue. Ultrasound waves have the advantage over X-rays of having lower energy and so are less likely to damage healthy cells.

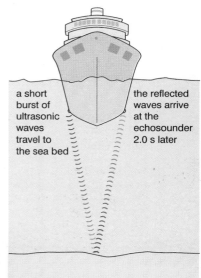

a short burst of ultrasonic waves travel to the sea bed

the reflected waves arrive at the echosounder 2.0 s later

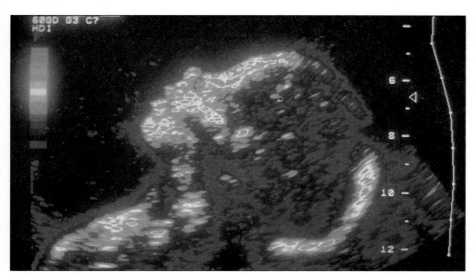

The fetus reflects the ultrasound waves more than the surrounding amniotic fluid.

Check yourself

QUESTIONS

Q1 In a thunderstorm why do we see the lightning before we hear the thunder?

Q2
a) (i) What causes a sound? (ii) Explain how sound travels through the air.
b) Astronauts in space cannot talk directly to each other. They have to speak to each other by radio. Explain why this is so.
c) Explain why sound travels faster through water than through air.

Q3 Ultrasonic waves are longitudinal waves.
a) What does the word 'ultrasonic' mean?
b) What does the word 'longitudinal' mean?
c) The waves travel through carbon dioxide more slowly than through air. How do the frequency and wavelength change when ultrasonic waves pass from air to carbon dioxide?

Q4 A fishing boat was using echo sounding to detect a shoal of fish. Short pulses of ultrasound were sent out from the boat. The echo from the shoal was detected 0.5 seconds later. How far away from the boat was the shoal of fish? (Sound waves travel through water at a speed of 1500 m/s.)

REMEMBER! Cover the answers if you want to.

ANSWERS

A1 Light waves travels faster than sound waves.

A2
a) (i) Vibrations. (ii) Compressions and rarefactions travel through the air.
b) Sound waves cannot travel through a vacuum, radio waves don't require a medium.
c) Particles in water are closer together so the vibrations are passed on more quickly.

A3
a) Above the limit of audible hearing.
b) The motion of the medium is parallel to the direction of movement of the wave.
c) The wavelength decreases, the frequency remains the same.

A4 375 m.

TUTORIAL

T1 *Only when the lightning flash is directly overhead will there be no noticeable gap between the flash and the crack of thunder.*

T2
a) *These are both common question so don't forget. Sound is a longitudinal wave. The compressions and rarefactions are small differences in air pressure.*
b) *This is a big difference between sound waves and electromagnetic waves.*
c) *The denser the material the greater the speed of sound. In concrete, for example, the speed of sound is 5000 m/s.*

T3
a) *The frequency of ultrasonic waves is too high for the human ear to hear.*
b) *This contrasts with transverse waves where the medium moves at right angles to the movement of the wave.*
c) *You should remember that $v = f \times \lambda$; v decreases, f stays the same so λ must decrease.*

T4 $s = v \times t \times \frac{1}{2} = 1500 \times 0.5 \times 0.5 = 375\,m.$

EXAMINATION CHECKLIST

The facts and ideas that you should understand by studying this topic.

To be successful at Foundation GCSE Tier you should be able to:

- recall the difference between a transverse and longitudinal wave
- recall the main features of transverse waves including crest, trough, amplitude, wavelength and frequency
- recall that waves can be reflected and refracted
- recall that high frequencies mean short wavelengths and that low frequencies mean long wavelengths
- recall that waves transfer energy without transferring matter
- recall the order of the parts of the electromagnetic spectrum in terms of increasing wavelength: gamma rays, X-rays, ultraviolet, visible light, infrared, microwaves and radio waves
- recall the uses and problems of each part of the electromagnetic spectrum
- explain how the reflection of light enables objects to be seen
- describe some common uses of mirrors and lenses
- draw simple ray diagrams to show reflection and refraction
- recall that light travels faster than sound
- recall that the frequency of ultrasound is higher than the upper threshold of human hearing
- describe some of the medical applications of ultrasound
- recall that waves can be transmitted through the Earth.

In addition, to be successful at Higher GCSE Tier you should be able to:

- describe the features of longitudinal and transverse waves
- explain how diffraction occurs
- use the formula $v = f\lambda$
- explain the uses and effects of the main areas of the electromagnetic spectrum in terms of their physical properties
- explain how refraction occurs
- explain total internal reflection and its application in fibre optics
- explain how waves transmitted through the Earth can be used to provide evidence for its structure
- recall that longitudinal P-waves travel through solid and liquid whereas transverse S-waves cannot travel through liquid.

Key Words

Tick each word when you are sure of its meaning.

amplitude
compression
critical angle
decibels
diffraction
electromagnetic spectrum
frequency
lateral inversion
longitudinal
medium
pitch
rarefaction
real image
reflection
refraction
seismic waves
total internal reflection
transverse
virtual image
visible spectrum
wavelength

EXAM PRACTICE

Sample Student's Answers & Examiner's Comments

a) The positions of the UV and IR regions of the electromagnetic spectrum have been correctly identified.

b) Again the correct selections have been made.

c) (i) The candidate appeared to be confident in tackling this question but has made two mistakes. The correct formula for speed in terms of distance and time has been given and 1 mark is scored for this. However, the distance has been converted into metres but the velocity has not. Note that the figures could have been used as they were given, i.e. distance in km and speed in km/s. One mark has been lost for incorrect use of units. The final mark has been lost for not appreciating that the radar waves actually travelled 100 km (to the plane and back).
(ii) The first mark is for correctly explaining why radar waves are better than sound waves for detecting aircraft. The second mark is for explaining why sound waves are better for use in water. A common mistake with this type of question is to only answer the first part.

● A mark of 8/10 corresponds to a grade A on this higher tier question.

---

1 Part of the electromagnetic spectrum is shown below:

gamma rays	R	S	visible light	T	micro-waves	radio waves

a) Which letter shows the position of:

(i) ultraviolet radiation? letter S ✓ (1)

(ii) infra-red radiation? letter T ✓ (1)

b) Name the electromagnetic radiations which are described by the following sentences.

(i) This radiation **cannot** be detected by the eye but is emitted by hot objects.
    infra-red ✓ (1)

(ii) These are the shortest radio waves – they are used in communications and to produce heat.
    microwaves ✓ (1)

(iii) This radiation has a short wavelength and is emitted by unstable nuclei.
    gamma rays ✓ (1)

c) (i) A radar station detects a jet aircraft when it is at a distance of 50 km. This is shown in the diagram below.

How long does it take for the radar waves to travel from the transmitter to the aircraft and back to the receiver? Show how you obtain your answer.

Note: All electromagnetic radiation has a velocity of 3×10^5 km/s.

$$speed = \frac{distance}{time} \checkmark \quad or \quad time = \frac{distance}{speed}$$

$$time = \frac{50,000 \text{ m.}}{3 \times 10^5} \text{ ✗} = 0.167 \text{ sec.}$$ (3)

(ii) Explain **one** reason why radar waves are used to detect aircraft while sound waves are used to detect submerged vessels.

radar waves travel much faster through the air than sound ✓ waves do but radar waves are absorbed by water particles whereas sound waves are not. ✓ (2)

(Total 10 marks)

8/10

Southern Examination Group

Questions to Answer

Answers to questions 2 and 3 can be found on page 373.

2 Diagrams A, B, C and D show oscilloscope traces of four different sounds.

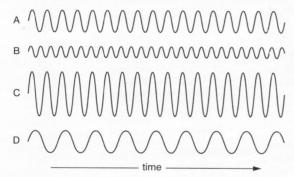

time

a) Which diagram, A, B, C or D, represents:

   (i) the loudest sound? (1)

   (ii) the sound with the highest pitch? (1)

b) The sound labelled C has a frequency of 500 Hz and a speed of 340 m/s.

   Calculate the wavelength of this sound. Show your working. (3)

c) The drawing shows how the speed of P-waves change as they travel through the Earth.

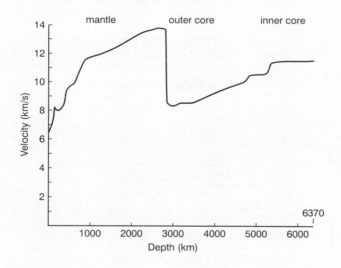

   (i) Explain why the speed of the waves falls at a depth of 3000 km. (2)

   (ii) Explain what evidence the speed of the P-waves through the Earth has given scientists about its internal structure. (4)

Northern Examination and Assessment Board

3 a) Look at the diagram. Some of the parts of the electromagnetic spectrum have been named.

gamma waves	X-rays	A	visible light	B	radio waves

    (i)   Write down the name of part A.   (1)

    (ii)   Write down the name of part B.   (1)

b) The diagram shows two plane mirrors held at 90° to each other. A ray of light shines onto one of the mirrors. It makes an angle of 45° with the mirror.

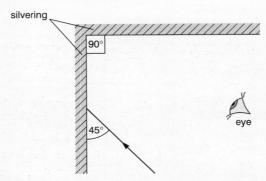

    (i)   Draw and label the normal line at the point where the ray hits the mirror.   (1)

    (ii)   The ray of light is reflected by the mirrors and reaches the eye.

          Carefully draw the path that the ray takes. Label the sizes of the angles.   (1)

c) The diagram shows a ray of light shining onto the end of a fibre optic cable.

    (i)   Continue the path of the ray of light. Put your answer on the diagram above.   (1)

    (ii)   Fibre optics can be used for telecommunications. They are replacing telephone wiring. Explain the advantages of fibre optics.   (3)

Midland Examining Group

CHAPTER 19

THE EARTH AND BEYOND

The Earth is just one of the planets orbiting the small star that we call the Sun. The Sun is just one star situated near the end of one of the spiral arms of the Milky Way galaxy – a galaxy swarming with billions of stars. The Milky Way is just one of the billions of galaxies that make up the Universe.

Even though we occupy a tiny corner of the Universe, scientists can still study it on a grand scale. The everyday force of gravity plays a big part in the movement of the planets, the life cycle of stars and the future of the Universe.

THE SOLAR SYSTEM

The **solar system** is made up of the Sun and its planets. The planets are **satellites** of the Sun and are kept in **orbit** by the gravitational pull from the Sun. The orbits of the planets are slightly oval or elliptical.

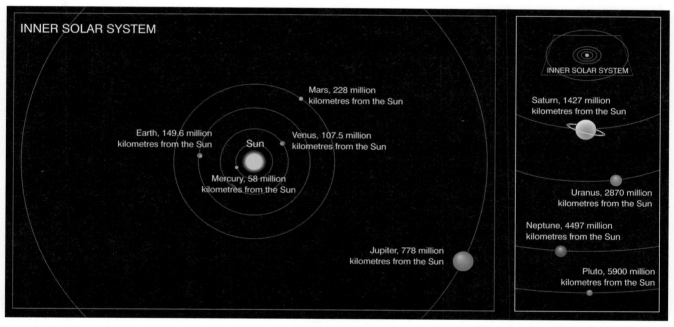

Nine planets orbit the Sun.

Asteroids are fragments of rock, up to 1000 km in diameter, that orbit the Sun between the four inner planets and the five outer planets. The asteroids were formed at the same time as the Solar System.

Comets go round the Sun like planets but their orbits are much more elliptical. Comets spend much of their time too far away from the Sun to be seen. They are thought to be made from frozen ice and rock. When they get close to the Sun some of the solid turns into gas, forming a 'tail' which points away from the Sun.

A comet moves round the Sun in an elliptical orbit.

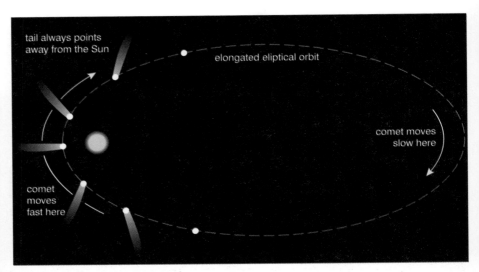

The Halle Bopp comet was last easily visible from the Earth in March 1997.

Centripetal force

Once you start something moving, it will carry on in a *straight line* until another force acts on it. This is more obvious in deep space where there is no gravity or friction. (Outside the Earth's atmosphere there is, of course, no air resistance.) So how can natural and artificial satellites orbit the Earth in a *circle*? For the same reason that the planets orbit the Sun – gravitational pull.

To keep an object moving in a circle, you have to constantly pull it towards the centre of that circle. A force that constantly pulls towards a centre is called a **centripetal force**. Near the Earth, gravity exerts a force that always acts towards its centre. Gravity keeps satellites like the Moon moving in a near-circular orbit around the Earth. Artificial satellites orbit the Earth for the same reason, and as they don't need motors or rockets to stay in such orbits, they can stay there for decades.

At the right distance from the Earth, an unpowered orbit takes exactly 24 hours. This means that an artificial satellite orbiting above the Earth's equator and moving in the same direction as the Earth's spin, will appear to be flying over the same spot

GALAXIES

The Sun is part of a groups of stars called a **galaxy**. Our Sun is part of the Milky Way galaxy. The Milky Way is in the shape of a spiral. We can see other galaxies through telescopes. These other galaxies are so far away that it is difficult to see the individual stars, so earlier astronomers called them **nebulae** (singular: nebula) meaning 'a bright cloudy spot'. These days, we can see these galaxies properly, and the word nebula now usually means a cloud of gas where stars are being born. Many nebulas are formed by stars exploding.

The Milky Way galaxy from the 'side' and from 'above'. Most stars are in the galactic centre but our solar system is near to the end of one of the spiral arms.

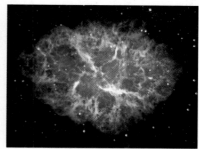

The Crab Nebula is the remains of a supernova. It is made of material blown away when a massive star exploded.

Life-cycle of a star

Birth (Thousands of millions of years)	Life (Tens of millions of years)	Death (Millions of years)
● The star starts as a huge cloud of hydrogen gas and dust.	● Huge gravitational forces pull the outer parts of the star towards its core.	● The hydrogen in the star's core gets used up and the core starts to cool.
● Gravity pulls the hydrogen atoms closer together.	● The energy produced by the nuclear fusion ensures that the gases on the outside are very hot.	● The star then starts to collapse and the core gets hotter again very quickly enabling other nuclear fusion reactions to start.
● The gas gets hotter as the cloud gets smaller and more concentrated.	● The pressure exerted by the hot gases exactly balances the force of gravity.	● A sudden surge of radiation is produced and the star expands to form a red giant or red supergiant.
● In the very high temperatures created, the hydrogen atoms join together to form helium atoms in a process called nuclear fusion. Large amounts of energy are released in this process.	● This balance continues for millions of years until all the hydrogen has been used up and the nuclear reactions stop.	● In a medium-sized star like the Sun, the nuclear fusion reactions finish and the core collapses under gravity and the red giant forms a white dwarf star.
● The core sends out light and other radiation and the cool, outer layers are blown away, sometimes forming planets around the new star.	● Some stars appear red-orange but hotter stars appear blue-white.	● In a large star, when the core collapses there is a sudden explosion and the red supergiant forms a supernova. A supernova is brighter than a whole galaxy of stars. A very dense core is left behind becoming a neutron star and, if it is very small and dense, a black hole.

Check yourself

QUESTIONS

Q1 The same area of sky was observed at the same time on two successive nights.

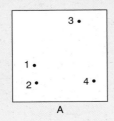 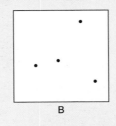

A   B

One of the 'stars' in diagram A is a planet. Which one is it? Explain your choice.

Q2 How does a normal star produce energy?

Q3 The list contains the stages in the life cycle of a star. Arrange them in the correct order.

**Supernova Clouds of hydrogen gas
Red supergiant Blue star**

Q4 a) A satellite moves around the Earth at a constant speed in a circular orbit. (i) What force is acting on the satellite and (ii) what is the direction of this force?
  b) Some satellites travel in geostationary orbits. Explain what is meant by a geostationary orbit.

REMEMBER! Cover the answers if you want to.

ANSWERS

A1 2. Its position changes.

A2 Hydrogen nuclei join together to make helium nuclei in a process know as nuclear fusion.

A3 Clouds of hydrogen gas → blue star → red supergiant → supernova.

A4 a) (i) The force due to gravity, its weight, which (ii) acts towards the centre of the Earth.
  b) The satellite is always at the same position above the surface of the Earth.

TUTORIAL

T1 *The word 'planet' comes from an old word for 'wanderer'.*

T2 *Don't get confused with nuclear fission which is the process used to obtain energy in nuclear power stations. Very high temperatures are required before nuclear fusion will take place.*

T3 *The sequence is given in the table on page 341. A supernova is only produced from a very large star. A star such as our Sun will eventually form a white dwarf.*

T4 a) *Gravity always pulls towards the centre of the Earth. The satellite's weight is providing the centripetal force required for it to move in a circle.*
  b) *The ring directly above the equator at the height of geostationary orbits is full of communication satellites.*

THE ORIGIN OF THE UNIVERSE

Evidence for the way that the **Universe** formed has been obtained by analysing the frequency of light received from distant galaxies.

When a police car is travelling towards you the pitch or **frequency** of the siren is higher than when the police car is moving away from you. This effect is known as the **Doppler Effect**. In a similar way the frequency of light we obtain from a star is affected by the movement of the star. If the star is moving away from the Earth, light waves reach the Earth with a lower frequency than that emitted by the star. This change in frequency is called a **red shift**. Light from all the distant galaxies shows a red shift, so it follows that all these galaxies are moving away from us. In fact, all galaxies are moving away from each other. The Universe is expanding.

Galaxies are moving away from each other. The more distant galaxies are moving at the greatest speed.

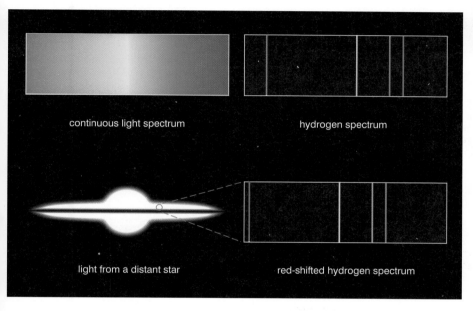

continuous light spectrum

hydrogen spectrum

light from a distant star

red-shifted hydrogen spectrum

The top picture shows the spectrum of white light. The second picture shows the spectrum produced by hydrogen when it glows. The bottom picture shows the spectrum of hydrogen from a star seen through a telescope. The pattern is shifted towards the red end. This red shift shows that the star is moving away from us.

THE BIG BANG THEORY

Galaxies are moving apart from each other. One explanation for this is that everything in the Universe originated from the same point and then exploded apart. This theory is known as the **Big Bang**. Since this explosion, the Universe has been expanding and cooling. Calculations suggest that the Big Bang could have occurred between 11 and 18 billion years ago.

Astronomers have been able to make predictions about the future of the Universe. These predictions depend on the mass of the Universe (which isn't known with any certainty) and the speed at which it is expanding. There are three main possibilities:

- **The Universe will keep on expanding**. The galaxies move fast enough to overcome the forces of gravity acting between them.

- **The expansion will slow down and stop**. The galaxies will remain in a fixed position. The gravitational forces will exactly balance the 'expansion forces'.

- **The Universe will start to contract**. The force of gravity acting between the galaxies will overcome the forces causing expansion and pull them back together again.

Check yourself

QUESTIONS

Q1 What is the Doppler Effect?

Q2 What is the evidence that supports the big bang theory?

·········· **REMEMBER! Cover the answers if you want to.** ··········

ANSWERS

A1 The change in frequency of a wave due to the relative motion between observer and emitter.

A2 Red shift evidence shows that all galaxies are moving apart from each other. Reversing the direction of movement of the galaxies shows that they could all have come from the same starting point.

TUTORIAL

T1 *If the emitter is moving away from the observer the frequency is reduced – if the emitter is moving towards the object the frequency is increased. In the case of light waves, a reduction in frequency results in a shift to the lower frequency (red) end of the visible spectrum.*

T2 *It was the famous astronomer Edwin Hubble who first observed the red shift and then proposed the big bang theory.*

Key Words

Tick each word when you are sure of its meaning.

asteroid
Big Bang
black hole
centripetal force
comet
Doppler Effect
galaxy
nebula
neutron star
nuclear fusion
red giant
red shift
red supergiant
satellite
solar system
supernova
Universe
white dwarf

EXAMINATION CHECKLIST

The facts and ideas that you should understand by studying this topic.

To be successful at Foundation GCSE Tier you should be able to:

- recall that the Earth is one of a number of planets that orbit the Sun
- describe the relative positions of the Earth, Moon, Sun and planets
- recall that gravitational forces determine the motion of planets, comets and satellites
- recall that stars are very hot and give off light
- recall that the Universe consists of stars, planets and large groups of stars called galaxies
- describe the relative sizes and distances of stars, planets and galaxies.

In addition, to be successful at Higher GCSE Tier you should be able to:

- recall that circular motion requires a centripetal force
- recall that gravity provides the centripetal force for orbital motion
- describe the life history of stars
- recall ideas about the Universe expanding and the Big Bang theory.

EXAM PRACTICE

Sample Student's Answers & Examiner's Comments

a) The candidate has correctly chosen 'red shift'.

b) **(i)** There are two marks for this part. One mark for appreciating that the shift has occurred to longer wavelength or lower frequency, the other mark for stating that this shows the galaxy is moving away from the Earth.
(ii) One mark has been gained for stating that the greater shift shows that the distant galaxy is moving at a faster speed than the nearer one.

c) The answer lacks the detail required for 3 marks. A mark has been scored for mentioning 'gravity'. Additional marks are awarded for stating that the gravity could pull the galaxies together (1 mark) so that they move apart more slowly or actually start to move towards each other (1 mark).

- A mark of 5/7 corresponds to a grade B on this higher tier question.

---

1 If light from a galaxy is analysed in a spectrometer, the spectrum is found to contain lines at certain frequencies. This general pattern of lines is similar from all galaxies and can be matched by light from a source on the Earth.

The diagrams show spectra from a stationary Earth source, a nearby galaxy and a distant galaxy.

Earth

near galaxy

distant galaxy

violet light

wavelength getting longer

red light

a) What is this apparent change in frequency called?
   red shift ✓ .... (1)

b) Explain what this information shows about the movement, relative to the Earth, of:

   (i)   the nearby galaxy;
         the shift is to longer wavelength (lower frequency) ✓ so
         galaxy moving away from Earth ✓ .... (2)

   (ii) the distant galaxy.
         greater shift therefore distant galaxy moving away even
         faster ✓ .... (1)

c) Some theories suggest that the rate of expansion of the Universe is slowing down. What could cause the rate of expansion to slow down? Explain briefly why it would slow down the rate.
   the force of gravity ✓ ....
   ....
   ....
   .... (3)

   5/7

   (Total 7 marks)

   Midland Examining Group

Questions to Answer

Answers to this question can be found on page 374.

2 a) The following table gives information about some of the planets.

Planet	Distance from the Sun (million kilometres)	Radius (kilometres)	Average Density (g/cm^3)
Earth	149	6350	5.52
Jupiter	773	70960	1.33
Mars	227	3360	3.94
Mercury	58	2400	5.43
Uranus	2886	25275	1.3
Venus	108	6025	5.24
Neptune	4469	25200	1.76

Use the information given in the table to answer the following questions.

(i) On which of these planets is the surface temperature likely to be the lowest? (1)

(ii) Which of the planets is closest to Earth? (1)

(iii) What pattern is there between size and density of the planets? (1)

(iv) What pattern is there between the density of the planets and their position in the Solar System? (1)

b) (i) Explain how stars form from large gas clouds. (3)

(ii) Use words from the following list to show the Sun's likely evolution.

neutron star   red giant   supernova   white dwarf (2)

(iii) Name and explain the process by which energy is produced in the core of the Sun. (3)

London Examinations

Radioactivity was discovered at the end of the nineteenth century when a scientist called Henri Becquerel noticed that uranium blackened photographic film even in the dark. Today the debate about nuclear radiation and nuclear energy continues. Knowing more about radioactivity, its benefits and drawbacks, can help to inform this nuclear debate.

UNSTABLE ATOMS

By now, you should be very familiar with the structure of atoms: a positively charged central **nucleus** surrounded by shells, or orbits, of moving **electrons**. The nucleus is made of positively charged **protons** and neutral **neutrons**. Getting all those positive charges packed into the nucleus is not always easy – remember that like charges repel! Still, most nuclei are very stable and stick together. But some nuclei are unstable, they 'decay' and break apart into more stable nuclei. This breaking apart is called **nuclear fission**.

When unstable nuclei break apart, they give out particles, radiation and energy. Atoms whose nuclei do this are said to be **radioactive**. There are three main types of things emitted from a nucleus. They are called **alpha** (α), **beta** (β) and **gamma** (γ) rays. A stream of these rays is referred to as **ionising radiation** (often called nuclear radiation, or just 'radiation' for short).

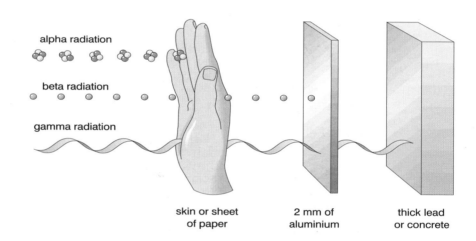

alpha radiation

beta radiation

gamma radiation

| skin or sheet of paper | 2 mm of aluminium | thick lead or concrete |

	alpha (α)	**beta (β)**	**gamma (γ)**
Description	A positively charged particle, identical with a helium nucleus (two protons and two neutrons)	A negatively charged particle, identical to an electron.	Electromagnetic radiation. Uncharged.
Penetration	4–10 cm of air. Stopped by a sheet of paper.	About 1 m of air. Stopped by a few mm of aluminium	No limit in air. Stopped by several cms of lead or several metres of concrete.
Effect of electric and magnetic fields	Deflected.	Deflected considerably.	Unaffected – not deflected.

Evidence for the 'nuclear' atom

Alpha particles were used in a famous experiment, back in 1911, that was the first evidence for the familiar 'nuclear' structure of the atom. In the last century, many scientists believed that atoms were just spheres of positive charge embedded with specks of negative charge (electrons).

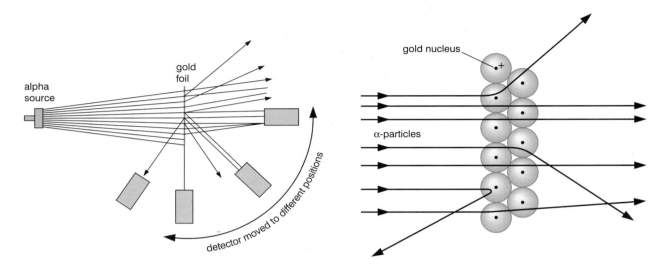

In the crucuial experiment, a stream of alpha particles was aimed at a piece of gold foil a few atoms thick. Most of the alpha particles passed through the foil undeflected. This showed that most of the atom is empty space. Some alpha particles passed through but were deflected and a few bounced back as if they had done a U-turn.

The deflections meant that the atom's mass and charge isn't spread out but must be clustered together. The very small number of U-turns must have been caused by 'head-on' collisions with a concentration of mass and positive charge. This meant that such clusters were very small. Altogether, this showed that an atom has a tiny cluster of positive charge – a nucleus.

BACKGROUND RADIATION

Radioactivity is measured using a Geiger–Müller tube linked to a counter.

All ionising radiation is invisible to the naked eye, but it affects photographic plates. Individual particles of ionising radiation can be detected by a Geiger–Müller tube. If you link a Geiger–Müller tube to a counter you can measure the number of particles of radiation that enter the tube.

Even without a radioactive source nearby, you will always find ionising radiation wherever you go. There is always ionising radiation present. This is called **background radiation**. Background radiation is caused by:

- radioactivity in soil, rocks and materials like concrete
- radioactive gases in the atmosphere
- cosmic rays from the Sun.

HALF-LIFE

The **activity** of a radioactive source is the number of ionising particles it emits each second. Over time, fewer nuclei are left in the source to decay, so the activity drops. The time taken for half the radioactive atoms to decay is called the **half-life**.

Starting with a pure sample of radioactive atoms, after one half-life half the atoms will have decayed. The remaining undecayed atoms still have the same chance of decaying as before, so after a second half-life, half of the remaining atoms will have decayed. So a quarter of the atoms will remain undecayed after two half-lives.

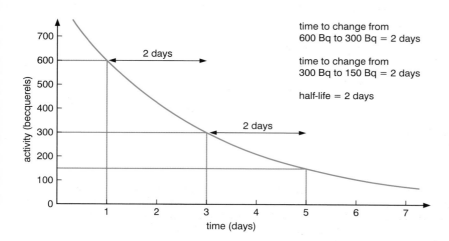

time to change from
600 Bq to 300 Bq = 2 days

time to change from
300 Bq to 150 Bq = 2 days

half-life = 2 days

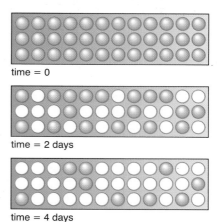

The half-life is 2 days. Half the number of radioactive atoms decays in 2 days.

Worked example

A radioactive element is detected by a Geiger–Müller tube and counter as having an activity of 400 counts per minute. Three hours later the count is 50 counts per minute. What is the half-life of the radioactive element?

Write down the activity and progressively halve it. Each halving of the activity is one half-life:

0	400 counts
1 half-life	200 counts
2 half-lives	100 counts
3 half-lives	50 counts

3 hours therefore corresponds to 3 half-lives. 1 hour therefore corresponds to 1 half-life.

NUCLEAR EQUATIONS

In the same way that chemical changes can be written as chemical equations, you can write down nuclear changes as nuclear equations. Each nucleus is represented by its chemical symbol with two extra numbers written before it. Here is the symbol for Radium-226:

the top number is the **mass number**
(the total number of protons and neutrons)

the bottom number is the **atomic number**
(the number of protons)

$$^{226}_{88}\text{Ra}$$

The mass numbers and atomic numbers must balance on both sides of a nuclear equation.

Alpha decay – the nucleus emits an α-particle
(2 protons and 2 neutrons)

radium-226 (parent nucleus)   radon-222 (daughter nucleus)

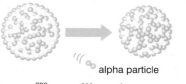

alpha particle

$$^{226}_{88}\text{Ra} \longrightarrow ^{222}_{86}\text{Rn} + ^{4}_{2}\alpha$$

Beta decay – a neutron changes into a proton in the the nucleus

polonium-218 (parent nucleus)   astatine-218 (daughter nucleus)

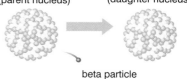

beta particle

$$^{218}_{84}\text{Po} \longrightarrow ^{218}_{85}\text{At} + ^{0}_{-1}\beta$$

Check yourself

QUESTIONS

Q1 List three causes of natural background radiation.

Q2
a) Which type of nuclear radiation is in the form of a helium nucleus?
b) Describe the composition of a helium nucleus.
c) Use the helium nucleus to explain the difference between atomic number and mass number.

Q3 The diagram shows some of the results of an experiment using some very thin gold foil that was carried out by the scientist Sir Ernest Rutherford.

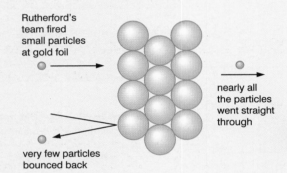

Rutherford's team fired small particles at gold foil

nearly all the particles went straight through

very few particles bounced back

a) What type of radioactive particle did Rutherford use?
b) What did Rutherford deduce from the observation that most of the particles passed straight through the gold foil?
c) What did Rutherford deduce from the observation that some particles bounced back from the foil?

Q4 The following graph shows how the activity of a sample of sodium-24 changes with time. Activity is measured in 'becquerels' (Bq).

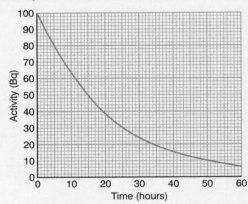

a) Sodium-24 has an atomic number of 11 and a mass number of 24. What is the composition of the nucleus of a sodium-24 atom?
b) Use the graph to work out the half-life of the sodium-24.

Q5 At 9 a.m., a radioactive atom has a count rate of 600 counts per minute. At 10 a.m., the count rate is 300 counts per minute. What will the count rate be at 12 noon?

Q6 The following equation shows what happens when a nucleus of sodium-24 decays.

$$^{24}_{11}\text{Na} \rightarrow ^{x}_{y}\text{Mg} + ^{0}_{-1}\beta$$

a) What type of nuclear radiation is produced?
b) What are the numerical values of x and y?

REMEMBER! Cover the answers if you want to.

ANSWERS

A1 Radioactivity in rocks, concrete and soil; cosmic rays from the Sun; radioactive gases in the atmosphere.

TUTORIAL

T1 *Background radiation varies depending on which part of the country is tested. Cornwall tends to have high background counts due to the large amounts of granite present in the Earth's surface in Cornwall.*

ANSWERS

A2
a) Alpha radiation.
b) 2 protons and 2 neutrons.
c) Atomic number is the number of protons (in the case of helium this equals 2). Mass number is the number of protons and neutrons (in the case of helium this equals 4).

A3
a) Alpha particles.
b) Most of the atom is empty space.
c) The atom must have a very high concentration of mass and positive charge. Rutherford called this the nucleus.

A4
a) 11 protons, 13 neutrons.
b) 15 hours.

A5
75 counts per minute.

A6
a) beta particle.
b) $x = 24$; $y = 12$.

TUTORIAL

T2
More details on atomic structure are given in chapter 12. Atoms are ordered in the periodic table in terms of their atomic numbers.

T3
a) Gamma particles would not have deflected at all as they are uncharged. Beta particles might work, but they deflect very easily because they have such a small mass.
c) The neutron wasn't discovered until 1932.

T4
a) The atomic number gives the number of protons. The difference between the mass number and the atomic number equals the number of neutrons.
b) The count falls from 100 Bq to 50 Bq in 15 hours. It also falls from 50 Bq to 25 Bq in 15 hours.

T5
The count rate has halved in 1 hour so the half-life must be 1 hour. The count rate will halve again to 150 in another hour (11 a.m.) and halve again to 75 in yet another hour (12 noon).

T6
a) A beta particle is an electron. The 'beta' symbol can also be written as an electron 'e'. The electron is shown with an atomic number of −1 and a mass number of 0.
b) The mass numbers must balance on the left-hand and right-hand sides of the equation (24 = 24 + 0). The atomic numbers must balance on the left hand and right hand sides of the equation (11 = 12 − 1).

THE USES AND DANGERS OF RADIOACTIVITY

Radioactive materials can be dangerous. Alpha, beta and gamma radiation can all damage living cells. Alpha particles, due to their strong ability to ionise other particles, are particularly dangerous to human tissue if they get inside your body. Gamma radiation is dangerous because of its high penetrating power. Nevertheless, radiation can be very useful – it just needs to be used safely.

- Sterilising – Gamma rays can be used to kill bacteria. This is used in sterilising medical equipment and in preserving food. The food can be treated after it has been packaged.

- Smoke detectors – A smoke alarm includes a small radioactive source that emits alpha radiation. The radiation produces ions in the air which conduct a small electric current. If a smoke particle absorbs the alpha particles, it reduces the number of ions in the air, and the current drops. This sets off the alarm.

Some smoke detectors contain a radioactive source.

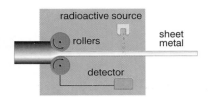

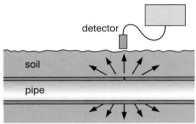

Sheet thickness control.

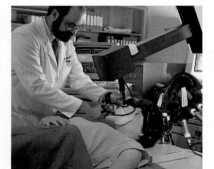

Tracers detect leaks.

- Thickness measurement – Beta particles are used to monitor the thickness of paper or metal. The number of beta particles passing through the material is related to the thickness of the material. The detector can automatically move the rollers closer together or further apart.

- Checking welds – A gamma source is placed on one side of a weld and a photographic plate on the other side. Weaknesses in the weld will show up on the photographic plate.

- Medical – In radiotherapy high doses of radiation are fired at cancer cells to kill them.

- **Tracers** – Radioactive substances with half-lives and radiation types that suit the job they are used for. The half-life must be long enough for the tracer to spread out and to be detected after use but not so long that it stays in the system and causes damage.

    Medical – detecting blockages in vital organs. A gamma camera is used to monitor the passage of the tracer through the body.

    Agricultural – monitoring the flow of nutrients through a plant.

    Industrial – monitoring the flow of liquid and gases through pipes to identify leakages.

Inhaling a medical tracer.

RADIOACTIVE DATING

Igneous rocks are made when molten rock from inside the Earth reaches the surface and cools down. The rock contains small quantities of uranium-238 – a type of uranium that decays with a half-life of 4500 million years, eventually forming lead. The ratio of lead to uranium in a rock sample can therefore be used to calculate the age of the rock. For example, a piece of rock with equal numbers of uranium and lead atoms in it must be 4500 million years old. This would be unlikely as the Earth is only 4000 million years old.

A similar method is used to find out the age of plant or animal remains. This method is called radioactive **carbon dating**. Carbon in living material contains a constant, small amount of the radioactive isotope carbon-14 which has a half life of 5700 years. When the living material dies the carbon-14 atoms slowly decay. The ratio of carbon-14 atoms to the non-radioactive carbon-12 atoms can be used to calculate the age of the plant or animal material.

NUCLEAR FISSION

The energy produced in a nuclear power station also comes from the decay of unstable nuclei, that is, nuclear fission. Atoms of uranium-235 are bombarded with neutrons. This causes the uranium atoms to split into two smaller atoms (barium and krypton). This fission of the atoms is accompanied by the release of energy, other neutrons and gamma rays. The products of the fission are radioactive and so need to be disposed of very carefully. However, the amount of energy produced is considerable. A single kilogram of uranium-235 produces approximately the same amount of energy as 2 000 000 kg of coal!

Don't confuse this process with the nuclear reactions taking place in the Sun. Fission describes nuclei *splitting* up. In the Sun, small atoms *combine* together to form larger atoms in a process known as **nuclear fusion**. Considerable amounts of energy are produced in this process too.

This woman's body was preserved in a bog in Denmark. Radioactive carbon-14 dating showed that she had been there for over 2000 years.

Check yourself

QUESTIONS

Q1 What type of radiation is used in (i) smoke detectors, (ii) thickness measurement, (iii) weld checking?

Q2 This question is about tracers.
a) What is a tracer?
b) The table below shows the half-life of some radioactive isotopes.

Radioactive Isotope	Half-life
lawrencium-257	8 seconds
sodium-24	15 hours
sulphur-35	87 days
carbon-14	5700 years

Using the information in the table only, state which one of the isotopes is most suitable to be used as a tracer in medicine. Give a reason for your choice.

Q3 a) When uranium-238 in a rock sample decays what element is eventually produced?
b) Explain how the production of this new element enables the age of the rock sample to be determined.

REMEMBER! Cover the answers if you want to.

ANSWERS

A1 (i) Alpha particles, (ii) beta particles, (iii) gamma rays.

A2 a) A tracer is a radioactive isotope used in detection.
b) Sodium-24. The lawrencium-257 has too short a half-life; the sulphur-35 and carbon-14 have half-lives which are too long.

A3 a) Lead.
b) The ratio of uranium-238 to lead-207 enables the age of the rock to be determined.

TUTORIAL

T1 Details of these uses are given on pages 351 and 352. Alpha particles are the most ionising, gamma rays are the most penetrating.

T2 a) Tracers are widely used to detect leaks and blockages.
b) The tracer must be radioactive for long enough for it to be detected after injection into the body but must not remain radioactive in the body for longer than necessary.

T3 a) In fact the uranium-238 decays through a chain of short-lived intermediate elements before forming lead.
b) In a similar way the carbon-14 : carbon-12 ratio is important in finding out the age of previously living material.

Key Words

Tick each word when you are sure of its meaning.

alpha particles

background radiation

beta particles

carbon dating

gamma rays

half-life

nuclear fission

nuclear fusion

radioactive

tracers

EXAMINATION CHECKLIST

The facts and ideas that you should understand by studying this topic.

To be successful at Foundation *GCSE Tier you should be able to*:

- recall that radioactivity results from the nucleus of an atom breaking up
- recall the main sources of background radiation
- recall the relative penetrating power of alpha (α), beta (β) and gamma (γ) radiation
- recall that an alpha (α) particle is a positively charged particle (a helium nucleus)
- recall that a beta (β) particle is a negatively charged particle (an electron)
- recall that a gamma (γ) ray is an uncharged electromagnetic wave
- recall that the emission from a radioactive substance decreases with time
- recall that nuclear radiation can damage healthy cells in the body
- recall some of the uses of nuclear radiation.

In addition, to be successful at Higher *GCSE Tier you should be able to*:

- explain the concept of half-life
- explain how the radioactive dating of rocks depends on the calculation of the uranium/lead ratio
- explain the medical uses of radioactivity including the use of tracers.

EXAM PRACTICE

Sample Student's Answers & Examiner's Comments

1 A radioactivity detector is placed near a radioactive source. The detector is used to take measurements. It is moved further away for each reading. Look at the diagram.

source   detector

a)   The detector can detect three kinds of radiation:

alpha   beta   gamma

(i)   Which kind of radiation is stopped by 5 cm of air?

alpha ✓ .... (1)

(ii)   Which radiation will pass through paper but not 3 mm of aluminium?

beta ✓ .... (1)

(iii)   Which kind of radiation is the most penetrative?

gamma ✓ .... (1)

b)   Nirvisha tries a different experiment with the radioactive detector.

She uses a different radiation source. She takes readings from the detector every 10 s.

Look at the graph of her results.

(i)   Use the graph to estimate the half life of the radiation source.

30 secs ✓ .... (1)

(ii)   Uranium is found in rock samples.

The uranium changes slowly into lead by radioactive decay.

The approximate age of a rock sample containing uranium can be determined. Explain how.

Analyse rock to find out the ratio of lead and uranium. ✓✓ Use the half-life ✓ to estimate how much uranium has decayed. The more lead there is the older the rock is. ✓

....

.... (4)

a)   *The correct answers have been given to parts (i), (ii) and (iii). The correct order of penetration of alpha, beta and gamma has been given.*

b)   *(i)   Only an estimate can be given, the graph is not accurate enough to give a precise value. The time taken for the reading to fall from 250 to 125 is approximately 30 seconds. An answer from about 25 to 40 s would be acceptable.*

(ii)   An extended answer is required here and there are 4 scoring points. The candidate has written a good answer. Reference to the amount of lead scores 1; reference to the ratio of lead to uranium scores a second mark. Appreciation that as more decay occurs, more lead is produced, thus showing the sample to be older also scores 1 mark. Reference to half-life scores the final mark.

EXAMINER'S COMMENTS

c) **(i)** *The candidate has calculated the correct answer. The step-wise halving of the activity is a good strategy and clearly shows how many half-lives are needed.*

(ii) *A mark has been scored for stating that a half-life of 30 years is too long. The second mark has been missed. Noting that the radioactive source would stay in the body too long is correct, but why is it too long? The consequence of it being too long has to be given, in this case the possible damage to healthy cells.*

● *A mark of 10/11 corresponds to a grade A on this higher tier question.*

c) (i) One type of radioactive iodine is used in medicine. It can be injected into the body to help diagnosis. Its half-life is 8 days. Its reading immediately after injection is 400 counts per minute.

What will the reading be after 24 days?

$400 \xrightarrow{8} 200 \xrightarrow{16} 100 \xrightarrow{24} 50$ .... 50 units ✓ .... (1)

(ii) Radioactive caesium has a half life of 30 years. Explain why this is not used to help medical diagnosis.

The half-life is too long ✓ and the radioactive source would stay in the body too long. .... (2)

⟨10/11⟩ (Total 11 marks)

Midland Examining Group

Questions to Answer

The answers to this question can be found on page 374.

2 a) Finish the table about alpha, beta and gamma radiations.

Type of radiation	What is it?	What type of charge does it carry?
alpha		
beta	electron	negative
gamma		

(4)

b) Technetium-99 is a radioactive isotope which is often used by hospitals in the examination of patients. It emits only gamma rays; it has a half-life of 6 hours.

A doctor wishes to examine the action of a patient's heart. He injects a substance containing technetium-99 into the blood. It mixes with the blood. As the substance passes through the heart, the radiation from it is detected outside the patient's body. Measurements are made while 500 heart beats take place.

(i) Why does the doctor choose a gamma emitter with a half-life of 6 hours rather than one with a half-life of 6 minutes or 6 days? (5)

(ii) A sample of technetium-99 has a count-rate of 6400 counts per minute. After 24 hours the count rate was calculated as 400 counts per minute. Explain how this calculation was made. (2)

(iii) When a single measurement of the count-rate was taken, it was 413 counts per minute. Suggest two reasons why the measured count-rate is different from the calculated value. (2)

Midland Examining Group

Marks scored are shown in ordinary brackets. Alternative answers that score the same mark are indicated by a slash (/), separate answers that score marks each are separated by semi-colons (;).

CHAPTER 1 LIFE PROCESSES AND CELLS

Answers

2 a) A = (cell) membrane (1), B = nucleus (1), C = cytoplasm (1).
   b) B/nucleus (1).

3 a) D/nucleus (1).
   b) C/cytoplasm (1).
   c) B/cell membrane (1).

4 1 = chloroplast (1), 2 = vacuole (1), 3 = cell wall (1), 4 = nucleus (1).

5 a) 54 (1).
   b) Cylinders in water absorbed water and cells swelled (1); cylinders in sugar solution lost water and cells shrank (1); loss or gain of water is related to the surrounding solution concentration (1).
   c) **Two** from: volume; flexibility; diameter; mass. (2)

Examiner's Comments

2 a) Almost every cell will contain these three parts. There are only a few exceptions, for example red blood cells do not contain a nucleus.
   b) Chromosomes or genes would also be correct answers although they are not included as parts on the diagrams.

3 a) You must learn the jobs of the parts of cells.
   b) Some reactions take place in the nucleus but the question asked where do **most** take place.
   c) Substances pass through the fully permeable cell wall, but it is the membrane that **controls** whether substances enter or leave.

4 This is a straightforward question if you have learned the parts of cells.

5 a) $(52 + 54 + 56 + 54 + 55) \div 5 = 54.2$. As the other figures are given to the nearest whole number this is rounded down to 54. There is no mark for the unit in this question, but you may sometimes lose marks for not including them so ensure that you do. Questions often include some maths and graph plotting so prepare for this when you sit your exam.
   b) The potato cells contain a solution. If these cells are surrounded by a weaker solution (such as water) they will absorb water by osmosis. The sugar solution must have been stronger than the cell solution, as those cells have lost water. Animal cells burst if they take in too much water – plant cells do not burst because of the inelastic cell wall.
   c) Note that changes in width would have been too small to measure. Acceptable answers show that you have understood what is happening – that the cells have absorbed or lost water.

CHAPTER 2 HUMAN BODY SYSTEMS

Answers

2 a) Left diagram: arrow from ventricle through valve A (1).
     Right diagram: arrow from left atrium to left ventricle (1).
   b) An explanation to include the following points: stop blood flowing (1); in the wrong direction / control direction of blood flowing (1).
   c) Pulmonary vein (1).
   d) Needs to produce higher pressure / to pump blood around body (1).

3 a) Any **two** from: water; hormones; carbon dioxide; amino acids; sugar; protein; antibodies; urea. (2)

b) Reacts with / carried by haemoglobin (1), as oxyhaemoglobin (1).

4 a) Make it faster (1).

b) More difficult for oxygen to be absorbed / carbon dioxide to be removed / reduced air volume (1).

c) Bronchus (1).

5 a) X (1).

b) Diffusion (1).

c) To be carried round the body / to all body cells (1).

d) (i) Enzymes (1). (ii) Converts large molecules to small molecules (1).

6 a) **Three** from: wave of muscular contractions; called peristalsis; fibre provides bulk; allows muscles to grip. (3)

b) (i) Starch (1).

(ii) Add biuret solution (or components) (1). Turns purple / lilac (1).

c) The enzyme / amylase is present in saliva and is specific to the reaction involving starch (1). The enzyme / amylase will not function at the pH of the digestive juice from the stomach (1).

d) Waste products from reactions within the body are excreted (1). Undigested food has not been used within the body so is egested (1).

Examiner's Comments

2 a) Blood flows from the atria to the ventricles. Whether the valves are open or closed tells you if blood is flowing into or out of the ventricle.

b) This answer applies to any valve whether in the heart or in the veins.

c) One you'll just have to learn. The left atrium receives blood from the lungs. The vessels to and from the lungs are the **pulmonary** artery and vein. This one is taking blood INto the heart so it is a veIN.

d) A common question is why the left ventricle is more muscular than the right ventricle, but this question is comparing the ventricle to the atrium. The atria fill up with blood from the veins so when the ventricles expand they can rapidly fill with blood. It is the ventricles that actually do the pumping.

3 a) You should know these.

4 b) If the lungs are filled with fluid there is less room for air to enter. This means you would have to breathe faster to still get enough oxygen.

c) A common mistake would be to give bronchiole or trachea. A significant number of candidates would miss this one out because there wouldn't be a space after the question. Make sure you read every line.

5 a) Smaller molecules are absorbed more easily.

b) Active transport would also be a correct answer.

c) Although you may have studied science by studying different topics at different times, in exam questions you can expect to refer to different topics in the same question. This question is really about the job of the blood.

d) Common mistakes would be to give acid or bile. These help enzymes to work but it is the enzymes that digest the food. Part (ii) is really just asking what is meant by digestion.

6 a) If you have forgotten the scientific word for something then you may still pick up marks if you answer in such a way that it is clear you understand the topic. In this example you can still get full marks without using the word 'peristalsis'. Of course if you do know the word then use it!

b) Another one to learn. If you have forgotten then as a last resort make a guess – you might be lucky. You certainly won't get a mark by leaving it blank. Although there is only one sentence for the question you are asked two things. Make sure you answer both parts.

c) Remember each enzyme has its own optimum pH.

d) This is asking the difference between egestion and excretion, each mark is for a definition.

CHAPTER 3 BODY MAINTENANCE

Answers

2 a) A = axon (1), B = (myelin) sheath (1).
   b) Synapses / nerve endings close together (1).
   c) Paralyses (1), signals blocked to motor neurones (1). Alternatively: stops
      sensation (1), blocks signals from sensory neurones (1).

3 a) Days 18, 19 (1), day 28 (1).
   b) Growth/build up/thickening (1), of uterus wall (1).
   c) (i) Progesterone (1). (ii) Ovary / corpus luteum (1). (iii) Menstruation
      (1). (iv) Pregnancy (1).

4 **Five** from: (high levels of glucose cause) pancreas to release insulin; insulin
   travels in blood to liver; changes excess glucose to glycogen; glycogen stored
   in liver; (when level of glucose in blood drops) pancreas releases glucagon;
   glucagon changes glycogen to glucose. (5)

5 a) The more cigarettes smoked the greater the risk of bronchitis (or
      reverse) (1).
   b) (i) Stops mucus moving (1).
      (ii) Dust and chemicals remain (in trachea) (1), cough to bring up
           irritants/mucus/dust/chemicals (1). No marks for infection /
           inflammation / blockage.
      (iii) Line alveoli/lungs/air sacs with dust and chemicals/tar (1).
           Stops/reduces exchange of oxygen/carbon dioxide, or carbon
           monoxide effect on red blood cells (1).

6 a) **Three** from: slow action / speed; produced in one place and acts at
      another; carried in blood; cause / control changes in body; chemical
      messengers. (3)
   b) (i) Reduces blood clots (1), which block arteries in heart / would cut off
           oxygen / blood supply (to heart muscle) (1).
      (ii) Disprin (1), second mark is for a reason, such as: no side effects / not
           using paracetamol which causes liver damage (1).

Examiner's Comments

2 a) Again these just need to be learned. Questions often start with a
      straightforward question testing your knowledge and then go on to test
      your understanding.
   b) Although ideally you should remember scientific words you may still get
      marks if you describe the answer clearly, as this example shows.
   c) When a question says 'suggest' it means that there isn't simply one
      correct answer that you will have been expected to learn. Instead your
      understanding of the topic is being tested and a scientifically reasonable
      answer should get the marks.

3 a) Even if you forget everything about hormones when you sit your exam,
      you could still answer this question by using the graph.
   b) Again, even if you cannot remember exactly what oestrogen does there
      are clues in the question. For example you can see that day 9 to 14 is not
      long after menstruation (bleeding) occurs so it would be reasonable to
      assume that this is when the uterus lining is re-growing.
   c) The answers to parts (i), (ii) and (iv) just have to be learned although
      you could work out the answer to (iii) by looking at the graph. Note that
      even if you couldn't remember the name 'progesterone' this need not
      stop you answering the other parts correctly.

4 You can see from the fact that there are 5 marks allocated that a detailed
   answer is required. In this case a lot of information has already been given
   in the question and marks will not be given for simply repeating that
   information.

5 a) Another straightforward start to a question where the answer can be gained by reading the information given.

b) (i) If you did not know that cilia 'wave' moving the mucus along there is enough in the question to let you work out the answer.

(ii) The key point is that as the cilia aren't removing the dirt and microbes then coughing is the only way left. Giving correct information that does not answer the question, does not gain any marks.

(iii) 'Short of breath' means that oxygen is not entering into the blood as well as it should. To get both marks here does not need two answers, as one point described in some detail is sufficient.

6 When you are given a lot of information don't forget to use it. It may give you at least part of the answer.

b) (ii) You are not expected to know the answer to this already, so don't be put off because some of the words may be unfamiliar. You are being asked to make a reasoned judgement and explain it.

CHAPTER 4 PLANTS

Answers

2 a) An explanation to include **two** of the following points: no light / needed light; for photosynthesis; existing starch removed. (2)

b) (i) Chlorophyll (1). (ii) Oxygen (1).

c) **Three** from: P3 has CO_2; Q3 has no CO_2; P2 has light; Q2 has no light. (3)

3 a) Indication on diagram that water enters through roots, passes up the stem and out through the leaves (1).

b) (i) Osmosis (1).

(ii) Water passes through semi-permeable membrane (1), from region of low concentration to region of high concentration of ions (1).

c) (i) Loses water through leaves / transpiration (1), changes size of stomata / holes in leaves / pores (1).

(ii) **Two** correct reasons, e.g. temperature control by evaporation, prevents drying out / dehydration / cell damage / plasmolysis. (2)

4 a) Increased growth / high yield (1).

b) Potassium, phosphate (1).

c) Active transport (1), from low to high concentration (1).

d) Make proteins (1), (cell) growth / enzymes (1).

Examiner's Comments

2 a) This is an example of an extended question where you could get full marks by giving two answers, (the first and third points on the mark scheme), or by giving an answer with some further explanation (the first and second points). Either is valid as the question has not asked for a specific number of answers.

b) Although you have not been told that the whole question is about photosynthesis, you are expected to realise that. In which case these answers are straightforward if you know the (word) equation for photosynthesis.

c) Looking at the diagram it should be clear that some parts of the leaves do not have access to the factors necessary for photosynthesis and the production of starch.

3 a) You have been asked to draw 'arrows' as you have to show the **direction** of the water movement. Some candidates would lose the mark by simply drawing a line, even if it follows the correct route.

b) If you forget the correct term, osmosis, you can still gain marks if your explanation is correct. Many candidates would make the mistake of not attempting the second part if they could not give the first answer.

4 a) The answer 'make them grow' would not gain a mark as the crop would grow anyway. The point is that fertilisers would *increase* the growth.

   b) If you are asked for, in this case, two answers, don't write down more than that or you could lose marks.

   c) There are two marks available for this question so you are expected to give more than a one word or phrase answer. It may not always be clear to you what the examiner is looking for so, in this type of question (which is different to part b) you can write at more length if you have several valid points to make. Do bear in mind that you will be able to get full marks on the lines given on a real exam paper, so if you extend your answer by much more you are probably going off the point.

   d) Another extended answer. The simple answer is 'make proteins' but the question is for two marks, so you need to give a more detailed explanation of why proteins are needed to gain the second mark.

CHAPTER 5   ECOLOGY AND THE ENVIRONMENT

Answers

2 a) (i) Marks for reasons, not for saying 'yes'. Two specific changes e.g. light, soil, nutrients, air, moisture, temperature. 1 mark each (2). Not hedge position.

   (ii) **Either:** Grow different dandelions (1) in same conditions (1) **or** same plants (1) in different conditions (1).

   b) Dandelions more even height (1), because (named) condition changes or conditions now same (1).

3 a) (i) **Three** from: water logged soil contains no air / spaces filled with water; therefore bacteria are active / favoured; nitrates broken down / nitrogen lost; less nitrate in soil; nitrate shortage limits photosynthesis. (3)

   (ii) Plants need nitrogen / nitrates (1). Shortage limits photosynthesis / growth (1).

   b) **Two** from: cereal crops / wheat need lots of nitrate; would soon exhaust natural supply; extra N_2 promotes photosynthesis / growth. (2). Plus **two** from: clover has nitrogen-fixing bacteria; returns nitrates to soil; therefore doesn't need extra supply. (2)

   c) Nitrates needed for growth in spring (1); artificial nitrate available immediately (1); manure breaks down slowly (1); therefore must be added well in advance (1).

4 a) **Two** from: persist; toxic to other animals; passed along food chain. (2)

   b) (i) **Either:** control agent increasing rapidly (1) because lots of food (1), **or** pest decreasing slowly (1) because agent still small numbers (1).

   (ii) Pest decreasing rapidly below harmful level; agent numbers levelling off because food supply decreases (1).

   (iii) Agent and pest below harmful levels; agent rises when pest rises / reference to oscillation. (2)

   (iv) Both pest and agent below harmful level; pest staying low. (2)

Examiner's Comments

2 a) (i) You are expected to use your knowledge and judgement about which factors could be different for the different dandelions. There are two marks for the question which should guide you in to giving at least two answers. (You do not get a mark for the first part as it could be a 50:50 guess.) That it may be warmer and more moist (due to less wind) are probably the best answers, but other answers would be allowed as there is not enough information to disallow them.

   (ii) This is not an answer you will have learned. Expect to get questions like this where you have to think about the particular situation.

b) Expect to be asked to *predict* what might happen in different situations. Just look at what's happening and use those ideas. Here for example if the hedge was removed conditions would be more like those of the smaller dandelions in the picture. In this case you do get a mark for each part of the question as neither is a simple yes or no.

3 a) The diagram reminds you that denitrifying bacteria convert nitrates in the soil into nitrogen gas in the air. You should know that 'anaerobic' means that the bacteria do not need oxygen (see chapter 2). You are expected to make the link that waterlogged soil will contain little oxygen so only the denitrifying bacteria will survive in it and these will be removing the nitrates that would otherwise be available to the plants. Three marks indicates that a detailed explanation is needed for part (i).

b) You should know that clover is a leguminous plant and contains nitrogen-fixing bacteria but here you are given a big help in the diagram which shows clover taking nitrogen from the air. Spell out all the steps in your line of reasoning to make sure you get full marks.

c) You may not have learned this point but you should know that manure decays and releases nitrates slowly (remember this is why it causes less water pollution than artificial fertilisers). Follow that line of reasoning and you are on your way to the answer.

4 a) A straightforward factual recall question to get you started.

b) (i) You have not been told that the 'biological control agent' is some kind of organism, probably a predator that feeds off the pest, you are expected to realise this. In this example you could get both marks by explaining fully what is happening to *either* of the organisms. However it is recommended in an exam, when you do not know the mark scheme, to answer as fully as you can and in this case explain *both* graphs.

(ii) Make sure you confine your comments to what is happening to the graph only between the times asked in the question. To do otherwise indicates you do not fully understand the question or cannot read the graph properly. In this case you can only get full marks by explaining both lines on the graph.

(iii) The situation now is that the populations are oscillating (going up and down) in a similar way to the lynxes and snowshoe hares earlier in the chapter.

(iv) The key reason here is that the pest numbers *stay* low. If pesticide was used the numbers would probably eventually increase again unless repeated treatments were used and even then some of the pests would probably become resistant. The point about the control agent numbers staying low is also important as you would want to minimise the effects of introducing this organism on the rest of the ecosystem.

CHAPTER 6 GENETICS AND EVOLUTION

Answers

2 a) **Four** from: select parents large and tasty **not** tall; cross plants (and grow seeds); select best plants ; repeat process. (4)

b) (i) Vegetative / asexual / cloning (1).
(ii) Clones / identical copies / all same. **Not** clones if cloning in (i) (1).

3 a) Hh (1).
b) All / 100% (1).
c) 3 tall : 1 short (1). There is a 1 in 4 chance of receiving two recessive genes (1).

4 a) Webbed feet (1), have greater area to 'push' against the water (1); small wings (1), can be 'tucked in' to reduce resistance (1).
b) (i) From fossils (1). (ii) Changes to fossils over time (1).

5 a) Natural population has a wide range of variations; because it has a large number of alleles; selective breeding reduces the number of alleles; cloning perpetuates this reduced number of alleles. (4)

b) **Four** from: cut genes for disease resistance; from chromosomes of resistant variety; introduce into chromosomes of 'ordinary banana'; tissue culture to produce disease resistant plants / clone; enzymes cut chromosomes. (4)

Examiner's Comments

2 a) Exam question about selective breeding cover many examples, but the basic idea of choosing individuals with the best features, crossing them, selecting the best offspring and so on, will be the same.

b) (i) In case you are unfamiliar with the term 'runner', the picture shows clearly that it has grown from the original plant as an example of asexual reproduction.

(ii) The essential point about asexual reproduction is that the offspring are *genetically* identical to the parent. If the plant had been crossed with another, the desired features might have been 'lost'. Note that the mark scheme does not allow you to get two marks for repeating the same point. This is common to all mark schemes.

3 a) A common mistake would be the answer HHhh. Don't forget each parent only passes on *one* of each gene pair to the offspring.

b) A straightforward question. This is testing that you appreciate that H is dominant to h.

c) When individuals that are heterozygous for a characteristic with a dominant allele and a recessive allele are crossed, the probabilities of the offspring will *always* be in the ratio of 3 showing the dominant phenotype to 1 showing the recessive phenotype.

4 b) When answering exam questions look at how they are set out. These two questions are grouped together, which means that there is a good chance that the answers are linked. As part (i) is about fossils there is a good chance that (ii) will be as well.

5 a) When you are asked to look at this amount of information before a question you will almost certainly not be able to get full marks without using that information. This does not mean that you can simply find all the answers in the passage but that you will have to use it, *as well as your own knowledge*, to answer the question. Also note that in this kind of question it is usually easy to pick up some of the marks but the danger with having so much space to write in is that some candidates will try to write at length, often repeating a lot of information already given in the question, and not gain many marks because they have not answered the question.

b) Like questions about selective breeding the examples you will be asked about in genetic engineering questions will probably be new to you. This does not make them more difficult since the basic procedures in genetic engineering of transferring genes will be the same. As there are 4 marks, make sure you spell out the steps in detail.

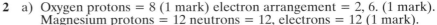

CHAPTER 8 STRUCTURE AND BONDING

Answers

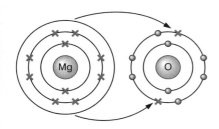

2 a) Oxygen protons = 8 (1 mark) electron arrangement = 2, 6. (1 mark). Magnesium protons = 12 neutrons = 12, electrons = 12 (1 mark).
  b) The magnesium atom transfers its two outermost electrons to the oxygen atom (1). Both atoms achieve a full outer electron shell (1). The magnesium forms the Mg^{2+} ion; oxygen forms the O^{2-} ion (1). Correct atom diagram (left) (1).
  c) When an ionic crystal melts energy is needed to break the strong electrostatic attractions between the ions (1). As the ions in magnesium oxide have charges of 2+ and 2– the strength of attraction will be greater than in sodium chloride (1+ and 1–) (1). More energy is needed therefore to overcome the attraction, so the melting point will be higher (1).

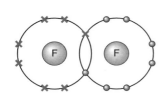

3 a) Diagram (left) shows electron arrangement as 2, 1 (1 mark).
  b) (i) 9 (1), (ii) 10 (1).
  c) Isotopes are atoms of the same element / have the same number of protons (1) with different numbers of neutrons/mass numbers (1).
  d) Hydrogen – covalent (1), magnesium chloride – $MgCl_2$ (1).

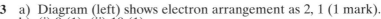

  e) Diagram (left) shows one shared pair of electrons (1).
  f) **Two** from: high melting point or boiling point; conducts electricity when molten or in solution / behaves as an electrolyte; often soluble. (2)

Examiner's Comments

2 a) The proton number equals the electron number and the numbers 24 and 16 are the mass numbers.
  b) A common mistake is to leave out the formulae of the ions formed. Always include a diagram when answering this type of question. The atom diagrams do not need to include the contents of the nucleus but you should include all the electrons, not just those in the outermost shell. A well drawn atom diagram will score all the marks except the mark for the formulae of the ions.
  c) You are looking for three different marking points. The word 'suggest' indicates that you are being asked to use your understanding to answer a question which you may not have covered in lessons. Try and visualise a model of an ionic structure and what must happen to it when it melts.

3 a) The diagram shows there are three protons therefore there must be three electrons. The first electron shell can only hold two electrons.
  b) Remember the lower number is the atomic number (equal to the number of protons or electrons). The difference between the two numbers is the number of neutrons.
  c) Candidates often confuse isotopes with allotropes. (An allotrope is a different physical form of the same element).
  d) Remember that the bonding between non-metal atoms is covalent. The formula of magnesium oxide can be worked out using the periodic table. Magnesium is in Group 2 (forms 2+ ions); chlorine is in Group 7 (forms 1– ions). Two chloride ions are needed for each magnesium ion.
  e) You were asked to show outer electrons only. You wouldn't lose a mark if you included inner electrons.
  f) Look for two different types of property. You would probably not get two marks for giving high melting point and high boiling point. A key property of ionic compounds is their behaviour as electrolytes. Don't forget this one.

CHAPTER 9 FUELS AND ENERGY

Answers

2 a) Kills fish in lakes / damages plants / damages metalwork / damages statues (1).
  b) **Three** from: more than one greenhouse gas; other gases have a greater greenhouse effect; global warming may be due to periodic climatic variation; extra carbon dioxide may be removed by more photosynthesis; or by dissolving in the oceans. (3)
  c) (i) $C_{10}H_{22}$ (1).
    (ii) Alkanes (1).
    (iii) Exothermic (1).
    (iv) Any of C–C, C–H or O=O (1).
    (v) Exothermic/ energy transferred to the surroundings (1).

3 a) A molecule containing carbon and hydrogen atoms only (1).
  b) C_5H_{12} (1).
  c) Fractional distillation: crude oil is evaporated (1); condensed (1); at different temperatures (1); producing fractions with similar numbers of carbon atoms (1). Cracking: large hydrocarbons broken down (1); into smaller alkanes (1); and alkenes also formed (1). (Maximum of 5)
  d) The molecule contains carbon–carbon double bonds. (1)
  e)

```
     H   H   H   H   H   H
     |   |   |   |   |   |
--- —C — C — C — C — C — C—---
     |   |   |   |   |   |
     H   H   H   H   H   H          (2)
```

  f) (i) Unreactive / air tight (1).
    (ii) There are strong covalent bonds between the polymer chains (1). On heating these bonds are not broken until the whole structure breaks down (1).

Examiner's Comments

2 a) The acid is sulphuric acid and so has a relatively low pH.
  c) (ii) To check that you haven't made a simple slip, the general formula for an alkane is C_nH_{2n+2} where n = number of carbon atoms.
    (iii) Energy transferred out (EXit) to the surroundings, so exothermic.
    (iv) All the bonds in the reactants, i.e. decane and oxygen are broken. Carbon dioxide and water are the products.
    (v) It is easy to get this the wrong way round. Bond breaking requires energy, energy is taken in, it is endothermic. (Imagine the effort of pulling a molecular model apart.) Bond making must be the reverse, it is exothermic.

3 a) Don't forget the word 'only'. Without it you will lose the mark.
  b) Remember the general formula for an alkane is C_nH_{2n+2}.
  c) This answer requires 'extended prose'. The mark allocation indicates that 5 different points are needed. The candidate has been asked to describe fractional distillation **and** cracking. It is important to cover both of these.
  d) This is a popular question. Try and learn the terms saturated and unsaturated and their meanings.
  e) Start with the left hand molecule. Link one of the C=C bonds to the molecule on the right. Do the same with the middle molecule. Redraw the molecules with single bonds and no double bonds.
  f) (i) There are a number of properties of poly(ethene), but few are appropriate to the use given in the question. Think carefully before writing down any old answer.
    (ii) This relates to the important concept of bonds within (intramolecular) and between (intermolecular) molecules. Unlike thermoplastics, thermosetting plastics have strong cross links between the polymer molecules. This means that they cannot be melted down and re-moulded.

CHAPTER 10 ROCKS AND METALS

Answers

2 a) (i) An igneous rock is formed when molten magma from the mantle
cools. (1)
(ii) A metamorphic rock is formed when a sedimentary or igneous
rock (1) is subjected to high temperatures and pressures (1).
   b) **Four** from: small pieces of rock are broken off the igneous rock due to
weathering and erosion; wind and rain break off particles; water freezes
in cracks and expands resulting in fragments breaking off as the ice
melts; the particles are transported by rain water into streams and rivers;
the particles then subjected to high pressures; over a period of millions
of years. (4)
   c) **Four** from: sedimentary rock can be forced into the Earth's mantle when
two tectonic plates collide; this process is called subduction; with the
high temperature and pressure in the mantle the rock melts and forms
magma; when plates move apart the magma is forced up to fill the gap;
either as a volcano; or a mid-ocean ridge; and the magma then cools and
forms igneous rock. (4)

3 a) (i) Aluminium oxide / alumina / bauxite (1).
(ii) Carbon (1).
(iii) The carbon anodes react with / are oxidised by, the oxygen produced
at the anode (1). Carbon dioxide is produced (1).
   b) (i) They could use bronze to make hunting weapons (spears) which were
sharper than those made from stone. (1) They could use bronze to
make cooking utensils which were stronger than pottery (1).
(ii) The more unreactive the metal the longer ago it was discovered (1).
(iii) Magnesium cannot be extracted from its ore by reduction using
carbon (1) as it is more reactive than carbon (1). Instead it is
extracted using electrolysis / electricity (1) which has only been
available over the last 125 years or so (1).
   c) (i) Coke (carbon) (1) and limestone (1). (ii) Slag (1).

Examiner's Comments

2 a) It would be easy to miss the second mark in part (ii). Always try and find
as many different points as there are marks.
   b) This is a piece of extended writing. You should be guided by the fact that
there are 4 marks and so 4 different ideas are needed. The key ideas are
those of weathering, erosion and transportation. Try and give examples
for each idea.
   c) Again a question that requires an extended answer. Details of the rock
cycle are needed. To be absolutely precise subduction occurs when a
continental plate collides with an oceanic plate. The oceanic plate being
more dense will be pushed down into the mantle.

3 a) Metals are nearly always extracted from their oxides. A high
temperature is needed to keep the aluminium oxide molten which only
serves to increase the rate of the reaction between carbon and oxygen.
   b) (i) This is not the time to write a history essay! Look for the clues in the
question. The key words are 'not brittle', 'sharpened' and also the
reference to 'stone' and 'pottery'.
(ii) Look at the table given in the question. There is only one mark so
look for an overall pattern.
(iii) You should recognise that magnesium is a reactive metal and so will
be extracted by electrolysis rather than by heating its oxide with
carbon.
   c) These are recall questions. If you haven't learned the details of the blast
furnace you won't be able to work out the answers.

CHAPTER 11 CHEMICAL REACTIONS

Answers

2 a) The mixture becomes warm (1).
   b) The paste is a catalyst (1). Without the paste the reaction takes a very long time. (1).
   c) **Two** from: setting time will decrease; reaction is faster; particles are moving faster. And **two** from: particles have more energy; a greater proportion of molecules have more energy than the activation energy (2 points); more effective collisions; greater collision frequency. (Maximum of 4.)

3 a) A reversible reaction (1).
   b) **Two** from: add a catalyst (iron); increase temperature; increase the concentration of the reactants; increase pressure. (2)
   c) **Four** from: (*economic*) provides income for local industry, community; provides income for the nation; increases the yield of food crops / provides more food; (*social*) provides employment; (*environmental*) factory could be an eyesore; increased transport close to factory; possible pollution risks. (Maximum of 4)
   d) (i) 32% ± 1% (1). (ii) Decreases (1). (iii) Increases (1).

Examiner's Comments

2 a) Remember that the transfer of energy is one of the important features of a chemical reaction.
   b) You should notice that when you compare experiments 1 and 3 only the volume of paste is changed and this results in experiment 3 having a shorter setting time (greater rate of reaction).
   c) Notice that more marks are obtained for greater detail and precision.

3 a) This is a very common question.
   b) In this reaction all the common ways of controlling a reaction can be used. In fact, the catalyst is finely ground to increase its surface area as well.
   c) The question asks for economic, social and environmental effects. You should try to give at least one example in each category.
   d) Always take care to understand the scale on a graph. It is a common mistake to assume that each 'square' is one unit.

CHAPTER 12 THE PERIODIC TABLE

Answers

2 a) (i) Atomic number (1).
      (ii) Te has a higher relative atomic mass than I / in the wrong order (1). This is due to isotopes (1). Te has more neutrons than expected and this accounts for the higher RAM (1).
   b) (i) Reactivity increases down Group 1 (1).
      (ii) The hydroxide (1) and hydrogen (1).
      (iii) The electrons are 'lost' or transferred to the substance they react with. (1)
   c) (i) F 2, 7 / Cl 2, 8, 7 / Br 2, 8, 8, 17 / I 2, 8, 8, 18, 17. (1)
      (ii) Reactivity decreases down the group (1).
      (iii) Atoms become larger (1) less easy to gain the electron needed (1).

3 a) (i) Group 1 / alkali metals (1).
      (ii) Rubidium hydroxide (1). Hydrogen (1).
      (iii) **Two** from: metal melts; flames; fizzing; metal floats; metal dissolves. (2)

b) (i) Any named acid (hydrochloric acid, nitric acid, sulphuric acid) (1).
(ii) Magnesium iodide (1) and hydrogen (1).
(iii) Magnesium dissolves (1) bubbles of gas / colourless gas formed (1).
c) (i) Acid effect / red / pH < 7 (1).
(ii) Sodium carbonate dissolves (1) fizzing / bubbles of gas (1).
(iii) Sodium selenate / water / carbon dioxide (1).

Examiner's Comments

2 a) Elements are arranged in the periodic table in order of increasing atomic number, i.e. in order of increasing numbers of protons. Atomic mass is made up from the number of protons and neutrons. It is the neutrons that cause the apparent incorrect order. There isn't a simple pattern in the number of neutrons in atoms.
   b) On the left side of the periodic table reactivity increases down the group. The hydroxide is the alkali, hence alkali metals. The gas given off is hydrogen.
   c) (i) Working out the electronic structure was covered in chapter 8. You may need to revise this. The first electron shell can hold 2 electrons, the 2nd can hold 8, the 3rd can hold 8, the 4th can hold 18, and the 5th can hold 18.
      (ii) In the groups on the right of the periodic table (non-metals) reactivity decreases down the group. Fluorine is the most reactive Group 7 element.
      (iii) Non-metals react by gaining electrons (in contrast to metals which lose electrons). The bigger the atom the less attraction an approaching electron feels and so the harder it is to gain that electron.

3 a) (i) The question stated that the periodic table could be used. The question is much harder if you don't use it.
      (ii) As the question states you are not required to know the chemistry of rubidium but you are expected to know that it will behave in a similar way to sodium. Therefore, as sodium forms sodium hydroxide and hydrogen the products from the rubidium reaction can be predicted.
      (iii) The question asks for observations. Candidates very often write about the products formed and not what they would see.
   b) (i) Once again you are not expected to know the chemistry of iodine but you should know that iodine behaves in a similar way to chlorine. Hydrogen iodide solution will resemble hydrogen chloride solution which is commonly called hydrochloric acid.
      (ii) Magnesium and hydrogen *chloride* would form magnesium *chloride* and hydrogen, therefore hydrogen *iodide* would form magnesium *iodide* and hydrogen.
      (iii) The question is still asking whether you can recall the reaction of magnesium and hydrochloric acid.
   c) (i) H_2SeO_4 will behave in exactly the same way as H_2SO_4, sulphuric acid, as selenium and sulphur are both in Group 6 of the periodic table. H_2SeO_4 must be acidic.
      (ii) A carbonate reacts with an acid to produce carbon dioxide and the carbonate will dissolve.
      (iii) The word equation for this type of reaction is:
         carbonate + acid → salt + carbon dioxide + water.
         In this case the salt is sodium selenate (like sodium sulphate).

CHAPTER 13 CHEMICAL CALCULATIONS

Answers

2 Symbols   Cu   Fe   S
   Mass   5.27   4.61   5.27 (1 mark)
   Moles $\frac{5.27}{64} = 0.082$ $\frac{4.61}{56} = 0.082$ $\frac{5.27}{32} = 0.165$ (1 mark)

   Ratio $\frac{0.082}{0.082} = 1$ $\frac{0.082}{0.082} = 1$ $\frac{0.165}{0.082} = 2$ (1 mark)
   This gives the formula $CuFeS_2$ (1 mark).

3 Equation H_2 + Cl_2 →   2HCl
   Moles   1 mole   1 mole   2 moles (1 mark)
   Mass ratio   2 tonnes   71 tonnes   73 tonnes (1 mark)
   Multiply by 2.05. To make 150 tonnes you need: 146 tonnes (1 mark).

4 a) Fe_2O_3 formula mass = 56 + 56 + 16 + 16 + 16 = 160 g
       A mark for getting two 56s, a mark for getting three 16s, and a final
       mark for stating the answer correctly in grams.
   b) Equation Fe_2O_3 +   3CO   →   2Fe   + $3CO_2$
       Moles   1 mole   2 moles (1 mark)
       Mass ratio   160 tonnes   112 tonnes (1 mark)
       Multiply by 200
       Mass   32 000 tonnes   22 400 tonnes (1 mark)
       Mass of iron = 22 400 tonnes

5 Equation $CaCO_3$ →   CaO   + CO_2
   Moles   1 mole   1 mole   1 mole (1 mark)
   Mass ratio   100 tonnes   56 tonnes   (1 mark)
   Multiply by 4 400 tonnes   224 tonnes   (1 mark)
   Mass of limestone = 400 tonnes (1 mark).

Examiner's Comments

2 Marks are given for the calculation, so show your working clearly.

3 The answer need only be given to the nearest tonne as this was how the
   quantity was given in the question. Notice that there is no need to work out
   the mass of hydrogen required.

CHAPTER 14 ELECTRICITY

Answers

2 a) There is no complete circuit / conducting path (1).
   b) A and B (1).
   c) C (1).
   d) C (1).
   e) A and B (1).
   f) (i) Ammeter (1). (ii) Amps or A (1).
   g) (i) 3 (V) (1). (ii) 6 (V) (1).

3 a) Fuse Y is in the immersion heater circuit whereas fuse X is in the
       lighting circuit. The immersion heater has a greater power (3000 W)
       than the lighting circuit (1). It therefore takes a larger current (1).
   b) A short-circuit, for example when a lamp filament breaks (1). This
       causes a large current to flow through the fuse (1).
   c) The lamps can each be switched independently of each other (1). If one
       lamp 'blows' the others remain alight (1).
   d) **Two** from: stops electrocution (1); fault makes casing become live (1);
       causes large current to flow to earth (1); melting the fuse (1).

4 a) (i) R = V/I (1), R = 4/1.25 = 3.2 Ω (1).
      (ii) I = V/R (1), I = 12/3.2 = 3.75 A (1).
   b) (i) Increase in temperature (1) causes an increase in resistance (1).
      (ii) From the graph, $R = 2\,\Omega$ (1),
           $V = IR = 0.5 \times 2$ (1 mark) = 1 (V) (1 mark).
      (iii)$P = VI = 1 \times 0.5$ (1 mark) = 0.5 W (1 mark).

Examiner's Comments

2 a) Switch 1 controls both lamps, as it is open neither lamp will work.
   b) The current can now flow from the battery through lamps A and B and
      return to the battery. A complete circuit has been made. The parallel
      circuit through lamp C is still broken and so lamp C will not light.
   c) This time the circuit through lamp C is now complete. Lamps A and B
      are isolated as switch 2 is open.
   d) In this parallel circuit the current will divide between the route involving
      lamps A and B and the route involving lamp C. As the resistance
      provided by two lamps in series is greater than that provided by one
      lamp, more current will flow through lamp C than through lamps A and
      B. Therefore lamp C will be the brightest.
   e) Lamps in the same route of the circuit are in series. In contrast lamp C
      is in parallel to lamps A and B.
   f) Remember that an ammeter is always part of the circuit - the current
      actually flows through it. The unit of current is the ampere or amp (A).
   g) The potential difference across the battery is 6 V and this must equal the
      potential difference in the circuit. (i) As the lamps are identical the
      potential difference across each lamp will be the same, i.e. 3 V.
      A common error is to think that the p.d. is 6 volts across each
      component in the circuit. (ii) The lamps are in parallel. The p.d. across
      the two lamps must be the same as the p.d. across the battery.

3 a) Remember that power = current × voltage. As the voltage is the same
      (mains voltage) the current in the heating circuit will be greater than the
      current in the lighting circuit.
   b) The fuse wire is designed to only allow a certain current to flow before
      the heating effect of the current is sufficient to melt the fuse wire. Any
      sudden surge of current will cause the fuse to 'blow' or melt.
   c) In a question like this where there are two marks try and look for two
      separate points. The fact that switches are shown in the diagram is a clue
      to the first point. The second point is the perhaps the more obvious
      advantage of a parallel circuit over a series circuit.
   d) This is a very common question. Sometimes it is set in the context of
      double insulated appliances. The way the earth wire and the fuse work
      together to ensure safety is a very important idea. If you are still in
      doubt look back at page 259.

4 a) Always write down the formula you intend to use. In both these cases
      the Ohm's law equation needs to be rearranged. If you are not sure how
      to do this look back at page 253. In this question the units are given. The
      second question depends on the first. You can still get full marks for the
      second part even if you use the incorrect value for the resistance.
      Providing you show your working you will get marks for showing the
      approach you are using.
   b) (i) The first important point is to say how resistance changed with
           increasing current. The other point is to appreciate that when
           resistance is not constant it is due to the effect of temperature. This
           has a big effect in the case of a filament lamp.
      (iii) You must remember the equation for working out electrical power.
           In this question the units of power are not given (watts) and so to
           obtain the mark for the answer you must also give the correct unit.

CHAPTER 15 ELECTROMAGNETISM

Answers

2 a) (i) $V_p/V_s = N_p/N_s$ (1 mark), $24/V_s = 5/100$, $V_s = 480$ V (1 mark).
(ii) To reduce energy loss (1). Higher voltage results in lower current (1). The lower the current the less the heating effect (1).
b) **Two** from: key connects circuit; causing the electromagnet to pull contacts; thus completing the circuit to the starter motor. (2)
c) The wires have a magnetic field around them (1), they are in the magnetic field produced by the permanent magnet (1), the two magnetic fields interact (1), Fleming's Left Hand rule (1) predicts the movement will be to the right (1).

Examiner's Comments

2 a) (i) Don't forget to write down the equation. You will notice that 1 mark has been given for this.
(ii) The use of high voltages for transmitting electricity is a common question on higher tier papers.
b) Always try and follow the sequence logically. In questions about relays remember that there will be two circuits. The one with the electromagnet in will be used to switch on the other one.
c) A very detailed answer is required to obtain 5 marks. Again it is very important to think the situation through logically. You will notice only one mark is given for the movement of the wire. Fleming's left hand rule: **f**irst finger = **f**ield, se**c**ond finger = **c**urrent, thu**m**b = **m**ovement.

CHAPTER 16 **FORCES AND MOTION**

Answers

2 a) (i) L is on the first horizontal section (1) and H is on the shallow sloping section (1). See diagram (right).
(ii) 1000 m. (1)
(iii) Speed = distance/time (1 mark) = 400/150 = 2.67 m/s (1 mark).
b) (i) 20 cm² (1).
(ii) Pressure = force/area (1 mark) = 300/20 = 15 N/m² (1).

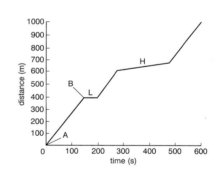

3 a) (i) Speed = distance/ time (1 mark) = 240/30 = 8 m/s (1 mark).
(ii) The speed will have been slower than the average at the start (1) and so must be above average at its fastest (1).
b) (i) See diagram (right). (1)
(ii) See diagram (right). (1)
(iii) 2 s.
c) (i) From 5.5 s to 10.0 s approx. (1).
(ii) The athlete is accelerating (1) and has not yet reached a constant speed (1).
d) (i) See diagram (right). (1)
(ii) Distance = speed × time = 2 × 10 = 20 m (1). Distance from finish line = 100 – 20 = 80 m (1).

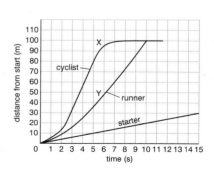

Examiner's Comments

2 a) (i) A horizontal part of the graph indicates that Peter has stopped.
      (iii) Notice again how there is a point for showing the equation.
   b) The area is calculated from the length × width, i.e. $10 \times 2 = 20 \, cm^2$.
      Remember that the force is spread equally between the two tyres.
      Always remember to start with the formula you are going to use.

3 a) The athlete cannot run at a constant speed right from the start of the
      race. He must accelerate at the beginning.
   b) (i) This is where the positive gradient of the curve starts to decrease.
      (ii) This is where the curve becomes straight indicating constant speed.
      (iii) Cyclist reaches 100 m mark after 8 s, runner reaches it after 10 s.
   c) (i) A steady speed is indicated by a straight line with a positive gradient
           in a distance-time graph.
      (ii) If the line is not straight the speed is not constant and so indicates
           acceleration or deceleration.
   d) After 15 s the starter will have walked 30 m. As the speed was constant
      the line will be straight.

CHAPTER 17 **ENERGY**

Answers

2 a) (i) Radiation (1). (ii) Conduction (1).
   b) (i) Copper painted black (1).
      (ii) Matte black is a good absorber of radiation (1), copper is a good
           conductor of heat (1).
   c) Advantage: day to day cost of energy is low (1). Disadvantage: high
      installation costs / not effective on a dull day (1).

3 a) P.E. = $m \, g \, h$ (1), P.E. = $1 \times 10 \times 400 = 4000 \, J$ (1).
   b) Less than 4000 J, some energy will be wasted due to friction (1).
   c) Electrical → potential (1).
   d) Not 100% efficient (1), friction in the pumps wastes energy (1).
   e) Good use can be made of 'off-peak' electricity in preparation for when
      demand is high (1).

4 a) (i) Work done = force × distance (1 mark) = $50 \times 10 \times 10$ (1 mark) =
           5000 J (1 mark).
      (ii) Power = Work done / time taken (1 mark) = 5000/6 (1 mark) =
           833.3 W (1 mark).
   b) It is transferred to the surroundings (1) as thermal energy (1) due to
      friction between the boy and the slide, and the boy and the water (1).

Examiner's Comments

2 a) Energy is obtained from the Sun by radiation. The energy will be
      conducted from the particles in the solid to the particles in the liquid.
      The energy will be transferred through the water by convection.
   b) Metals conduct well. Dull black surfaces absorb radiation better.
   c) Renewable energy sources such as solar, wind and wave have the
      limitation of inconsistent supply.

3 a) Note: the vertical height of 400 m is shown on the diagram.
   b) Although energy is conserved not all the energy is transferred into useful
      energy – friction is a common cause of 'wasted' energy.
   c) This is the reverse of the transfer used to generate electricity.
   d) This is the same point as mentioned in part ii) above.
   e) Electricity cannot be stored on the National Grid. Using electricity to
      pump the water when there is little demand for electricity is an indirect
      way of storing electrical power.

4 a) Always remember to include the unit.
   b) There are three scoring points so look for three separate ideas.

CHAPTER 18 WAVES

Answers

2 a) (i) C (1), (ii) B (1).
   b) $v = f \times \lambda$ or $\lambda = v / f$ (1 mark). $\lambda = 340/500$ (1 mark), $= 0.68$ m (1)
   c) (i) At 3000 km the solid rock mantle (1) changes to the liquid rock (1) outer core.
       (ii) The velocity increases quickly – the Earth has a thin crust (1). The velocity increases within the mantle – the density (viscosity) of the mantle increases with depth. (1) The velocity drops appreciably at the beginning of the outer core – the outer core is liquid (1). The velocity increases from the outer core to the inner core – the inner core is solid (1).

3 a) (i) Ultraviolet (1)(ii) Infrared (1).
   b) (i) Normal drawn correctly and labelled as in diagram (right). (1)
       (ii) Ray drawn with 45° angles marked as in diagram (right). (1)
   c) (i) Correctly drawn with equal angles of incidence and reflection as in the diagram (below). (1)

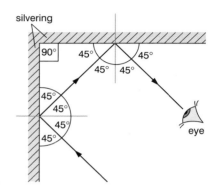

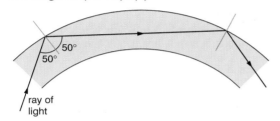

ray of light

       (ii) **Three** from: no interference / better quality transmissions (1); cables thinner / more signals carried (1); more security (1); less expensive than copper wiring (1). (3)

Examiner's Comments

2 a) The wave with the loudest sound has the greatest amplitude. The wave with the highest pitch has the highest frequency.
   b) Always write down the equation you are going to use and don't forget the units.
   c) (i) Always refer back to the graph. The clues given are the labels 'mantle' and 'outer core'.
       (ii) This question requires an extended prose answer. You must look for at least 4 scoring points. You can see that the scoring points refer to the four key areas of the graph.
3 a) It is vital that you know the order of waves in the electromagnetic spectrum.
   b) The angles should be drawn using a protractor. Forgetting to label the angles will result in the loss of the second mark.
   c) Again it is advisable to use a protractor when drawing ray diagrams. Note that 50° is above the critical angle so total internal reflection will occur.

CHAPTER 19   THE EARTH AND BEYOND

Answers

2 a) (i)   Neptune (1).
      (ii) Venus (1).
      (iii) Generally the larger the size the lower the density (1).
      (iv) Generally the higher density planets are nearer to the Sun (1).
   b) (i)   Gas clouds drawn together by gravity (1), core becomes very hot (1),
            nuclear fusion begins (1).
      (ii) Red giant (1) → white dwarf (1).
      (iii) Nuclear fusion (1), hydrogen atoms join to form helium atoms (1), a
            considerable amount of energy is released (1).

Examiner's Comments

2 a) (i)   The planet with the lowest surface temperature is likely to be the
            one furthest from the Sun. (iii) and (iv) Only 1 mark is given so it is
            not necessary to look for an exact mathematical relationship (if, in
            fact, there is one). In each case a simple pattern is all that is
            required.
   b) (i)   There are three scoring points so look for three separate ideas. With
            this type of question it is important to get the separate processes in
            the right order.
      (ii) A supernova is formed from the collapse of a very large star, much
            larger than the Sun.

CHAPTER 20   RADIOACTIVITY

Answers

2 a) Alpha is a helium nucleus(1) with a positive charge (1).
      Gamma is electromagnetic radiation (1) with no charge (1).
   b) (i)   **Five** marks from: Gamma rays are the only ones that penetrate the
            skin (1), alpha and beta particles would be absorbed (1) causing cell
            damage (1). 6 minutes would be too short (1) and time is needed for
            the tracer to mix with the blood (1), but 6 days would expose the
            patient to too much radiation (1).
      (ii) 6400 (time = 0), 3200 (time = 6 hours), 1600 (time = 12 hours), 800
            (time = 18 hours), 400 (time = 24 hours). (2)
      (iii) Radioactivity is a random process (1) effect of background
            radiation (1).

Examiner's Comments

2 a) 2 protons + 2 neutrons would be an acceptable alternative to a
      helium nucleus.
   b) (i)   There are a lot of marks available here and so a detailed answer is
            required. There are really two things to explain: firstly, why is a
            gamma emitter used (rather than alpha or beta); secondly, why is a
            half-life of 6 hours chosen. Unless you read the question carefully it
            is easy to miss the first part.
      (ii) You get one mark for realising there are four half-lives in 24 hours,
            and another mark to show the count rate halving four times.
      (iii) Adjustments always have to be made to eliminate the background
            count. Radioactive decay is based on the probability of an atom
            decaying. As a result of this there are often small variations and
            smooth radioactive decay curves are rare.

Published by HarperCollins*Publishers* Ltd
77–85 Fulham Palace Road
London W6 8JB

www.**Collins**Education.com
On-line support for schools and colleges

© HarperCollins*Publishers* Ltd 2001

First published 2001
Reprinted 2001 (twice)

ISBN 0 00 711201 7

Chris Sunley and Mike Smith assert the moral right to be identified as the authors of this work.

British Library Cataloguing in Publication Data

A catalogue record for this book is available from the British Library.

Edited by Alan Trewartha

Production by Kathryn Botterill

Cover design by Susi Martin-Taylor

Book design by Rupert Purcell and produced by Gecko Limited

Index compiled by Julie Rimington

Printed and bound by Bath Press

Acknowledgements
The Authors and Publishers are grateful to the following for permission to reproduce copyright material:

London Examinations, a division of Edexcel Foundation
Edexcel Foundation, London Examinations accepts no responsibility whatsoever for the accuracy or method of working in the answers given.

Midland Examining Group
The Midland Examining Group bears no responsibility for the example answers to questions taken from its past question papers which are contained in this publication.

Northern Examinations and Assessment Board
The authors are responsible for the possible answers/solutions . and the commentaries on the past questions from the Northern Examinations and Assessment Board. They may not constitute the only possible solutions.

Southern Examining Group
Answers to questions taken from past examination papers are entirely the responsibility of the authors and have neither been provided nor approved by the Southern Examining Group.

Photographs
Allsport 13, 283, 285B, 313; John Birdsall Photography 258, 327L; Frank Lane Picture Agency 46; Geoscience Features 303L, 303R, 304; Peter Gould 130; Holt Studios International 87, 91; Mark Jordan 327R; Andrew Lambert 136, 105, 113T, 129B, 177, 207, 222, 224, 255, 259, 277, 351; Natural History Museum 148; Natural History Photographic Agency 112, 117; Science Photo Library 2, 16, 21, 25, 39, 43, 45, 51, 61, 71, 94, 103, 104, 113B, 129T, 285T, 324, 333, 340, 341, 352T, 352B; Tony Stone 79

Illustrations
Roger Bastow, Harvey Collind, Richard Deverell, Gecko Ltd, Ian Law, Mike Parsons, Dave Poole, Chris Rothero and Tony Warne

Every effort has been made to contact the holders of copyright material, but if any have been inadvertently overlooked, the Publishers will be pleased to make the necessary arrangements at the first opportunity.

You might also like to visit:
www.**fire**and**water**.com
The book lover's website

INDEX

absorption 27, 28–9, 69
a.c. 250, 275, 277
acceleration 285, 288–94
acid rain 93, 160
acids 181, 219
activation energy 167, 168, 197
    and catalysts 203
active sites on enzymes 206
active transport 6–7, 69
activity 349
adaptation 78, 117
addiction 51
addition reactions 162
adrenal glands 41
aerobic respiration 13, 14
agrochemicals 89–91
air 22, 306
    metal reactions with 180, 222
air resistance 284, 285
alimentary canal 27
alkali metals 221–3
alkalis 219
alkanes 155–60, 161–2, 166–7
alkenes 161–2
alleles 107–11
allotropes 148
alpha particles 347–8, 349, 351
alternating current see a.c.
aluminium, extraction of 188
alveoli 20, 22
ammonia, Haber Process for 209
amplitude 320
anaerobic respiration 13–14
animal cells 2, 3
    blood cells 15, 16, 50–1
    neurones 36–8
    sex cells see eggs; sperms
anodes 187, 188, 189
antibodies 50
antigens 51
arteries 17–18
artificial selection 112
asexual reproduction 104–5
asteroids 339
atomic mass 235
atomic number 139, 349
atomic structure 138–42, 348
atoms 139, 347–8
    moles of 235–6
atria 17
auxin 71
axons 37

bases 103, 219
batteries 250
beating trays 97
beta particles 347, 349, 352
Big Bang theory 343
biodegradable materials 163
biological controls 90
biomass 82, 303
blood 15–18, 50–1
boiling 134, 135, 155–6
bonds 142–9, 155–6, 163
    in chemical reactions 166–7
Boyle's law 296
brains 39
breathing 20–3, 48
bronchi 20
Brownian Motion 136

capillaries 17–18

carbohydrates 24, 25
    see also starch; sugars
carbon 148, 207, 352
    in metal extraction 185, 186–7
carbon cycle 86, 161, 164
carbon dioxide 13, 22, 145
    in carbon cycle 86, 161, 164
    in greenhouse effect 92, 160–1
    in photosynthesis 60, 62–3
carnivores 81
catalysts 203
    enzymes 26, 205–7
cathodes 187, 188, 189
cells 1–3, 4, 5–8
    genes in 103, 104–6
    see also animal cells; plant cells
cells (electrical) 182
central nervous system 36
centripetal force 340
CFCs 94
charge 247
chemical changes 196–7
chemical digestion 25–7
chemical equations 128–33, 239–40
chlorine 222, 226–8
chlorophyll 2, 60, 61
chloroplasts 2, 61
chromosomes 2, 103, 104–6
    in sex determination 109–10
cilia 20
circuits 246–52
    power in 257–61
    relay switches in 271–2
    resistance in 252, 253–5
circular motion 340
circulatory system 15–19
clones 104, 113
coal 154
collision theory 196–7, 200–1, 202
colon 27
combining power 125–8
combustion 159–61, 166–7
comets 339–40
community 78
competition 78
compounds 125–8
compressing gases 296
compression 319
concentration 5–6
    and reactions 200–1, 208–9
concentration gradients 5, 22
condensation 134
conduction 306
conductors 253, 254, 306
consumers 81
Contact Process 210
continental plates 175–6
contraception 43
convection 174–5, 307
copper, purification of 188–9
core, Earth's 174–5
costs, energy 258, 308–9
covalent bonding 144–7, 148
cracking of oil fractions 156–7
Crick, Francis 103
critical angle 330
crude oil 154–9
crust, Earth's 174, 175
current 246–7, 255, 274–5
cycles 86–8, 176–7
    carbon 86, 161, 164
cytoplasm 1, 2

Darwin, Charles 116–17
dating, radioactive 352
d.c. (direct current) 250
deceleration 285
decibels 333
decomposers 87
denatured enzymes 26, 206
diabetes 43, 109
diamond 148
diaphragm, thoracic 21
diatomic substances 129, 227
diffraction 322
diffusion 5–6, 136
digestion 24–30
diodes 255
direct current (d.c.) 250
diseases 43, 50–1, 109
displacement reactions 181, 227
dissolving 136, 196
distances 289–90, 291
distillation 154–6
DNA 103, 104, 114
dominant alleles 107
Doppler effect 343
dot-and-cross diagrams 143–4
drugs 51–2
duodenum 27
dynamos 275

Earth 174–9, 322, 339
    magnetic poles 270
earthquakes 175, 322
ecology 78–102
ecosystems, defined 78
effectors 38
efficiency 309
egestion 27, 83
eggs, human 42–3, 105–6
elastic materials 284
electric motors 272
electrical energy 246–9, 272
electricity 246–69
    generation of 274–5
    transmission of 276–7
    see also electromagnetism
electrolysis 131, 185, 187–9
electrolytes 182, 187–8
electromagnetic induction 274–8
electromagnetic spectrum 319, 324–6
    see also gamma rays; ultraviolet radiation
electromagnetism 270–82
electronic structure 140, 228
    of halogens 226
    of metals 221, 223
electrons 138–42
    in bonding 142–4, 145
    and conductivity 253, 254, 255
    and current 246–7
    in static electricity 261–4
electrostatic induction 262
elements 125
emulsifying fats 26–7
endocrine system 41–4
endothermic reactions 165–6, 167
energy 303–18
    activation 167, 168, 197
    and catalysts 203
    in chemical reactions 165–8, 196
    efficiency 309
    electrical 246–9, 272
    in food chains 81, 82–4
    kinetic 272, 312, 313